COPY #2

W9-CAD-643

DISCARDED

DISCARDED

COPY #2

LAB 257

THE DISTURBING
STORY OF THE
GOVERNMENT'S SECRET
PLUM ISLAND GERM
LABORATORY

LAB 257

MICHAEL CHRISTOPHER CARROLL

Mattituck-Laurel Library

WM

WILLIAM MORROW
An Imprint of HarperCollins*Publishers*

Map of Plum Island on pp. xii–xiii © by William H. Wheeler. Reprinted by permission.

LAB 257. Copyright © 2004 by Michael Christopher Carroll. All rights reserved. Printed in the United States of America. No part of this book may be used or reproduced in any manner whatsoever without written permission except in the case of brief quotations embodied in critical articles and reviews. For information address HarperCollins Publishers Inc., 10 East 53rd Street, New York, NY 10022.

HarperCollins books may be purchased for educational, business, or sales promotional use. For information please write: Special Markets Department, HarperCollins Publishers Inc., 10 East 53rd Street, New York, NY 10022.

FIRST EDITION

Designed by Renato Stanisic

Printed on acid-free paper

Library of Congress Cataloging-in-Publication Data
Carroll, Michael C.
    Lab 257 : the disturbing story of the government's secret Plum Island germ laboratory / Michael Christopher Carroll.
       p. cm.
    Includes bibliographical references.
    ISBN 0-06-001141-6
    1. Plum Island Animal Disease Laboratory—History. 2. Virology—Research—New York (State)—Plum Island—History. 3. Laboratories—New York (State)—Plum Island—History. 4. Animals—Diseases—Research—New York (State)—Plum Island—History. 5. Animals as carriers of disease—Research—New York (State)—Plum Island—History. I. Title: Lab two fifty-seven. II. Title: Disturbing story of the government's secret Plum Island germ laboratory. III. Title.
QR359.5.U6C37 2004
614.5'8'0973—dc22

                                                                    2003044279

---

04 05 06 07 08 WBC/RRD 10 9 8 7 6 5 4

To my mother,
Audrey Joan Carroll

—M.C.C.

For a quiet outing where the air is a real tonic to a run-down man or woman; where you could have your bluefish and blackfish, sea bass, and lobsters fresh out of the water, cooked to a turn; milk, eggs, and chicken in abundance, Plum Island was the ideal place. What more could one ask?
—PLUM ISLAND ADVENTURER (1909)

Plum Island is a biological Three Mile Island.
—PLUM ISLAND EMPLOYEE (2003)

CONTENTS

To the Reader

Statements made during interviews and direct quotes from primary materials are contained in quotation marks, and I altered some of them slightly for grammar. Fictitious names are used to protect the identity of sources and they too are contained in quotes. Scenarios and dialogue I constructed are based upon recollections of multiple sources and primary and secondary research, and as such are placed in italics. The title *Lab 257* refers to one of the two laboratory buildings that exist on Plum Island, the other being Laboratory 101.

Based largely upon interviews, official documents, and detailed research, this book is also the product of personal visits to Plum Island, after the sixth of which I was abruptly denied further access by the U.S. Department of Agriculture, on the grounds of "national security."

Further information for the reader can be found in the Source Notes at the end of the book.

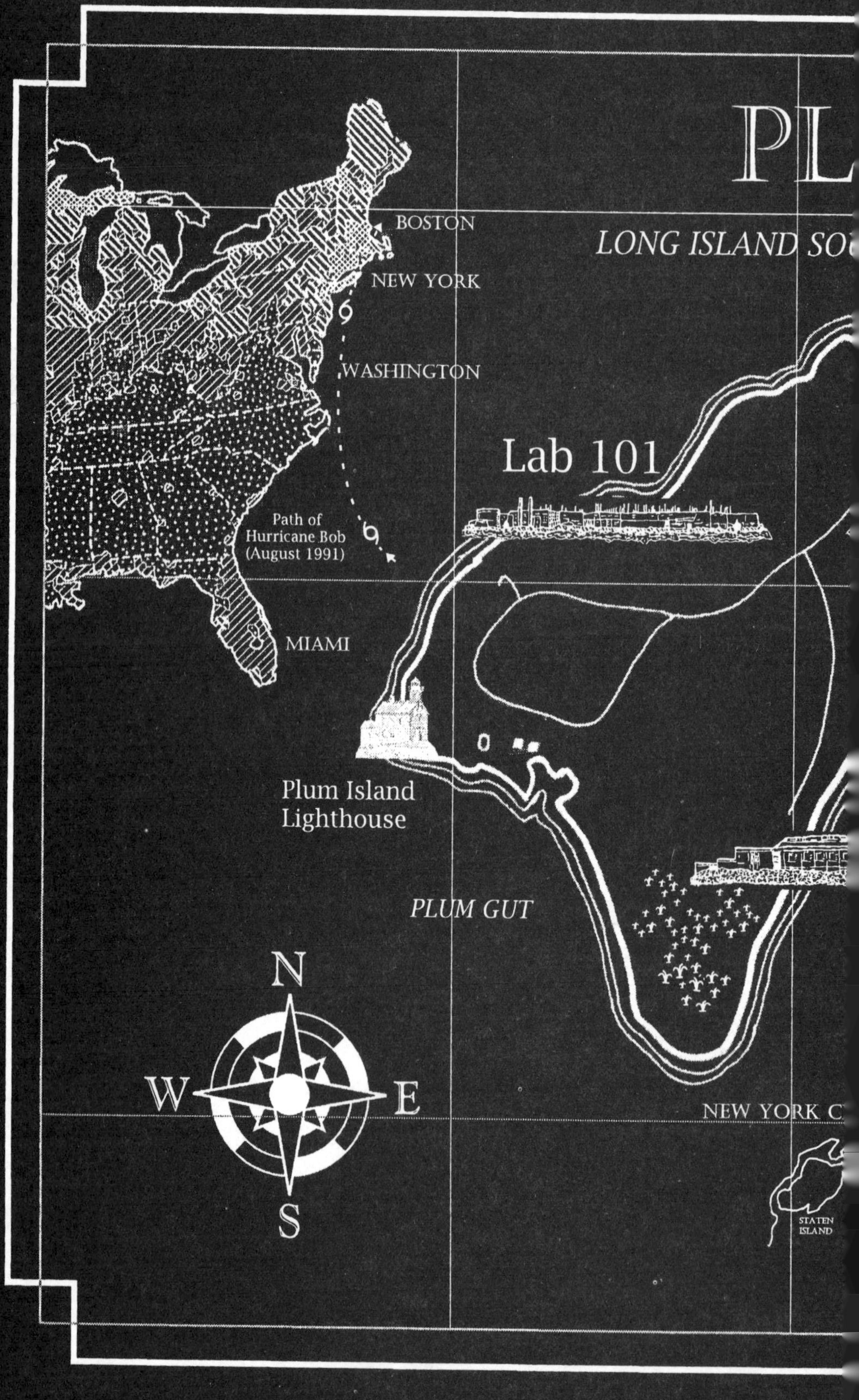

BOSTON

NEW YORK

WASHINGTON

Path of
Hurricane Bob
(August 1991)

MIAMI

PL

LONG ISLAND SO

Lab 101

Plum Island
Lighthouse

PLUM GUT

N

W   E

S

NEW YORK C

STATEN
ISLAND

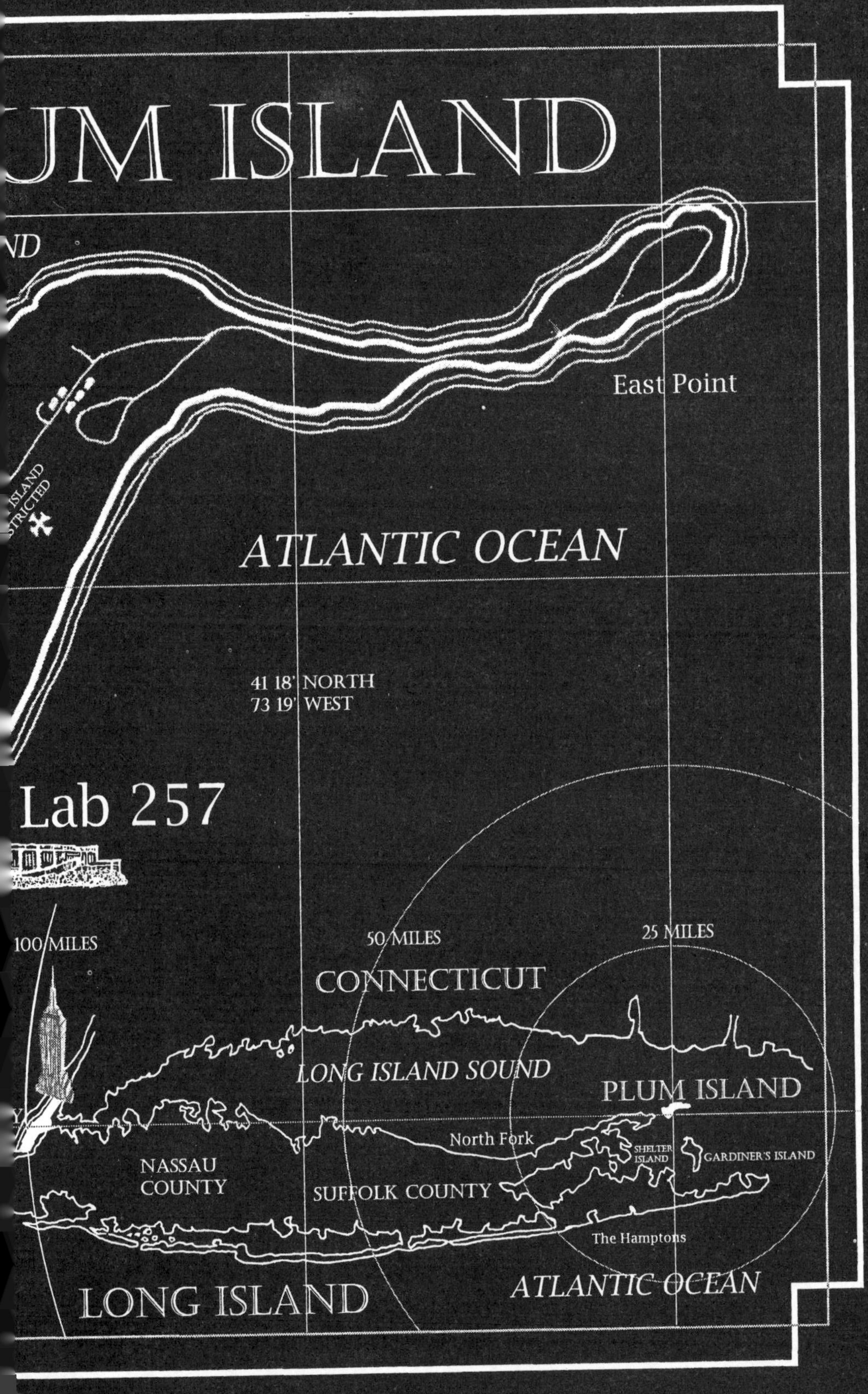

UM ISLAND

East Point

ATLANTIC OCEAN

41 18' NORTH
73 19' WEST

Lab 257

100 MILES

50 MILES

25 MILES

CONNECTICUT

LONG ISLAND SOUND

PLUM ISLAND

North Fork

SHELTER
ISLAND   GARDINER'S ISLAND

NASSAU
COUNTY

SUFFOLK COUNTY

The Hamptons

LONG ISLAND

ATLANTIC OCEAN

Preface

In May 1998, Sultan Bashiruddin Mahmood, a Western-educated nuclear physicist, was cheered by throngs gathered in the narrow cobblestone streets of Islamabad. They set off fireworks and wielded large placards emblazoned with his likeness. As the powerful chairman of the Pakistan Atomic Energy Commission, Mahmood developed nuclear power plants with the help of Canada and other Western nations. At the direction of the Pakistani government, Mahmood and others co-opted this technology and developed nuclear weapons; and that May their first atom bomb was successfully tested to counter a nuclear test by their archenemy, neighboring India. A joyous national celebration in Mahmood's and the other nuclear scientists' names ensued, a celebration rivaling Pakistan's independence day holiday.

Mahmood was a national hero.

In 2002, he was under twenty-four-hour house arrest and his assets were frozen.

The clean-shaven, moderately religious nuclear engineer, educated in Britain in the early 1970s, had drifted toward hard-line Islamic views by the mid-1990s. The snow-white-haired Mahmood became a bona fide religious extremist. He began fortune telling by reading people's palms and grew a long, unkempt beard. Urging Pakistan to adopt neighboring Afghanistan's Taliban way of life, his writings stressed the sacredness of keeping beards, the correlation between natural catastrophes and places of immoral behavior, and hidden scientific knowledge from Allah contained in the Koran. After he obstreperously opposed the Nuclear Test Ban Treaty in favor of a vigorous bomb-testing program, an apprehensive Pakistani leadership forced Mahmood out of the atomic energy program. He then relocated to Kabul, Afghanistan, where he set up a relief organization called Tameer-e-Ummah, or "Islamic Reconstruction."

One month after the September 11, 2001, terrorist attacks on America, U.S. Army intelligence and the CIA intercepted telephone communications between Mahmood and Taliban leader Mullah Mohammed Omar, whose oppressive, terrorist-sponsoring regime was on the verge of being crushed by American forces.[1] The United States government declared Tameer-e-Ummah, his Islamic "charity," a terrorist front organization. Its assets were seized by Pakistan's central bank, and Pakistani troops arrested their nuclear hero and turned him over to American authorities. CIA Director George J. Tenet rushed to Islamabad to take personal control of the Mahmood affair.

During his interrogation about the phone calls, Mahmood claimed to have discussed only Omar's personal safety and a flour mill that the charity wanted to build in Kandahar. But he failed six separate lie detector tests and disputed the results, calling the technology flawed. Then he admitted meeting on multiple occasions not only with Omar but also with a rogue Saudi millionaire named Osama bin Laden (and his top deputy, Al Zawahiri) as recently as August 2001. He described meeting with bin Laden and Omar, perched atop fluffy pillows, drinking green tea and praising Allah, in a scene reminiscent of the video footage aired worldwide featuring Osama bin Laden plotting destruction with his cohorts. Mahmood told interrogators that bin Laden had more in mind than schools and flour mills—bin Laden said he had radioactive material and wanted to know how to build a "dirty bomb" nuclear device (and "other things," according to Mahmood's son Azim, who later confirmed the meeting).

Fearing further embarrassment to its already tarnished nuclear energy program, Pakistan decided in early 2002 not to put Mahmood—whose "noble cause" was featured on the cover of the December 2001 issue of the radical glossy *Kashmir Jihad*—on public trial. The Pakistani government instead charged him with violating national secrecy oaths. Today, he remains under house arrest (and a media gag order) in a gated, two-story home.

After Pakistan arrested Sultan Bashiruddin Mahmood and U.S. forces secured the Kabul perimeter as part of the Afghanistan phase of America's war on terrorism, CIA and U.S. Army commandos stormed Mahmood's Kabul residence and the offices of Tameer-e-Ummah. Inside, they found copies of books he authored, with disturbing titles such as *Mechanics of the*

---

[1]The contents of the Mahmood–Mullah Omar conversations have not been made public.

Doomsday and Life After Death, and *Cosmology and Human Destiny,* which contained the following prediction: "During the autumn of 2000–2001, due to the high energy solar cycle and its impact on people, certain crazy actions are likely to happen." Then they uncovered some very incriminating raw material. One item was a diagram of a helium balloon designed to release large quantities of anthrax spores into the atmosphere. Another was a large document titled "Bacteria: What You Need to Know." And a hefty file folder contained a bundle of papers from an Internet search on anthrax vaccines.

Then there was a dossier. It contained information on a place in New York called the Plum Island Animal Disease Center. There was even a *New York Times* article in the dossier about this island.

Why would an associate of Osama bin Laden be so interested in some obscure New York island?

If Plum Island isn't exactly a household name in America, it apparently holds a prominent place in the minds of people like Sultan Bashiruddin Mahmood. And it's not the sandy beaches and swaying plum trees they're after.

Less than 2 miles off the east end of Long Island and 85 miles from New York City lies an unimposing 840-acre island, unidentified on most maps. On the few where it can be found, Plum Island is marked red or yellow, stamped U.S. GOVERNMENT—RESTRICTED DANGEROUS ANIMAL DISEASES.

Six huge ferries, each carrying 1,000 people and 120 cars, pass within one quarter mile of it every day, making the journey from Orient Point—the tip of Long Island's North Fork—to New London, Connecticut. More than one million people live and spend their summers in the Hamptons, within a scant mile or two from its shores, yet few know the name of this pork chop–shaped island that lies on the periphery of the largest population center in the United States. Even fewer can say whether it is inhabited or why it doesn't exist on the map.

I've often been asked why I decided to write about Plum Island. In the summer of 1992, as part of a ritual before picking up a friend from the Connecticut-to-New York ferry, I drove out to Orient Point, the end of the narrow strip of rural land that lies on either side of Route 25. Abandoning my car at the end of the road, I hiked through a mile's worth of tall beach grass, making my way to the very tip of Long Island. I climbed a high rocky bluff that sloped off to a sandbar that stretched far into the water, until it stopped abruptly at an ancient lighthouse that looked more like a cast-iron coffeepot than a warning beacon. Crouching on the bluff, I gazed at the waves from the Long Island Sound as they met the current of Gardiner's Bay—the precipice of the great Atlantic Ocean—crashing together, spitting spindrift high into the air, and falling against the shore. Through the light haze, ten miles off to the north, was the long coastline of Connecticut; between us, the ferry slowly chugged its way toward me. Out past the lighthouse was a wide green landmass. It looked deserted except for a powder blue water tower that sprouted above a green canopy. The island triggered a series of thoughts—rumors of biological warfare tests, news stories about deadly virus experiments, talk about Lyme disease being hatched there, stories about a man who worked on Plum Island and contracted some strange, undiagnosed ailment during a storm. *But it doesn't make sense—it looks so pristine! What is happening out there? And why?*

The long white ferry charged into the foreground, sailing through Plum Gut, the deep, narrow strait between where I sat and Plum Island. As I returned to the car, I resolved to uncover one day exactly what Plum Island is.

Those were my thoughts over a decade ago. Years later, fresh out of law school, I began to revisit that haunted place, trying to put the pieces of the puzzle together. To begin unraveling the *real* story—what happened (and still happens) behind the island's locked gates—I needed to educate myself in a variety of divergent topics. I immersed myself in research on a wide range of far-reaching subjects—geology, colonial history, animal disease, human disease, animal psychology, microbiology, biological warfare, coastal artillery, terrorism, lighthouses, and American Indians. I spoke to scientists, government officials, local residents, historians, and past and present employees of Plum Island, collecting firsthand accounts of this inexplicably anonymous place. Friends thought I had become a bit fanatical about the topic. They were right—the passion of "Plum" (as it's called by those in the know) had consumed me.

Research turned up disquieting descriptions of Plum Island, like "Right Out of *The Andromeda Strain*" and "Uncle Sam's Island of No Escape" and "The Setting of an Ian Fleming Novel." And my personal favorite: "Asking for directions to Plum Island is like asking for directions to Frankenstein's castle." One local newspaper took an opposing viewpoint, glibly describing Plum Island as a place where "white-coated scientists work on animal viruses before they wash up, climb aboard a ferry and then eat dinner with their families. It's no more exciting or diabolical than that." As I delved deep into researching and writing this story, I harkened back to the science fiction novels I enjoyed as a child, and came to a numbing realization: scientific truth can indeed be stranger than science fiction.

I came to realize that the USDA is far more than wholesome Grade A eggs, and that veterinarians are not all "doggy docs" who tend gingerly to the well-being of golden retrievers and calicos. Once the story began to take shape, those mom-and-apple-pie feelings quickly dissipated. I found Plum Island to be more than a nearby atoll covered in a blanket of trees and secrecy. The island's vibrant history stretches back 350 years: it was once an ancient Indian fishing outpost, later colonized by early English settlers, then a sheep and cattle farm, a Revolutionary War battleground, a rendezvous for the British in 1812, a coastal defense fort, a submarine mine factory, and an Army biological warfare laboratory. Since 1954, the ostensible mission of Plum Island's Animal Disease Center has been to protect America's $100 billion livestock industry and defend it from foreign viruses, like the foot-and-mouth disease virus epidemic that ravaged Europe in 2001. After September 11, 2001, its mission returned to biological warfare.

Today, it is home to virginal beaches, cliffs, forests, ponds, bogs, trails, paths, roads, buildings, and people—and the deadliest germs that have ever roamed the Earth.

On the last of my six voyages to Plum Island, I toured the island grounds with retired draftsman Ben Robbins. He carried a flashlight in his rear jeans pocket, and wielded a two-foot machete in his left hand. We climbed through the Army bunkers and coastal defense installations of Spanish-American War vintage, while thorns and brush from everywhere snapped into my face. Ben knew every inch of Plum Island because he had to—for decades it was his job to draw the maps and blueprints. I quickly realized Plum Island is like a preserve that hasn't known much orderly planting, pruning, or occasional TLC. "We have industrial grade poison ivy," said Ben, pointing out pea-sized gray clusters reaching out to touch someone along the overgrown dirt trails. It's a dense jungle on the eastern seaboard—perhaps the most untouched, uncared for, undeveloped, and unnoticed spit of land around.

We reached a high bluff on the island's east end and entered an old Army weather station. Up a narrow crumbling staircase was a lookout room with a wind-speed detector and wind cups twirling in the breeze, hand-painted signal flags, and a Morse-code alphabet key. The room was littered with glass from broken thin slat windows and bird dung. Heavy brush concealed a breathtaking view of the Atlantic.

On the way out, I noticed a weathered gray metal ammunition storage box behind the door on the first floor. Brushing off decades of dust, I read faded stenciling, printed sideways along the box: SPRAYER—CHEMC—ENGI-

NEERS, TROOP SUPPLY. I wondered what it was used for. It is the highest point on the island, so outdoor tests spraying pathogens or insect vectors would have been run from here. But Ben was outside calling my name. I dashed out to meet him and continued the tour.

In the van returning to the lab for lunch, I looked down at my khakis. A burst of heat flashed up through my legs straight to my gut as I saw about eight or so tiny black dots, a few on each pant leg—immediately I realized I was covered in ticks. I carefully brushed them off onto the floor of the minivan, wondering why I'd been trekking through these overgrown woods with a short-sleeved shirt and no hat. Looking at the front passenger seat, where Ben was chatting amiably with our driver, I noticed with new interest his red mesh baseball cap and long sleeves. I tried to keep cool, but I was perspiring like crazy.

It didn't help that an awful burning smell joined us in the van at that point, hard to describe except to say the stench reminded me of something noxious that shouldn't be burned or smelled. Then we saw a thick cloud creeping slowly to the southwest. Ben whirled around on me and chirped, "Look—it's time to burn the animals!" The rolling black cloud was heading for Gardiner's Island, on its way over to greet Sag Harbor and the Hamptons.

When we stopped for lunch back at the laboratory, I excused myself and went to the men's bathroom. There I swiped off seven ticks attached to my shirt, and four more on my the inside of my pant leg. I ripped my fingers through my scalp, tilting my head toward the mirror in desperation, not knowing what to do should I see a tiny black dot attached to my skin.

Nervous as hell, I went into the stall, peeled off my clothes, and examined every inch of fabric and every inch of my body. Three more ticks met their demise in the toilet bowl. Thankfully, I found nothing, and I would contract nothing. The rest of the day, I flicked more black spots off my clothing and covered my head with my hands as we hiked under the trees.

The uncomfortable connections between West Nile virus, Lyme disease, and Plum Island unfortunately proved difficult to dismiss as my research deepened. Of course you cannot coax an island to speak up for itself. To chart the story of Plum Island, then, you must see the people that "touch" the place. They may have been there for a month or for thirty-five years, or they may never have set foot on its soil, but in some way they've shaped Plum Island. Through the lives of these individuals and the one common thread they share, the story unfolds. Among the key people in this book is Dr. Jerry Callis, a Georgia farm boy who devoted his life to animal disease research and rose through government ranks to command Plum Island for more than thirty years. There is Dr. Callis's successor, Dr. Roger

Breeze, an Englishman enraptured by the American meritocracy, seeking the grail of scientific glory. There are those caught in the cross fire—people like Phillip Piegari, a support employee mired in a biological meltdown during a violent hurricane, a catastrophe that would have easily been avoided were it not for management's recklessness, and Frances Demorest, one of Plum Island's oldest veterans, who paid a terrible price for speaking out. And while there aren't any three-headed chickens or five-legged cows as the USDA is quick to point out with a chuckle, you'll be introduced to Nazi germ warfare scientists, ancient American Indians, germ warriors, germ hunters, Mexican cowboys, the father of anthrax, virus outbreaks in New York's backyard, and a biological and environmental muddle about to boil over.

Plum Island demands a carefully written, unbiased look, not from both sides, but from all sides. It needs to be probed by someone who isn't obsessed with rooting out government waste and corruption, or wed to conspiracy theories. The purpose of this book, then, is to explore the last half century on Plum Island in depth—a half century of biological experimentation and scientific breakthroughs, darkened by upheaval, concealment, and astonishingly careless management. Plum Island's inhabitants are in many ways the gang that couldn't shoot straight. Except here, this gang's ammunition is the deadliest germs known to mankind.

It is my hope that the work presented here will shed some light on the fascinating story of Plum Island. May it prompt a frank and open discussion between the public and its government on protecting present and future generations from preventable catastrophes lurking in our midst—before the preventable becomes the inevitable.

LAB 257

PART 1

OUTBREAKS

1975: The Lyme Connection

> *Dear Ann,*
>
> *Have you ever heard of Lyme disease? I am writing this letter because I know you can help thousands of people by warning them about this awful sickness. I have been battling it for 18 months. Frankly I am not doing well.*
>
> *It would be impossible for me to describe the emotional and physical pain that I have been through. I am a 42-year-old man, married nearly 20 years, and have a family. The days of slinging a 100-pound sack of bird-seed over my shoulder and walking to the backyard are over.*
>
> *Today I can't even lift a five-pound sack of flour. There was a time when I could play nine musical instruments. I sang in the church choir and ran my own small business. Today, I do none of the above. I am saving all my energy to fight Lyme disease.*
>
> *The treatment costs are staggering. IV antibiotic therapy runs from $150 to $475 a treatment. . . . We have already taken out a third mortgage on our home. Had I been aware of the symptoms from the beginning, I could have had $15 worth of oral antibiotics and that would have done the job.*
>
> *Thank you, Ann, for allowing me to try to help others.*
> *—S.J.N., Mattituck, N.Y.*

Protecting a nuclear power plant is no small task. When it opened in the 1980s, the Shoreham nuclear power plant on Long Island's North Shore boasted a 175-man militia equipped with Uzi 9-millimeters,

AR-15 assault rifles, and 12-gauge shotguns. This elite paramilitary unit patrolled the "protected area," a dense forest hundreds of acres deep that buffered the "controlled area," a huge concrete dome sheltering the uranium nuclear reactor. Every eight hours, a fresh detachment of fifty men, armed to the teeth and clad in steel-toed boots, tan pants, and khaki shirts, marched in lockstep through the protected area along dirt paths and through marshes, their watchful eyes and ears continually scanning for intruders. One Shoreham security officer, a short, blond-bearded, barrel-chested man, remembers the scene during the 3:00 P.M. to 11:00 P.M. shift in October 1987. His platoon had just moved out, marching into a field where they often spotted herds of thirty or forty wild deer darting ahead of them into the wooded glen. He felt a brief pinch on his left ankle and thought it was the stiff new Army boots he was breaking in. Later that night, he went home and showered. Pulling off his white tube socks, he noticed a small red mark on his ankle. *Those damn boots,* he thought, and went to bed.

When he awoke the next morning, the nagging blister had grown, so he grabbed tweezers from the bathroom vanity and poked at the area. Suddenly, something started to move, and he realized it wasn't a blister at all. It was a live bug. Panicked, he frantically dug into it. As he extracted the critter, it broke in two, spilling its insides into the microscopic holes it punched into his body.

Seventy-two hours later, he thought he had caught the flu. Within a week, his joints began to ache.

Most people don't think of deer as swimmers. But swim they do. Indigenous to most of the United States and Canada, white-tailed deer can swim distances as long as four miles.

Their natural predators—wolf, bear, mountain lion, and coyote—are long extinct from the northeastern landscape, but one tiny foe remains. Poised atop a blade of grass, the deer tick waits patiently for anything warm-blooded to brush by, feeding on deer as well as smaller creatures like birds and mice. The tick jumps aboard and pierces its sharp mouth hooks into the skin of its unlucky host. A tiny glutton with a king-sized appetite, the tick sucks the blood of its host in a feast that can last up to two whole days, while it swells to a bubble over three times its original size. At the same time, the little parasite deposits its own fluids into the host, fluids that sometimes prove fatal.

The feeding habits of ticks and the swimming abilities of deer were of little concern to the residents of Old Lyme, Connecticut, in July 1975. This quaint New England town is, for the most part, an upper-crust community with tree-lined streets and fine colonial and Federal-style homes. As one of

America's oldest towns, founded by English Puritans, Old Lyme was enjoying its tricentennial as the nation prepared for a bicentennial. But a strange set of occurrences that year would forever change its reputation from a warm, charming enclave to a place of fear and despair.

Old Lyme, nestled on the banks of the Connecticut River, sits just a shade north of the Long Island Sound. The midsummer weather in 1975 was typical for coastal Connecticut—hot, sticky, and humid. As little ones frolicked in the sun, ignoring the blistering heat, and grown-ups sought refuge on their porches by night, grateful for a balmy summer breeze, Polly Murray and Judith Mensch noticed something unusual about their children. Seemingly out of nowhere, they were showing signs of strange physical and mental ailments. Alarmed, the two mothers quickly phoned their neighbors, who were observing strikingly similar conditions in their own children. Many of the kids in the neighborhood—and some adults—were suffering from the same skin rashes, throbbing headaches, and painful swollen joints.

Together, Polly and Judith brought their concerns to the Connecticut Department of Health, which immediately appointed physicians from Yale University to investigate. Initially, the doctors misdiagnosed thirty-nine children and twelve adults with juvenile rheumatoid arthritis, a condition they named "Lyme arthritis," after the town where the strange outbreak occurred. Two years later, scientists linked Lyme arthritis to the bite of a deer tick. And in 1981, Dr. Wally Burgdorfer, a researcher at the National Institutes of Health, discovered a thin spiral bacteria—in technical terms, a spirochete—immersed in the fluid of a deer tick. He proved that the new spirochete was to blame—not for a Lyme arthritis, but for an entirely new ailment: Lyme disease.

Borrelia burgdorferi (Bb), named in honor of its discoverer, attacks humans in a number of ways, which is one reason why it remains difficult to diagnose. Characterized by symptoms such as facial paralysis and stiff swelling in the neck and joints, *Bb* also causes maladies like meningitis and encephalitis—both swellings of the brain—and cardiac problems, including atrioventricular block, myopericarditis, and cardiomegaly. Because *Bb* attacks the body's central nervous system, additional symptoms of Lyme disease include acute headaches, general fatigue, fever, moodiness, and depression.

That brief pinch the nuclear power plant trooper felt on his ankle that afternoon was the bite of an enemy no larger than the period at the end of this sentence. The chance of finding something that size, even had it attached to his exposed forearm, was pretty slim. The foe was either a eight-legged nymph deer tick or a Lone Star tick, swelling up to one hundred times its size with his blood. And while it sipped away, the tick regurgitated hundreds of spiral-shaped *Bb* bacteria into the victim's blood.

The tick is the perfect germ vector, which is why it has long been fancied as a germ weapon by early biowarriors from Nazi Germany and the Empire of Japan to the Soviet Union and the United States. Fixing its target by sensing exhaled carbon dioxide, the creature grabs onto a mammal's skin with its legs and digs in with its mouth hooks. The tick secretes saliva that helps glue it to its host, making it difficult to separate. A special hormone in the tick counteracts antibodies sent by the host to fight it off, and the crafty tick secretes an anti-inflammatory to prevent itching—so the host hardly knows it's there.

If the tick is the perfect germ messenger, then *Bb* is an incredibly clever germ. Because its outside wall is hard to destroy, the bacterium can fight off immune responses and antibiotic drugs. *Bb* finds a home in the mouth and salivary glands of larvae and nymph ticks, and infects females' ovaries and the thousands of eggs they will lay after breeding while attached to deer (upon which *Bb* has little effect). Common in mice and birds as well, today there are five subspecies of *Bb* and over one hundred mutated substrains in the United States.

The question that experts haven't been able to answer is why this disease suddenly surfaced in Old Lyme, Connecticut, in the summer of 1975.

PROJECT PAPERCLIP MEETS PLUM ISLAND

> *"I do not believe that we should offer any guarantees to protection in the post-hostilities period to Germans. . . . Among them may be some who should properly be tried for war crimes or at least arrested for active participation in Nazi activities. . . ."*
> —PRESIDENT FRANKLIN DELANO ROOSEVELT (1944)

> *"To the victors belong the spoils of the enemy."*
> —U.S. SENATOR WILLIAM L. MARCY (1832)

Dr. C. A. Mitchell began his remarks at the 1956 Plum Island dedication day by reminiscing on the late world war:

> I often think and almost tremble at what could have taken place had our Teutonic enemies been more alive to this. It is said that some of their scientists pointed out the advantages to be obtained from the artificial sowing of disease agents that attack domestic animals. Fortunately blunders existed in the Teutonic camp as in our own. Consequently, this means of attack was looked upon as a scientific poppy dream. . . . If [as much] time and money were invested in biologic agent dispersion as in one bomber plane, the Free World would have almost certainly gone down to defeat.

The audience murmured in acknowledgment, but one dedication day VIP stirred uncomfortably—the director of the new virus laboratory in Tübingen, West Germany, personally invited by Plum Island Director Maurice S. "Doc" Shahan. The mind of the brown-haired man with the scar on his face and upper lip held a dark secret. He sat there perspiring, staring at Dr. Mitchell through his gray-brown eyes, wondering how many people knew his past.

For he—Dr. Erich Traub—was that "Teutonic enemy."

Strangely enough, he had every right to be there. He was one of Plum Island's founding fathers.

Nearing the end of World War II, the United States and the Soviet Union raced to recruit German scientists for postwar purposes. Under a top-secret program code-named Project PAPERCLIP, the U.S. military pursued Nazi scientistific talent "like forbidden fruit," bringing them to America under employment contracts and offering them full U.S. citizenship. The recruits were supposed to be nominal participants in Nazi activities. But the zealous military recruited more than two thousand scientists, many of whom had dark Nazi party pasts.[1]

American scientists viewed these Germans as peers, and quickly forgot they were on opposite sides of a ghastly global war in which millions perished. Fearing brutal retaliation from the Soviets for the Nazis' vicious treatment of them, some scientists cooperated with the Americans to earn amnesty. Others played the two nations off each other to get the best financial deal in exchange for their services. Dr. Erich Traub was troubleing on the Soviet side of the Iron Curtain after the war, and ordered to research germ warfare viruses for the Russians. He pulled off a daring escape with his family to West Berlin in 1949. Applying for Project PAPERCLIP employment, Traub affirmed he wanted to "do scientific work in the U.S.A., become an American citizen, and be protected from Russian reprisals."

As lab chief of Insel Riems—a secret Nazi biological warfare laboratory on a crescent-shaped island nestled in the Baltic Sea—Traub worked

---

[1] The best known PAPERCLIP recruit was Wernher von Braun, the brains behind the Saturn V rocket that brought the Apollo 11 astronauts to the moon, the visionary architect of Disneyland's fabled Tomorrowland exhibit, one of the founders of NASA, and the fatherly host of network television specials on outer space. The American public knew him as a warm, affable man with a thick German accent. They didn't know that the U.S. missile program was based upon von Braun's revolutionary V-2 rocket he designed for Hitler, a 50-foot-long, 13-ton intercontinental ballistic missile. And they didn't know that during World War II, von Braun was a major in Heinrich Himmler's SS, and that his V-1 and V-2 rockets—built by some 20,000 slave laborers in his Mittlewerk SS munitions factory—rained destruction upon Europe in Hitler's futile attempt to turn the tide near the end of the war.

directly for Adolf Hitler's second-in-charge, SS Reichsführer Heinrich Himmler, on live germ trials. He packaged weaponized foot-and-mouth disease virus, which was dispersed from a Luftwaffe bomber onto cattle and reindeer in occupied Russia. At Himmler's request, Traub personally journeyed to the Black Sea coast of Turkey. There, amid the lush Anatolian terrain, he searched for a lethal strain of rinderpest virus for use against the Allies. Earlier in the war he had been a captain in the German Army, working as an expert on infectious animal diseases, particularly in horses. His veterinary corps led the germ warfare attacks on horses in the United States and Romania in World War I with a bacteria called glanders. He was also a member of NSKK, the Nazi Motorists Corps, a powerful Nazi organization that ranked directly behind the SA (Storm Troopers) and the SS (Elite Corps). In fact, NSKK's first member, joining in April 1930, was Adolf Hitler himself. Traub also listed his 1930s membership in Amerika-Deutscher Volksbund, a German-American "club" also known as Camp Sigfried. Just thirty miles west of Plum Island in Yaphank, Long Island, Camp Sigfried was the national headquarters of the American Nazi movement. Over forty thousand people throughout the New York region arrived by train, bus, and car to participate in Nuremberg-like rallies. Each weekend they marched in lockstep divisions, carrying swastika flags, burning Jewish U.S. congressmen in effigy, and singing anti-Semitic songs. Above all, they solemnly pledged their allegiance to Hitler and the Third Reich.

Ironically, Traub spent the prewar period of his scientific career on a fellowship at the Rockefeller Institute in Princeton, New Jersey, perfecting his skills in viruses and bacteria under the tutelage of American experts before returning to Nazi Germany on the eve of war. Despite Traub's troubling war record, the U.S. Navy recruited him for its scientific designs, and stationed him at the Naval Medical Research Institute in Bethesda, Maryland.[2]

Just months into his PAPERCLIP contract, the germ warriors of Fort Detrick, the Army's biological warfare headquarters in Frederick, Maryland, and CIA operatives invited Traub in for a talk, later reported in a declassified top-secret summary:

---

[2]Another PAPERCLIP recruit who worked alongside Traub at the Naval Medical Research Institute was Theodur Benzinger, an "aviation doctor." Benzinger was invited by Heinrich Himmler to view a film on high-altitude simulations at Dachau using prisoners as human guinea pigs (a "37-year-old Jew in good condition who lasted 30 minutes—he began to perspire, wriggle his head, developed cramps, became breathless, and [with] foam collecting around his mouth became unconscious and died"). Arrested for war crimes by Nuremberg prosecutors, he denied experimenting with any prisoners. He evaded prosecution at the Nuremberg Doctors' Trial and rushed into the waiting arms of Project PAPERCLIP.

Dr. Traub is a noted authority on viruses and diseases in Germany and Europe. This interrogation revealed much information of value to the animal disease program from a Biological Warfare point of view. Dr. Traub discussed work done at a German animal disease station during World War II and subsequent to the war when the station was under Russian control.[3]

Traub's detailed explanation of the secret operation on Insel Riems, and his activities there during the war and for the Soviets, laid the groundwork for Fort Detrick's offshore germ warfare animal disease lab on Plum Island. Traub was a founding father.

Little is publicly available about his clandestine activities for the U.S. military. The *names* of two studies, "Experiments with Chick Embryo Adapted Foot-and-mouth Disease" and "Studies on In-vitro Multiplication of Newcastle Disease Virus in Chicken Blood," were made available under the Freedom of Information Act, but the research reports themselves (and many others) were withheld. With his "laboratory assistant" Anne Burger, who came over in 1951, Traub experimented with over forty lethal viruses on large test animals.[4]

Traub also spent time at the USDA laboratories in Beltsville, Maryland, where he isolated a new weapons-grade virus strain in the USDA lab. Studying a virulent strain of a new virus that caused human infections, Traub showed how it adapted "neurotropically" in humans by voraciously attacking nerve and brain tissues. This was the same potent virus that infected a human in Plum Island's first-ever germ experiment one year later.

By 1953, West Germany recognized a need for its own Insel Riems and built a high-containment virus facility in Tübingen. They asked Dr. Erich Traub to return to the Fatherland and assume command. Permission was granted. But there was a catch. "In view of Dr. Traub's eminence as an international authority and the recognizable military potentialities in the possible application of his specialty, it is recommended that future surveillance in appropriate measure be maintained after the specialist's return to Germany." In other words, the CIA would be tailing him for years. As

---

[3]According to the document, a transcript of the meeting was prepared by the CIA. When I requested the transcript under the Freedom of Information Act, the CIA informed me that no such report ever existed, and if it did, it would be withheld for reasons of national security.

[4]Linda Hunt, author of *Secret Agenda*, the seminal book on Project PAPERCLIP, believes Burger may not have been Traub's "assistant," but rather his mistress. Apparently other PAPERCLIP recruits had imported their mistresses from Germany along with their families when they came to America. No additional information is available to the public on Anne Burger.

soon as the lab opened for business, he turned to Plum Island for starter strains of viruses, which were gladly shipped over. USDA officials traveled to West Germany and visited his laboratory often.

ERICH TRAUB AND PLUM ISLAND

Everybody seemed willing to forget about Erich Traub's dirty past—that he had played a crucial role in the Nazis' "Cancer Research Program," the cover name for their biological warfare program, and that he worked directly under SS Reichsführer Heinrich Himmler. They seemed willing to overlook that Traub in the 1930s faithfully attended Camp Sigfried. In fact, the USDA liked him so much, it glossed over his dubious past and offered him the top scientist job at the new Plum Island laboratory—not once, but *twice*. Just months after the 1952 public hearings on selecting Plum Island, Doc Shahan dialed Dr. Traub at the naval laboratory to discuss plans for establishing the germ laboratory and a position on Plum Island.

Six years later—and only two years after Traub squirmed in his seat at the Plum Island dedication ceremonies—senior scientist Dr. Jacob Traum retired. The USDA needed someone of "outstanding caliber, with a long established reputation, internationally as well as nationally," to fill Dr. Traum's shoes. But somehow it couldn't find a suitable American. "As a last resort it is now proposed that a foreigner be employed." The aggies' choice? Erich Traub, who was in their view "the most desirable candidate from any source." The 1958 secret USDA memorandum "Justification for Employment of Dr. Erich Traub" conveniently omitted his World War II activities; but it did emphasize that "his originality, scientific abilities, and general competence as an investigator" were developed at the Rockefeller Institute in New Jersey in the 1930s.

The letters supporting Traub to lead Plum Island came in from fellow Plum Island founders. "I hope that every effort will be made to get him. He has had long and productive experience in both prewar and postwar Germany," said Dr. William Hagan, dean of the Cornell University veterinary school, carefully dispensing with his *wartime* activities. The final word came from his dear American friend and old Rockefeller Institute boss Dr. Richard Shope, who described Traub as "careful, skillful, productive, and *very original*" and "one of this world's most outstanding virologists." Shope's sole reference to Traub at war: "During the war he was in Germany serving in the German Army."

Declining the USDA's offer, Traub continued his directorship of the Tübingen laboratory in West Germany, though he visited Plum Island frequently. In 1960, he was forced to resign as Tübingen's director under a dark cloud of financial embezzlement. Traub continued sporadic lab research for another three years, and then left Tübingen for good—a scandalous end to a checkered career. In the late 1970s, the esteemed virologist Dr. Robert Shope, on business in Munich, paid his father Richard's old Rockefeller Institute disciple a visit. The germ warrior had been in early retirement for about a decade by then. "I had dinner with Traub one day—out of old time's sake—and he was a pretty defeated man by then." On May 18, 1985, the Nazis' virus warrior Dr. Erich Traub died unexpectedly in his sleep in West Germany. He was seventy-eight years old.

A biological warfare mercenary who worked under three flags—Nazi Germany, the Soviet Union, and the United States—Traub was never investigated for war crimes. He escaped any inquiry into his wartime past. The full extent of his sordid endeavors went with him to his grave.

While America brought a handful of Nazi war criminals to justice, it safeguarded many others in exchange for verses to the new state religion—modern science and espionage. Records detailing a fraction of Erich Traub's activities are now available to the public, but most are withheld by Army intelligence and the CIA on grounds of national security. But there's enough of a glimpse to draw quite a sketch.

began to feel like a man made out of glass, like someone hit me with a baseball bat and shattered me from the top of my head to the balls of my feet," the nuclear power plant guard recalls. "Never in my life had I experienced such pain." His hands gnarled into contortions, and his vocal cords weakened and then became paralyzed, rendering him mute. The left side of his body went numb. A rheumatologist misdiagnosed him—just like doctors misdiagnosed the children of Old Lyme, Connecticut, in 1975—with rheumatoid arthritis. Then, the neurological symptoms set in. He experienced violent mood swings where he would be calm one moment and bawling silly the next. A newfound sensitivity to light made him a prisoner in his home, with the shades drawn and lights turned off. Noise was magnified a hundredfold, to the point that the vibrations from a person walking across the floor were excruciating. His incessant reflexive coughing was so powerful, it broke three of his ribs and brought up large globs of blood. When he told the doctor he suspected he had the long-misunderstood ailment Lyme disease, the doctor laughed. But with the help of his wife, a registered nurse, he diagnosed himself with thirty-eight of the forty symptoms of Lyme. Results showed he had some of the highest known titers of *Borrelia burgdorferi* (*Bb*) known in New York State. He was ordered to a hospital bed for intensive intravenous antibiotic treatment. The treatments for Lyme disease, which can range from oral antibiotics to massive weekly IV infusions, are like "trying to put out a forest fire with a watering can," according to another sufferer.

The symptoms subsided six months after the tick bite, but came back with new fury five months later. More IV antibiotics were prescribed. "I remember sitting in my doctor's office and saying, 'Doc, you know, I think I'm losing my mind.'" His heart, trying to cope with the large doses of chemicals, was failing him, but he figured he had nothing more to lose. He was dying. Teetering on the edge, the security guard mustered up what little strength he had left for one last hope. A deeply religious man, active in his church as a lay minister for some twenty-five years, he turned to a higher power.

"I could no longer speak. As the joint pain became unbearable, this one finger hurt so much I wanted it amputated—and I asked the doctor to amputate it. I prayed to God, 'How do you expect me to preach when I can't speak?' and He said, 'You have a typewriter—type.' And I had one crummy finger left that I could move, it was the little finger on my left hand and I'm right-handed." He began to pinkie-type, letter by letter, *click-clack-click*, whatever came to mind. In the very first sentence he punched the *I* key too hard and it broke off and fell on the floor. "Forty years old, and I began to cry like a baby—and no sound came out." He finished the letter, carefully penciling in the *I*'s. "It was the most humble letter I have ever written in my life." He did not think anyone would even read the pathetic-looking half-typed, half-scrawled epistle. But plucking Steven J. Nostrum's letter out of the thousands she received each day, the nationally syndicated advice columnist Ann Landers not only read it—she printed it.

What happened next was completely unexpected. Nostrum received hundreds of phone calls and thousands of letters from Lyme disease sufferers and their family members who shared his pain. "People were actually calling up telephone operators to get the exchange for Mattituck, randomly dialing numbers beginning with 298 and asking, 'Is there a man in your town with the initials S.J.N. who's involved with Lyme disease?'" Buoyed by the overwhelming response, he began to heal himself by helping others. He set up a makeshift command center in the basement of his home. Nostrum started one of the nation's first Lyme disease support groups; guests from around the country came to his monthly gatherings at the local library. Through his organization, the Lyme Borrelia Out-Reach Foundation (Lyme Borrelia is the technical name of Lyme disease), he published a newsletter, sent out audiotapes, distributed literature, and hosted a monthly cable television program seen across the country. He spoke at civic associations, churches, and schools, and testified before a special U.S. Senate committee on Lyme disease.

"I've been involved in Christian ministry for thirty-seven years—and I'm not going to tell you I saw a burning bush," says Nostrum. "But I can look you square in the eye and tell you I felt a *real calling* to get the information out, no political agendas, strictly from a point of education and prevention." Nostrum's education would lead him into some surprising territory.

A ttorney John Loftus was hired in 1979 by the Office of Special Investigations, a unit set up by the Justice Department to expose Nazi war crimes and unearth Nazis hiding in the United States. Given top-secret clearance to review files that had been sealed for thirty-five years, Loftus found a treasure trove of information on America's postwar Nazi recruiting. In 1982, publicly challenging the government's complacency with the wrongdoing, he told *60 Minutes* that top Nazi officers had been protected and harbored in America by the CIA and the State Department.

"They got the Emmy Award," Loftus wrote. "My family got the death threats."

Old spies reached out to him after the publication of his book, *The Belarus Secret*, encouraged that he—unlike other authors—submitted his manuscript to the government, agreeing to censor portions to protect national security. The spooks gave him copies of secret documents and told him stories of clandestine operations. From these leads, Loftus ferreted out the dubious Nazi past of Austrian president and U.N. secretary general Kurt Waldheim. Loftus revealed that during World War II, Waldheim had been an officer in a German Army unit that committed atrocities in Yugoslavia.[5] A disgraced Kurt Waldheim faded from the international scene soon thereafter.

In the preface of *The Belarus Secret*, Loftus laid out a striking piece of information gleaned from his spy network:

> Even more disturbing are the records of the Nazi germ warfare scientists who came to America. They experimented with poison ticks dropped from planes to spread rare diseases. I have received some information suggesting that the U.S. tested some of these poison ticks on the Plum Island artillery range off the coast of Connecticut during the early 1950s. . . . Most of the germ warfare records have been shredded, but there is a top secret U.S. document confirming that "clandestine attacks on crops and animals" took place at this time.

---

[5]The Office of Special Investigations found that Waldheim participated in the transfer of civilians to SS slave labor camps, the deportation of civilians to death camps, the use of anti-Semitic propaganda, and the mistreatment and execution of Allied POWs.

Erich Traub had been working for the American biological warfare program from his 1949 Soviet escape until 1953. We know he consulted with Fort Detrick scientists and CIA operatives; that he worked for the USDA for a brief stint; and that he spoke regularly with Plum Island director Doc Shahan in 1952. Traub can be physically placed on Plum Island at least three times—on dedication day in 1956 and two visits, once in 1957 and again in the spring of 1958. Shahan, who enforced an ultrastrict policy against outside visitors, each time received special clearance from the State Department to allow Traub on Plum Island soil.

Research unearthed three USDA files from the vault of the National Archives—two were labeled TICK RESEARCH and a third E. TRAUB. All three folders were empty. The caked-on dust confirms the file boxes hadn't been open since the moment before they were taped shut in the 1950s.

Preposterous as it sounds, clandestine outdoor germ warfare trials were almost routine during this period. In 1952, the Joint Chiefs of Staff called for a "vigorous, well-planned, large-scale [biological warfare] test program . . . with all interested agencies participating." A top-secret letter to the secretary of defense later that year stated, "Steps should be taken to make certain adequate facilities are available, including those at Fort Detrick, Dugway Proving Ground, Fort Terry (Plum Island) and an island field testing area." Was Plum Island the island field testing area? Indeed, when the Army first scouted Plum Island for its Cold War designs, they charted wind speeds and direction and found that, much to their liking, the prevailing winds blew out to sea.

One of the participating "interested agencies" was the USDA, which admittedly set up large plots of land throughout the Midwest for airborne anticrop germ spray tests. Fort Detrick's Special Operations Division ran "vulnerability tests" in which operatives walked around Washington, D.C., and San Francisco with suitcases holding *Serratia marcescens*—a bacteria recommended to Fort Detrick by Traub's nominal supervisor, Nazi germ czar and Nuremberg defendant Dr. Kurt Blome. Tiny perforations allowed the germs' release so they could trace the flow of the germs through airports and bus terminals. Shortly thereafter, eleven elderly men and women checked into hospitals with never-before-seen *Serratia marcescens* infections. One patient died. Decades later when the germ tests were disclosed, the Army denied responsibility. A Department of Defense report later admitted the germ was "an opportunistic pathogen . . . causing infections of the endocardium, blood, wounds, and urinary and respiratory tracts." In the summer of 1966, Special Operations men walked into three New York City subway stations and tossed lightbulbs filled with *Bacillus subtilis*, a benign bacteria, onto the tracks. The subway trains pushed the germs through the entire system and theoretically killed over a million passengers.

Tests were also run with live, virulent, anti-animal germ agents. Two hog-cholera bombs were exploded at an altitude of 1,500 feet over pigpens set up at Eglin Air Force Base in Florida. And turkey feathers laced with New-castle disease virus were dropped on animals grazing on a University of Wisconsin farm.[6]

The Army never fully withdrew its germ warfare efforts against food animals. Two years after the Army gave Plum Island to the USDA—and three years after it told President Eisenhower it had ended all biological warfare against food animals—the Joint Chiefs advised that "research on anti-animal agent-munition combinations should" continue, as well as "field testing of anti-food agent munition combinations. . . ." In November 1957, military intelligence examined the elimination of the food supply of the Sino-Soviet Bloc, right down to the *calories* required for victory:

> In order to have a crippling effect on the economy of the USSR, the food and animal crop resources of the USSR would have to be dam-aged within a single growing season to the extent necessary to reduce the present *average* daily caloric intake from 2,800 calories to 1,400 calories; i.e., the starvation level. Reduction of food resources to this level, if maintained for twelve months, would pro-duce 20 percent fatalities, and would decrease manual labor perfor-mance by 95 percent and clerical and light labor performance by 80 percent.

At least six outdoor stockyard tests occurred in 1964–65. Simulants were sprayed into stockyards in Fort Worth, Kansas City, St. Paul, Sioux Falls, and Omaha in tests determining how much foot-and-mouth disease virus would be required to destroy the food supply.

Had the Army commandeered Plum Island for an outdoor trial? Maybe the USDA lent a hand with the trial, as it had done out west by fur-nishing the large test fields. After all, the Plum Island agreement between the Army and the USDA allowed the Army to borrow the island from the USDA when necessary and in the national interest.

Traub might have monitored the tests. A source who worked on Plum Island in the 1950s recalls that animal handlers and a scientist released ticks outdoors on the island. "They called him the Nazi scientist, when they came in, in 1951—they were inoculating these ticks," and a picture he once saw "shows the animal handler pointing to the area on Plum where they

[6]Traub likely developed the standardized Newcastle disease virus placed on the biological cluster bomb by concentrating the germ in chicken blood while he was working at the Naval Medical Research Institute for Project PAPERCLIP.

released the ticks." Dr. Traub's World War II handiwork consisted of aerial virus sprays developed on Insel Riems and tested over occupied Russia, and of field work for Heinrich Himmler in Turkey. Indeed, his colleagues conducted bug trials by dropping live beetles from planes. An outdoor tick trial would have been de rigueur for Erich Traub.

Somebody gave Steve Nostrum a copy of John Loftus's *The Belarus Secret* at one of his support group meetings. Steve had long suspected that Plum Island played a role in the evolution of Lyme disease, given the nature of its business and its proximity to Old Lyme, Connecticut. But he never publicly voiced the hunch, fearing a loss of credibility; hard facts and statistics earned him a reputation as a leader in the Lyme disease field. Now, *in his hands*, he had a book written by a Justice Department attorney who not only had appeared on *60 Minutes* but also had brought down the secretary general of the United Nations. Nostrum disclosed the possible Plum–Lyme connection on his own television show. He invited local news reporter and Plum Island ombudsman Karl Grossman to help him explore the possibilities in light of the island's biological mishaps. Asked why he wrote about Loftus's book in his weekly newspaper column, Grossman says, "To let the theory rise or fall. To let the public consider it. And it seemed to me that the author was a Nazi hunter and a reputable attorney— this was not trivial information provided by some unreliable person."

In October 1995, Nostrum, fresh off nursing duty (having earned an RN degree to help Lyme disease patients), rushed to a rare public meeting held by the USDA. In a white nurse's coat, stethoscope still around his neck, Nostrum rose. Trembling, his blond beard now streaked with gray, he clutched his copy of *The Belarus Secret* as he read the damning passage out loud for the USDA and the public to hear. "I don't know whether this is true," he said, looking at the dais. "If it is true, there must be an investigation—if it's not true, then John Loftus needs to be prosecuted." People in the audience clapped, and some were astonished. A few gawked, thinking he was nuts. How did the USDA officials react? "If stares could kill, I would have been dead," remembers Nostrum.

Hiding behind the same aloof veil of secrecy they had employed for decades, the USDA brazenly cut him off. "There are those who think that little green men are hiding out there," the officials responded to Nostrum. "But trust us when we say there are no space aliens and no five-legged cows." A few laughs erupted in the crowd. "It did nothing but detract from what I was saying," says Nostrum. "But I said it, and I had the documentation to support it."

One person the USDA couldn't laugh off as easily was Congressman Michael Forbes. A concerned Forbes called newsman Karl Grossman after reading his column. "He gets hold of me late one night," says Grossman, "and tells me he's going to make a surprise visit—*a raid on Plum Island*—the very next morning." He wanted Grossman (and the power of his pen) to come along for the ride.

At 6:45 a.m. on a crisp, clear spring morning, just days after the 1995 Oklahoma City bombing, Forbes walked up to the ferry captain at Orient Point flanked by Grossman and John McDonald, an investigative reporter from *Newsday*. "Hi, I'm Congressman Mike Forbes."

The captain's response was one of "pure fear." He radioed over all kinds of distress calls to Plum Island. Waiting impatiently on the gangplank, Forbes was in no mood to fool around. He repeated, "Look, I'm Congressman Forbes and *I'm going on this ferry*." Minutes later, he was riding on the ferry with the two journalists. The self-invited guests were hurried onto the bus and whisked into the office of new Plum Island Director Dr. Harley Moon. Dr. Moon had been recruited from the USDA's Ames, Iowa laboratory, which studies less dangerous germs domestic to the United States. Though he had no interest in relocating to New York, he agreed to temporarily help move Plum Island past the troubles of past administrations before returning to his native Iowa. Forbes, who had never been on Plum Island before, immediately began peppering Moon with questions that revealed extensive homework. Nostrum had briefed Forbes beforehand and given him a list of questions, first of which was the Plum–Lyme connection unearthed in John Loftus's *The Belarus Secret*.

"Forbes wanted to get to the bottom of a bunch of issues," remembers Grossman. "Like the hurricane—the electricity going out in that hurricane years before. And all the stuff about Plum Island and biological warfare. A real laundry list of things.

"They were very nervous. Very nervous. There were no PR guys around. It was just these scientists—and no spin. Moon had six or seven scientists come in and explain what they were doing. Forbes was in no mood to be snowed or soft-balled. To bullshit him would have been difficult. Meanwhile, McDonald and I were shooting questions at them, too. It was their worst nightmare."

Forbes then asked Dr. Moon about the allegations in *The Belarus Secret*.

"There is a great concern about the prevalence of Lyme disease on eastern Long Island," Mike Forbes said, winding up. "And here we have the highest incidence of it in the world."

Dr. Moon replied to the inquiries with half-apologies. "I'm sorry, Congressman, I wasn't here then"—"I don't know, Congressman"—"We don't have any paperwork on that"—"I can't say what the Department of Defense might have been doing *before* Agriculture came here." Factually, the answers were all true. After all, how *would* Harley Moon know about outdoor tick trials in the 1950s? Dr. Moon had been on Plum Island two months (he'd only be there for ten more, before assuming an endowed research chair at Iowa State University). And if there were documents on Plum Island that addressed it, they were long gone, thanks to the recently ordered destruction of Central Files. Moon enjoyed complete plausible deniability.

"I agree with you that it looks like there's more of an incidence here of Lyme disease," said Moon, "but that might be due to improvements in diagnosis."

"Above all," Director Moon said to the trio on their way out, "do no harm—that's the first principle. Is it risky doing research on foreign animal diseases? I can't say there is no time, no way, a virus can get out of here. The possibility is so small—we take a calculated risk. And the risk-benefit ratio says, 'Let's do it.' "

Forbes was visibly perturbed on the ferry ride back to Long Island. The surprise visit hadn't yielded a smoking gun. Not finding what he was looking for, he left behind a long, detailed letter demanding answers. But the aggressive congressman and the two gumshoes had forewarned them. Dangers still lurked on Plum Island—Forbes was certain of it. "As long as I'm a member of Congress," he promised he would watch over the island like a hawk.

LYME ISLAND

By 1990, the east end of Long Island had, by leaps and bounds, the largest incidence of Lyme disease in the nation. But why?

The Geography. You can pinpoint cases of Lyme disease on a map of the United States by drawing a circle around the area of largest infection. Now you can tighten that circle until a single point is reached. That point? Plum Island. Spokes radiate outward from this point and pass through neighborhoods boasting the highest rates of Lyme disease contamination in the nation.

The Vectors. In the 1950s, the cocking of a rifle was often heard on Plum Island, portending the demise of deer that swam from the mainland to forage. Over time, fewer rifle shots were heard as the numbers of deer swimming to and fro increased, collecting ticks along the way. And while deer were sporadically shot, there was no stopping the wild birds. Retired scientists Jim and Carol House have been "birding" on Plum Island for over twenty years now, never missing an Audubon Society Christmas bird count, where they scout the terrain with binoculars, scribbling notes and snapping pictures. "Plum Island has a unique bird life," says Jim House. "It's got purple sandpipers, harlequin ducks, robins, eiders, osprey, warblers, and woodcocks," says Carol. Plum even hosts golden and bald eagles, who come in and dine on the baby Canadian geese. "It's one of their favorite stops in the springtime." There's no short supply of bald eagle food, because massive flocks of Canadian geese rule the island in droves. "We call them the Canadian Air Force," says one worker. "We made a list of a hundred and forty different species," says Carol. "One time I counted over two hundred brown creepers." The American Museum of Natural History runs a wild bird colony on seventeen-acre Great Gull Island, just east of Plum Island, and no doubt that has something to do with the plethora of fowl.

Plum Island, an untrammeled plot of wild nature, lies in the middle of the Atlantic flyway, the bird migration highway that runs between breeding grounds and winter homes from the Caribbean to the Florida coast, up the East Coast to the icy reaches of Greenland.

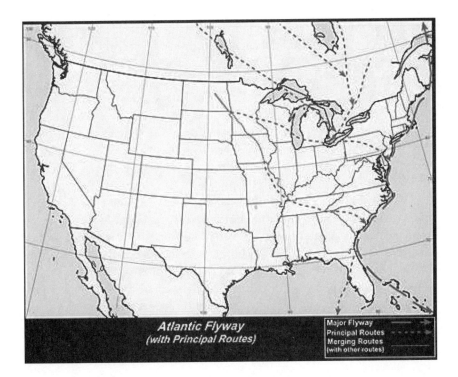

Atlantic Flyway
(with Principal Routes)

Major Flyway
Principal Routes
Merging Routes
(with other routes)

Plum Island breeders like Canadian geese, osprey, and seagulls nest on the island and make local trips to Connecticut and Long Island. Hundreds of thousands more, in all shapes and sizes, also gather on Plum Island, resting before crossing the Sound in the spring and flying along the Connecticut River valley toward Maine and Nova Scotia. Take swallows, for example. "One day you'd see a couple on the telephone wire," says Carol, "and then ten the next day, and then all of the sudden you'd see a hundred, then nothing. They'll wait until they get this critical mass, then go." Exactly where they settle is up to wind conditions at the time, which is often their first landfall.

That first landfall is Long Island and the Hamptons to the south and west, and coastal Connecticut, including Old Lyme, to the north. When the birds aren't migrating, they constantly travel locally between Long Island, Plum Island, and Connecticut for food and companionship. Thanks to all of this bird activity, Plum Island presents more vectors for the spread of infectious disease spread than perhaps anywhere else. Ticks have a long and varied menu: droves of small foraging birds (ticks find baby chicks irresistible), a tantalizing wild deer habitat, and thousands of mice and rats for tick larvae and nymphs to feed on. Plum Island is a Lyme disease tinderbox.

Because the wildlife vectors are beyond control, safety at Plum Island has to be controlled from the inside out. The only way, then, for Plum Island to be 100 percent fail-safe is to keep the biological blood samples and germs dormant, sealed and padlocked inside laboratory freezers.

But for the lab to do its work, these same germs have to be taken out, thawed, and uncorked on lab benches. And that's where the vulnerability begins, as Dr. Moon pointed out.

"Let's face it," Plum Island scientist Dr. Douglas Gregg once said to a reporter, "there can be no absolute guarantee of securing the island."

The Theories. Attempts by the scientific community to explain the origin of Lyme disease are far from convincing. One popular theory holds that ticks always had *Bb* bacteria—germs similar to *Bb* existed in Europe and Asia for three hundred years—and infections are the result of the human-altered habitat in which the pests live. A century ago, goes the theory, there were far fewer woods in the United States, and deer were near extinct. Because of the modern conservation movement, forests replaced farmlands and populations of deer, birds, and small animals surged. In this environment, ticks multiplied. Suburban developments did the same. As Dr. Ralph Tierno reasons, "By the mid-1970s, human beings' collective behavior had created circumstances that so favored the spread of Lyme disease that sooner or later it was bound to attract people's notice and demand a protective response. . . . [W]e made our own sickbed and then we had to lie in it."

This theory assumes that Lyme disease was a gradual problem that "attracted attention." Nothing is further from the truth. What occurred in Old Lyme in 1975 was the outbreak of an unknown illness, concentrated within a defined geographic location that infected thirty-nine children and twelve adults. It was a modest epidemic. Old Lyme's outbreak was a footprint of something that had *deposited* itself there and festered. Lyme disease cannot simply be ascribed to poor land-use patterns, when ten miles south of Old Lyme lies an untamed island teeming with ticks, birds, deer, and mice, hosting two high-hazard germ laboratories proven to be anything but reliable in containing foreign germs.

White-tailed deer often swim across Plum Gut, the two-mile-wide strait that separates Plum Island from Long Island. Countless birds—including seagulls, Canadian geese, and osprey—fly between coastal Connecticut, Long Island, and Plum Island. Old Lyme lies directly in the flight vector of birds that congregate on Plum Island and migrate north and south along the East Coast. When biological security was taken seriously in the early 1950s, deer were shot on sight by trained snipers. Even puppies and dogs, fatefully setting their paws on the island's beaches with their owners, would be euthanized. By 1975, germs on Plum Island increased in both numbers and virulence—but safety and security measures moved in the opposite direction.

Tests were supposed to be held in airtight laboratory rooms. Instead, internal government documents prove there were gaping holes in the lab roofs where air currents and insects freely came and went, depending upon the direction of the wind. What's more, the animals were held in outdoor pens, where they were injected with virus vaccines and fed out of open-air feeding troughs. Plum Island workers witnessed birds flying in and out of the pens, picking morsels from the troughs. Wild animals were shooed away, but not before birds swooped down and mingled with the test animals. One eyewitness reported seeing deer entering the animal pens to feed.

If Dr. Traub continued his outdoor germ experiments with the Army and experimented with ticks outdoors, the ticks would have made contact with mice, deer, and more than 140 species of wild birds known to frequent and nest on Plum Island. The birds spread their toxic cargo to resting and nesting perches atop the great elms and oaks of Old Lyme and elsewhere, just like they spread the West Nile virus throughout the United States.

Researchers trying to prove that Lyme disease existed before 1975 claim to have isolated *Bb* in ticks collected on nearby Shelter Island and Long Island *in the late 1940s*. That timing coincides with both Erich Traub's arrival in the United States on Project PAPERCLIP *and* the Army's selection of Plum Island as its offshore biological warfare laboratory.

The USDA's spokesperson, Sandy Miller Hays, is unconvinced about the possibility of a link between Lyme disease and Plum Island:

Lyme disease—well, the positive agent for Lyme disease was identified in 1948, which was about six years before Plum Island came into being. So then some people blame the Army. There is a part of me that says, "Let me get this straight, the U.S. Army, that had saved the world, came along after the war and said, 'Let's poison a bunch of ticks and turn them loose on people in Connecticut?'"

We kind of giggle around here about the stories, but some of them are just outlandish. I always laugh about Nazi scientists....

Do you want to hear about how [scientists] are keeping a cow from drooling or do you want to hear about poisoned ticks and the Nazi scientists? It's always more fun to tell those scary stories.

A PR expert, Hays had *Scientific American* eating out of her hand in June 2000, when they reported her as saying, "'We still get asked about the Nazi scientists,' ... [with] the slightest trace of weariness creeping into her voice." In their feature story on Plum Island, the prestigious magazine dubbed the intrigue surrounding the island as a "fanciful fictional tapestry."

But as much as Ms. Hays and *Scientific American* might like to laugh or shrug it off, hard facts are indeed facts: the Army and the USDA conducted numerous outdoor biological warfare experiments within United States borders; the Army and the USDA were cooperating in a germ warfare laboratory built on Plum Island; the U.S. recruited the key architect of Nazi Germany's germ warfare program who worked directly for Heinrich Himmler; after Fort Detrick and the CIA interrogated him, the Nazi scientist developed the idea to build Plum Island, modeled after his own germ warfare lab on Insel Riems; the USDA borrowed this Nazi scientist to work in its Washington, D.C.–area laboratories; and this very Nazi scientist is now confirmed to have been on Plum Island on at least three occasions.

These aren't "fictional tapestries" or "scary stories"—they are scary facts from which conclusions can and should be drawn.

While the Army and the USDA are quick to deny the Plum Island tick experiments ever occurred, every few years the public learns of a top-secret germ warfare test whose existence the U.S. government had long denied. Consider this 2002 Pentagon disclosure in the *New York Times* about a 1964 test:

The Defense Department sprayed live nerve and biological agents on ships and sailors in cold-war era experiments to test the Navy's vulnerability to toxic warfare, the Pentagon said today. [S]ix tests were carried out ... [h]undreds of sailors were exposed to poisons ... in some of the experiments, known as Project SHIPBOARD HAZARD, or SHAD. Of the six tests, three used sarin, a nerve agent, or VX, a nerve gas [agents used by Iraq's Saddam Hussein against his own people];

one used staphylococcal enterotoxin B, known as SEB, a biological toxin; one used a simulant [*Serratia marcescens*] believed to be harmless but subsequently found to be dangerous . . . 4,300 military personnel [were] identified as participants in Project SHAD.

PLAYING WITH FIRE

Former Plum Island Director Dr. Jerry Callis says the association between Lyme disease and Plum Island is absurd. "Not now or ever had we anything to do with Lyme disease," he says in measured impatience. "It's existed in Europe for years—I've had it three times." Asked if Plum Island ever worked with ticks on Plum Island, Dr. Callis gave a surprising answer.

"Plum Island experimented with ticks," he says, adding, "but never outside of containment. We had a tick colony, where you take them and feed them on the virus, and breed the ticks to see how many generations it would last, on and on, until it's diluted. Recently, they reinstated the tick colony."

Tick *colony*?

Journalist Karl Grossman pressed a Plum Island lab chief some years ago about John Loftus's claims in *The Belarus Secret*. "My impression is that there is no truth to this," said Dr. Charles Mebus. But like director Harley Moon, Mebus had not been there in the 1950s to form an "impression." He did tell the roving reporter what he knew firsthand: Plum Island previously worked on—and continued to work on—tick experiments on "soft ticks" that transmitted heartwater, bluetongue, and African swine fever viruses, but aren't normally known for spreading the *Bb* bacteria. But that wasn't the complete picture.

The lab chief failed to mention that Plum Island also worked on "hard ticks," a crucial distinction. A long overlooked document, obtained from the files of an investigation by the office of former Long Island Congressman Thomas Downey, sheds new light on the second, more damning connection to Lyme disease. A USDA 1978 internal research document titled "African Swine Fever" notes that in 1975 and 1976, contemporaneous with the strange outbreak in Old Lyme, Connecticut, "the adult and nymphal stages of *Abylomma americanum* and *Abylomma cajunense* were found to be incapable of harboring and transmitting African swine fever virus." In laymen's terms, Plum Island was experimenting with the Lone Star tick and the Cayenne tick—feeding them on viruses and testing them on pigs—during the ground zero year of Lyme disease. They did not transmit African swine fever to pigs, said the document, but they might have transmitted *Bb* to researchers or to the island's vectors. The Lone Star tick, named after the white star on the back of the female, is a hard tick; along with its cousin, the deer tick, it is a culprit in the spread of Lyme disease. Interestingly, at that time, the Lone Star tick's habitat was confined to Texas. Today, however, it is endemic throughout New York, Connecticut, and New Jersey. And no one can really explain how it migrated all the way from Texas.

Entomologist Dr. Richard Endris joined Plum Island in 1981 to spearhead increased tick research. Endris and the African swine fever team leader, Dr. William Hess, went to Cameroon and other parts of Africa on tick-hunting safaris. They stuck their arms deep inside burrows and were occasionally bitten by snakes and rats. They searched out wild warthog burrows in the brush using a "tick sucker"—a reversed leaf-blower with attached sieves and filters—to strain hundreds of tick specimens out of the moist sand. They set out little blocks of dry ice thirty feet apart and watched the ticks march to the smoking lumps (the carbon dioxide attracted them, fooling the ticks into thinking it was the exhale of mammalian hosts). Endris constructed two high-hazard "insectories," insect labs, one in the back corner of Laboratory 101 and another in the basement. Each insectory was equipped with sand-filled climate chamber incubators with lighting that simulated photoperiods, protective rims around the airlock doors covered in sticky glue, and seals across all windows and drains. The ticks were fed on the blood of hairless suckling baby mice, where they would attach and molt and breed. All told, he reared over 200,000 hard and soft ticks of multiple species.[7] Endris handled the tiny parasites with extreme care, using fine art brushes to move the minute nymphs into a transfer container—a urine specimen cup and a screen, glued together with globs of plaster of Paris. To test the ticks, the scientists first anesthetized diseased pigs, goats, mice, and calves, then placed the ticks on the sleeping animals. The ticks immediately attached and dug their mouth parts in. After a few hours of feeding, technicians detached the ticks with soft-tip forceps.

Endris set up quite an impressive tick colony when he arrived in 1981. But there was substantial earlier tick research. Dr. Hess's tick experiments in the 1960s and '70s with the Lone Star tick and others were conducted in unsafe conditions. Endris said the early tick research wasn't focused: "Plum Island was not set up to deal with ticks at all." In 1980, a Plum Island scientific oversight committee urged the USDA to hire "an *appropriately* trained medical entomologist," calling it a *"priority* item." The consultants also "strongly recommended" the construction of "a *modern, approved* insectory be undertaken for future research."[8] The advisers had serious concerns

---

[7]Dr. Endris also conducted experiments with sand flies on Plum Island in 1987 to test transmission of leishmaniasis, a bacterial ailment that if left untreated, has a human mortality rate of almost 100 percent. It is characterized by irregular bouts of fever, substantial weight loss, and swelling of the spleen and liver. The work was performed under contract for Fort Detrick, and serves as another example of a deadly germ warfare agent worked on at Plum Island for the Army, with no public knowledge or public safety precautions taken.

[8]Italics are added.

with the primitive tick colony then in operation under the veterinarian Dr. Hess, who had been with Plum Island since 1953.

Dr. Endris and his boss, Hess, were both fired in 1988 by incoming director Dr. Roger Breeze, who promptly closed down their precious tick labs. They wrapped up research, put the viruses back in the freezers, and dumped the ticks into the autoclave, which steamed them at over 100 degrees centigrade.[9] Endris, who went to work for Merck Pharmaceutical, scoffs at a Plum Island–Lyme disease connection. "Those kind of comments . . . indicate a gross ignorance of Lyme disease." Before being fired by Dr. Breeze, Endris served as the scientific member of Southampton's Joint Lyme Disease Task Force, and says with conviction he never heard of any Lyme disease relationship.

But Dr. Endris wasn't on Plum Island in 1975; his entomology expertise and the "modern, approved" tick insectory he built were a full six years away. Unfortunately, Dr. Hess, who could shed light on the old tick experiments, died in 1999 in New Hampshire. It is clear, though, that he was proud of and cherished his thirty-five-year scientific career there—his family scattered his ashes in Plum Gut among the trade winds of Plum Island.

Dr. Garth Nicolson, a national expert on immune system disorders, isn't satisfied with the ecological Lyme disease theory. "There's a high possibility," says Nicolson, who runs a California medical institute and has testified before Congress on Gulf War Syndrome, "that Lyme disease is a combination of infectious agents let loose from a laboratory, possibly from Plum Island by birds to the coast, causing multiple infections." Nicolson contends *Bb* is often found in tandem with mycoplasma bacteria, which causes many of Lyme's debilitating symptoms. Mycoplasmas found in foreign countries were studied on Plum Island since its inception; they may have been cross-contaminated with *Bb* and escaped the lab in the 1970s.

Dr. Wally Burgdorfer, who discovered the Lyme disease bacteria that bears his name, says, "The big question is where the ticks came from." He believes that imported deer from Europe brought the deer tick species, and with it the bacteria, to America, where all three proliferated. I ask Dr. Burgdorfer about the Lyme disease connection to Plum Island. "Touching on something like that may cause a hell of a lot of problems," he says. "You have to show a development in the 1960s and 1970s, and it seems impossible—

"*Unless*," he continues, "they cultivated the tick species on Plum Island, and unknowingly fed some ticks on animals or humans and a *Borrelia* spirochete [bacteria] accrued." Dr. Wally Burgdorfer isn't ready to prove

---

[9]"A good tick will last about one minute," says Endris. "Any living biological protein will coagulate at that temperature."

the link, but he's quick to point to the proven track record that helps make the case: "Plum Island is proof of the existence of breaks in biological safety." And of a proposed biosafety-level-four upgrade at Plum Island, the most dangerous, he admits, "Even if it's biosafety level four [the highest containment level], that doesn't mean it's safe."

The tidy, conventional thinking on Lyme disease overlooks what appears to be an inescapable truth: wild birds and deer contracted *Bb* from ticks impregnated with a myriad of exotic germs studied on Plum Island in helter-skelter, haphazard conditions. In theory, all it took was the single infected Lone Star tick that escaped from the laboratory, crawled up a blade of island grass, and dug its mouth parts into a small bird that passed by. Thought to be a mere nuisance and now carriers of disease, bird migration and swimming-deer traffic cycled ticks carrying *Bb* throughout mainland Connecticut and Long Island. The birds and deer (themselves now carrying *Bb* in their bloodstream) were bitten by more ticks on the mainland, including the ubiquitous deer tick, which in turn passed it along to more mice, birds, and deer. From this breeding ground the Lyme disease spiraled exponentially out of control, with over 150,000 documented cases to date, and tens of thousands going unreported.

Where *Bb* came from is as important as where it's gone. It exists in two-thirds of all ticks found in the eastern United States, and the disease has appeared in forty-five states and the District of Columbia. Thirteen thousand new cases are diagnosed each year. Untold numbers of infections go undetected, because the red bull's-eye ring—the infamous signature of Lyme disease—appears in only 60 to 80 percent of all cases. Some infections appear as red, orange, or purple rashes; some are oval in shape, others triangular, and still others are horizontal. Sometimes *Bb* infection doesn't even cause a rash.

At any rate, the apparent epicenter of Lyme disease seems dubiously close, too coincidentally close to Plum Island, a place that has raised far too many questions for the *Bb* link to be dismissed along with the three-headed chicken and five-legged cow.

You know, the worst, meanest, nastiest, ticks in the world are poli-*ticks*," says Steve Nostrum. He knows that Lyme disease is a hot potato few want to handle, especially when dealing with a secret island in the middle of the most profitable summer tourism spot and coveted real estate in America. Nostrum tried unsuccessfully for over a decade to get coverage in local newspapers warning people to take precautions. A reporter once took him aside and told him, "You want to know, Steve, why you can't get anything published in the area papers? Well, off the record, our major advertisers are telling us, 'If you run one more story on Lyme disease and ticks, we're pulling our ads.'" After all, fear of Lyme disease is bad business, and can devastate an economy that thrives off the fat wallets of summer vacationers. Nostrum's son came home from high school one day with a letter telling parents they were going to teach students about the AIDS virus. He called up the principal and said, "That's wonderful that you're doing this—but the chances of my son getting AIDS are pretty slim. Meanwhile, you have eighteen students in your school with IVs in their arms at this moment suffering from Lyme disease." He gave the school "several thousands" of brochures on Lyme disease, and how to protect against being bit by a tick. Not one pamphlet was handed out. "We live in an area that is endemic of Lyme disease, and there's not one brochure out here," says Nostrum. "What is wrong with this picture?"

Thanks to advocates like Nostrum, the public has become aware of the debilitating ailment and how best to prevent it. He was the first to reject GlaxoSmithKline's much-ballyhooed LYMErix, a genetically engineered Lyme disease vaccine that was pulled off the shelves in February 2002.[10] Nostrum isn't fully cured, either—you never can be. When he's run down, fatigue and deep joint pain set in. He'll never sing and play instruments again like he did back in the 1980s, when his quartet won a third place award from the Society for the Preservation and Encouragement of Barber Shop Quartet Singing in America. When he feels up to it, he spends evenings surf casting with his son, or stargazing from the telescope on his back porch. Most of all he likes to drive his car—New York State license plate MR LYME (his wife's, MRS LYME)—over to the beach and quietly walk the rocky shore, collecting sandglass and seashells.

He watches flocks of birds fly overhead, he listens to their faint squawks high up in the sky, gliding over the expanse of the Long Island Sound, toward Connecticut—and toward the epicenter of the epidemic of the strange bacteria that has claimed the better part of his life.

And he wonders.

---

[10]Though manufacturer GlaxoSmithKline generated $40 million in sales during its inaugural year, they cited poor sales when pulling it off the market. Some critics said it caused Lyme disease–like side effects such as arthritis and muscle pain, and created a false sense of security—it only reached 80 percent effectiveness after three doses were administered. It was also unsafe for children under fifteen years of age.

2

1999: East End Meets West Nile

> *When man domesticated certain lower animals for his personal comfort and gain, he assumed the obvious hazard of sharing their diseases.*
> —Dr. William S. Middelton (1956)

Birds were losing their minds at the Bronx Zoo. Some flew in perpetual circles. Others died in their cages. Veterinarian in charge Dr. Tracey McNamara was concerned about the twenty-four birds—ducks, owls, a bald eagle, a black-crowned heron, and magpies—suffering in perhaps the most famous zoo in the world in August 1999.

McNamara had recently attended Plum Island's foreign animal disease school, where vets were taught how to diagnose exotic disease outbreaks and respond. But when she dialed the emergency telephone number set up to report possible foreign animal diseases, she found the line disconnected. The program was presided over by new Plum Island acting director Dr. Lee Ann Thomas, who had recently replaced director Dr. Alfonso Torres, who had replaced director Dr. Harley Moon when he left to return to Iowa in 1996, who replaced director Dr. Roger Breeze. Since Dr. Breeze's departure, there had been no continuity in the laboratory's leadership. According to sources, no one wanted the job because Breeze pulled the lab's strings from afar.

In the last few days of that August, Dr. Deborah Asnis, a specialist in infectious disease at a small Queens hospital, noticed an unusual pattern. Two and then four hospital patients, all elderly, contracted fevers with headaches, muscle weakness, and mental ailments that progressed into comas. A fifth patient came in a few days later with the same symptoms, and neighboring hospitals admitted three more strikingly similar cases. Each victim had recently spent time outdoors in the evenings, and lived near Flushing Meadows–Corona Park, a marshy mosquito breeding ground on the Long Island Sound. Testing blood samples and fluid from spinal taps, the Centers for Disease Control (CDC) diagnosed the ailment as St. Louis encephalitis (SLE), a malady caused by a domestic arbovirus (a virus transmitted by airborne insects) found along the Ohio and Mississippi rivers. The pesky mosquito was suspected as the culprit. New York City immediately launched a massive $6 million aerial and ground pesticide campaign, spraying the pesticide malathion (a potent neurotoxin carcinogen) and distributing over 300,000 cans of DEET chemical insect repellant to city firehouses, which spawned a public fright of its own.

The determined forty-five-year-old McNamara kept calling other numbers until she reached someone, and sent her bird tissue samples to the USDA's domestic animal disease laboratory in Ames, Iowa. A trained pathologist, McNamara knew that SLE would attack the zoo's chickens; finding the chickens healthy, she suspected a different germ was sickening both her patients and Dr. Asnis's human ones. She called the CDC and told them about her freezer full of dead birds and about a possible animal–human link. A CDC scientist brushed her off, offering to send samples. "You're just dealing with some veterinary thing," he uttered with contempt. McNamara later told Madeline Drexler, author of *Secret Agents: The Menace of Emerging Infections*, that she felt the CDC treated her like a "dingbat, premenopausal female veterinarian in New York City." McNamara sent the CDC the virus samples anyway—infected tissues of a snowy owl, several rare Chilean flamingos, and a cormorant—but the agency paid them no mind.

McNamara then contacted the Army's germ labs at Fort Detrick. They requested samples be sent right away. Because of her persistence, Fort Detrick was able to determine that the germ from bird and human samples was the same virus—the "West Nile virus," a microbe never before seen in the Western Hemisphere.

The culprit spreading the virus was the same one that transmitted the Rift Valley fever virus and so many other deadly germs—the wily mosquito.

The public reacted with hysteria. Parents kept their children indoors. Television reporters blitzed viewers with neighborhood chemical spraying alerts and tips on avoiding the virus, which seemed to be affecting the elderly, young children, and those with weakened immune systems, such as patients undergoing cancer treatments. Officials in Greenwich, Connecticut, announced a 5:00 p.m. curfew on all outdoor activities. A story in *The New Yorker* by Richard Preston, author of *The Hot Zone*, the best-seller about the feared Ebola virus, helped stir the pot. He revealed that an Iraqi defector (Saddam Hussein's body double) told sources that West Nile virus strains were part of Iraq's biological warfare program. Worse, the virus strain isolated by Fort Detrick was found to be similar to strains held by the Russians. "It is really an epidemic," said a doctor from Mount Sinai Hospital in Manhattan, a few weeks after the initial diagnosis. "And this outbreak is still growing." By year's end, sixty-two people had been confirmed infected and five more died, raising the death toll to seven.

First discovered amid the swampy banks of the Nile River in Uganda in 1937, West Nile virus has increased its dominion far and wide in America, reaching forty-three states (as far west as Montana) via birds and mosquitoes. The casualties had shrunk to twenty-one infections and two deaths in 2000, but that was deceptive. Spreading with newfound fury in 2002, West Nile boasted 4,156 confirmed cases and 284 deaths in the United States. There were 329 cases in Louisiana alone, where the mosquito is half jokingly referred to as the state bird. Illinois topped all states with 884 confirmed infections and 64 deaths. An all-out chemical pesticide attack was waged from airplanes, pickup trucks, and handheld sprayers hopscotching from backyard to backyard.

News reporters interviewed doctors, who sought to play down the threat. During one recent *NBC Nightly News* segment, a doctor stressed that "only those over sixty are at risk." Another state health official said, "The chances of being infected are very, very minimal.... It's certainly nothing to be alarmed about." But there appears to be an increasing number of infections in children, and West Nile virus strains are now attacking younger adults—such as a fifty-three-year-old otherwise healthy man. However, even if the scourge preyed only upon senior citizens, that still amounts to some thirty-two million men and women, one out of every eight Americans. And they aren't just numbers—they are people's parents, children's grandparents, and America's "Greatest Generation." They are people like seventy-two-year-old Ernest Hunt from Louisiana, who succumbed three weeks after being bit by a mosquito while he and his wife, Becky, were enjoying a lakefront Fourth of July barbecue with family and friends. It is now a risk for people to tend to their gardens and take evening walks during the summer months, because the West Nile virus is lurking—everywhere.

By Labor Day 2003, over 5,000 human infections had been reported, and 95 deaths. The CDC predicted another 100 people would die by the end of the year. It is now estimated that since the initial August 1999 outbreak, 200,000 people have been exposed to the West Nile virus. The death toll is approaching 400 and rising. There is no known cure or human vaccine.

est Nile virus struck New York City in August 1999 because of an increase in international trade and travel, the scientists say. The same scientists who chalk up Lyme disease to changes in human land-use patterns say that West Nile virus came to the United States in the passenger cabin of a commercial airliner. One doctor lays out a scenario whereby mosquitoes board an 1998 El Al flight from Israel (where scientists say a similar strain existed in 1998) and bite passengers on board. After landing at John F. Kennedy International Airport in southern Queens, New York, some of the bitten passengers go home to the Flushing area in northern Queens, where they are bitten again by domestic mosquitoes, which then carry off the virus and lay eggs. Mosquitoes hitch a ride from the airport to Flushing, goes the theory. The marshy inlets of the Long Island Sound provide an optimal breeding area, and female mosquitoes pass along the virus to millions of their offspring for the 1999 mosquito season. This scenario is certainly plausible, but it is far from proven—and there is no scientific support to back it up. "Unless there's a molecular signature," notes one scientist, "you can't tell how it came in—you just can't."

Though a few researchers rule out the possibility that the West Nile virus outbreak was a bioterrorism attack waged by an unknown enemy, most likely Iraq, others aren't so sure. "Of course it could have been easily intentionally introduced," says one scientist familiar with bioterrorism. "Saddam Hussein had worked on it and threatened us with it. All someone had to do was come into JFK Airport with a inconspicuously small bottle of twenty or so infected mosquitoes, and go to the Bronx Zoo or to Flushing Bay and let them loose." The scientist rejects the conventional airplane theory because he says not one of the infected persons in 1999 was an airline passenger arriving from a foreign country.

Other than hypotheticals like these, the scientific community is at a loss to identify the origin of the 1999 West Nile virus outbreak.

For all their postulating, none of the seasoned virus hunters looked to New York–area germ laboratories. A spokesperson for the New York Department of Health said frankly, "It has not been on the top of our to-do list." The director of the CDC, James Hughes, said as interesting as the question of how was to entertain, "the answer may remain elusive—Mother Nature does not always reveal her secrets."

But was it Mother Nature's secret to keep?

he West Nile outbreak brings to light the kinship between animal and human virus diseases. It illustrates how zoonotic viruses like West Nile fever, Rift Valley fever, Ebola fever, anthrax, and influenza move easily between animals and humans to achieve devastating results. For example, horses, like humans, are particularly susceptible to West Nile fever virus. What the public doesn't know is that *at the same exact time* that humans in Queens were dying, dead horses were being quietly carted off Long Island farms to Plum Island. Where the El Al commercial jet scenario falls short is where the plight of the horses begins.

While Dr. McNamara tended to her Bronx Zoo bird flock and Dr. Asnis cared for her Queens patients, horse farm owners seventy-five miles east were placing frantic phone calls to Dr. John Andresen at the Mattituck-Laurel Veterinary Hospital. Owners of thirteen North Fork farms on the east end of Long Island phoned and each call sounded the same: their cherished horses were losing all sense of motor coordination. The horses' hindlegs were buckling, and they were stumbling, neighing, twitching, and convulsing. "It was a mystery," recalled Andresen, who normally received four horse-related phone calls in a year—not thirteen in a week. Going out to investigate, he found "neurological cases, which are uncommon in an equine practice." One horse had collapsed in the stable, thrashing, unable to right himself on his legs. "It was in bad shape," Dr. Andresen recalled. The poor stallion expired before Andresen could give its owner his prognosis. Alarmed by the rife similarities, he contacted New York State's head veterinarian, Dr. John Huntley, who in turn called in the USDA's emergency response team. It wasn't the first time this virus SWAT team had descended upon Long Island's North Fork; an earlier team spearheaded an investigation into a 1978 virus outbreak on Plum Island.

By August 26—just three days after Dr. Asnis phoned in her alert to the New York City Health Department—a strange disease was confirmed in eighteen cases on thirteen horse farms, all in the North Fork hamlets of Riverhead, Jamesport, and Mattituck. Ten horses died, either on their own or by lethal injection to ease their misery. Expert epidemiologists flew in from Kentucky, Iowa, and Wisconsin, and met with Dr. Andresen to get the lay of the land. The team collected samples from 146 horses in the area and found that an alarming 25 percent of them tested positive for West Nile virus. Each farm visited had large pools of standing water in watering areas, ripe for mosquito transmission.

All of the infections occurred in a five-mile radius. The epicenter of these mysterious horse deaths was less than twenty miles away from a faulty exotic virus disease laboratory.

Of all the counties polled in the New York City metropolitan area in 1999, Suffolk County's dead bird and infected mosquito counts were among the highest recorded. The story of the horses raises even greater suspicions. Infection records from the last few years indicate how West Nile virus spread. In 2000, there were 60 cases and 20 deaths in seven states. In 2001, 738 confirmed cases and 156 deaths occurred in 20 states. In the out-breal year, 1999, records charted all equines in New York, Connecticut, and Maryland (the range of the West Nile virus outbreak by the end of its first year) into account—horses, ponies, mules, burros, and donkeys. Strikingly, none of the 271,000 equines tested positive for the West Nile virus, with one exception—Long Island's North Fork.[1] The simultaneous equine outbreak of this zoonotic virus, with the human outbreak occurring over seventy-five miles away, reveals that the ground zero of the West Nile virus outbreak was the North Fork peninsula opposite Plum Island, not JFK International Airport in New York City.

Many of the stable owners bred and sold horses to make a living, so they were glad to keep the exotic virus infections in their stables a secret. As the emergency team finished its work that August, the USDA quietly gathered the horse carcasses from the stables. From the eighteen horses, they gathered pints of blood, spinal tap fluid, even whole brain hemispheres, and carefully whisked them to the Plum Island ferry. On Plum Island, a pathologist performed a necropsy (animal autopsy) on the carcasses, and the brain tissues were sliced up and examined under the microscope. Clinical signs were recorded; samples apportioned, sealed, labeled, and refrigerated; and the horse remains went down the chute to the incinerator charging room.

If Plum Island was uninterested in West Nile virus in the past, it seems to have acquired a taste for the germ. In October 1999, on the heels of the outbreak, Lab 101 held four healthy *Equus caballus*, the common domestic horse. They were given intravenous injections of a West Nile virus strain obtained from Dr. McNamara's Bronx Zoo collection of frozen infected birds. Each day, as the horses progressively sickened, animal handlers examined them. The scientists noted that the virus was not detectable in the blood until thirty days postinfection, but occurred within forty-five days; all four succumbed. Autopsies were performed, and four new "horse adapted" virus strains were extracted for the Plum Island virus library. If West Nile virus is not detectable in horses for thirty days, that means the North Fork horses became infected with the virus bug in July of 1999, and perhaps earlier.

This all begs the question: did Plum Island have West Nile fever virus at the time of the outbreak? Dr. Robert Shope's Yale Arbovirus Research Unit (YARU) across Long Island Sound held twenty-seven different strains of West Nile virus in its New Haven, Connecticut, freezers until 1995, when

---

[1] Two horses that tested positive for West Nile virus were stabled at Belmont Park racetrack in Nassau County, Long Island, west of the North Fork, which lies in Suffolk County. These horses were purportedly shipped from Suffolk.

he moved to the University of Texas and took his strains with him. YARU and Plum Island often trafficked in viruses, most notably the dangerous Rift Valley fever virus in 1977. Had Dr. Shope shared West Nile virus reference samples with his friend Plum Island director Dr. Roger Breeze—the island laboratory being the only official location where foreign animal germs like West Nile virus are supposed to be studied? When I ask him, former Plum Island director Jerry Callis says he didn't think it was in the virus repository in August 1999. "I never thought it was important enough. It's more of a mild virus, not among the serious animal virus diseases," he says, even though it appears fatal to birds. But Callis had left Plum Island in 1987 when Dr. Breeze came in. USDA official Wilda Martinez confirmed Dr. Callis's beliefs when she assured the public that West Nile wasn't studied there prior to the outbreak—but she declined to say whether the virus was in their freezers.

Former Plum Island scientist Jim House, a Cornell classmate of Dr. Andresen, thinks West Nile samples existed prior to 1999. "There were samples there, and it wasn't answered clearly to the public. They didn't honestly tell how many samples they had and that's when people started to get upset." It is certainly difficult to believe that the facility boasting the world's largest collection of animal viruses would not have the West Nile fever virus, especially given the fact that right across Long Island Sound, YARU had no less than twenty-seven strains of the germ. Too, the North Fork horse epidemic is a footprint; and the cluster of horse stables, a most inviting first rest stop for mosquitoes and birds carrying the virus and working their way west from Plum Island, down the narrow strip of the North Fork. The USDA dispatched an emergency team to Long Island to examine how many horses were infected and on which horse farms. Yet they never turned their magnifying glasses on themselves. Had the fox been guarding the henhouse?

Martinez told the *New York Post* in 1999 that "top security [at Plum Island] does not mean top-secret." But my requests under the federal Freedom of Information Act for a catalog of germs contained in the Plum Island virus library went denied on national security grounds.

3

1967: The Demise of the Ducks

> *You see, we do have an impact.*
> —Plum Island assistant director

West Nile virus and Lyme disease aren't the only suspicious germ outbreaks where Plum Island is concerned. There's also the demise of the Long Island duck industry.

In 1873, New York merchants Ed McGrath and James Palmer imported white Pekin ducks from China for American food consumption. The winding creeks, kettle-hole ponds, and sandy soils of eastern Long Island were prime for duck farms. By the turn of the century, no less than thirty separate commercial flocks were raised on Long Island. At the industry's height, eight million savory Pekins were consumed in the United States annually—and Long Island produced six and a half million of them. Pekin were reared on over seventy Long Island farms, "picked" at their necks with knives, plucked, packed six to a box, and stored in walk-in freezers to await shipment. Roasted, marinated "Long Island duck" recipes became a national fancy. The name endures to this day, but the Pekin ducks do not.

The thriving Long Island duck industry was dealt a sudden fatal blow in the winter of 1967. Scores of white breeders from three months to two years old died on three adjacent farms; other ducks were trembling, becoming listless, and unable to stand upright. Instead of gaily flapping and quacking, flocks quietly sat on the ground, wings outstretched and heads down, clinical signs of physical weakness and depression. Whatever the ailment, it was highly contagious. It soon killed off well over half the flocks. The USDA swept in and took charge, ordering some six thousand ducks immediately poisoned and over fifty thousand eggs destroyed, costing local duck farmers millions of dollars. The infection was found to be the virus that caused Dutch duck plague (also called duck virus enteritis), an avian disease endemic to Holland, Belgium, India, and China. It was the first outbreak of this foreign animal disease within the United States. Two independent animal doctors who discovered the outbreak postulated how it occurred:

> The detection of the disease on the American Continent invites some explanation as to origin. Had the disease been present for some time and remained undetected? Was the disease imported . . . ? Was it introduced indirectly by traffic in infected material . . . ? The possibility seems remote that the disease had been present on Long Island for any length of time. In the long history of the intensive duck industry on Long Island, the disease had been neither suspected nor reported.

One farm employed Dutch farmworkers and entertained Dutch tourists, but they were dubious of that connection. "It is difficult or impossible to evaluate the importance of such traffic." They were at a loss to explain such an explosive virus infection. No evidence could be found that Plum Island scientists investigated the nearby site of the outbreak, or whether they investigated a possible link between the island laboratory and the outbreak. But the USDA pamphlet "DUCK VIRUS ENTERITIS: An old world disease . . . in the new world" maintained, "In spite of careful investigations, the source of the original U.S. outbreak has not been determined."

Three years later, the USDA announced a complete eradication of the virus. The department showered the Long Island duck farmers with awards for the important role they played in the campaign. The Plum Island virus laboratory, coincidentally nearby, had developed a vaccine for the duck farms to protect against further incursion of this exotic virus. But the plaudits and awards rang terribly hollow. The farmers had essentially participated in two eradications—that of the duck virus and of their own livelihood. With the farmers' flocks crippled by Dutch duck plague virus, Suffolk County—under strong pressure from Hamptons-area real estate interests—crushed the remaining farms. The majestic white ducks and their baby ducklings were now unwanted pests quacking and defecating in the ponds and creeks of the new Hamptons social set. Invoking "protection of the environment," the government closed duck farms, family farms that for a century were the backbone of Long Island's east end economy. "Prompt enforcement of control measures confined the outbreak to New York State," crowed the pithy USDA pamphlet.

The USDA was wrong about its ballyhooed disease campaign results. A prescient article, buried deep within the *New York Times* on Christmas Eve 1967, hinted at what the future held: FATAL VIRUS FOUND IN WILD DUCKS ON LI. The tiny newsbrief reported that sixty wild black ducks were found dead in Flanders Bay, at the head of Long Island's North and South forks. A Department of the Interior official admitted the discovery posed a "more dangerous" problem than the duck-farm outbreak. He was right. Wild waterfowl, whose travel is impossible to control, picked up the virus from the Long Island duck farms and spread it north and south along the North American flyway to geese, swans, blue-winged teals, and mallards. In 1970 and 1971, the virus infected birds throughout Pennsylvania and Maryland. By 1973, it reached the Lake Andes National Wildlife Refuge in faraway South Dakota, where forty thousand mallards and Canadian geese died and had to be collected, piled into a mammoth heap, and set ablaze. The infection reached the Gulf of Mexico and Canada by 1975.

The virus continues to spread to this day.

After the infection at the three farms, two local veterinarians brought bird carcasses and tissue specimens to Plum Island to confirm the infection by Dutch duck plague. Just as they had during the West Nile virus outbreak years later, Plum Island officials told reporters there was no possible way their laboratory caused this outbreak, because they did not study the virus until *after* the outbreak occurred. However, it was they who first conducted "exhaustive" laboratory tests and confirmed the outbreak was a foreign duck virus, so it's safe to say they had diagnostic agents to detect the virus and conceivably the virus itself in their stores. After all, they had to match the virus against something to make a positive identification. The results, said the Plum Island scientists of the new virus unleashed in their backyard, were "particularly interesting . . . in that they point out that duck plague virus infection is, in most probability, a new infection on Long Island. It appears that the virus was introduced from outside areas to the Suffolk County duck farms." Could there have been a laboratory leak on the order of the government-acknowledged virus outbreak that occurred there a decade later? As early as 1970, Plum Island was again working on a strain of duck virus, and this one had infected a man suffering from infectious hepatitis virus. "The implications," wrote the scientists of their research, "offer some unpleasant possibilities: first as a mode of introducing a serious avian disease into this country, and second as a possible source of new human viruses by recombination between [the duck virus] and human influenza."

Today, vast tract-home developments and million-dollar Hamptons estates rise above the fields where sixty-four family-owned duck farms once proudly stood. Meanwhile, a few small duck farms on Long Island are left, eking out a meager existence. There are two reminders of the once grand Long Island duck industry: a twenty-foot-tall concrete duck, built by a farmer in 1931 to sell his ducks and eggs on the roadside, that now sits guarding the entrance to a county park, and the Long Island Ducks, a minor league baseball team established in 2000, its untactful name highlighting an extinct part of New York State's heritage. It's hardly an exaggeration to say that nine out of ten Long Island Duck fans couldn't explain the origin of their team's name.

Like West Nile fever and Lyme disease, Dutch duck plague is an emerging disease in the United States that continues to spread with no end in sight. And like the other two, no one knows to this day how the duck virus entered into the United States and ravaged the duck farms of eastern Long Island.

Speaking of the vaccine they prepared for the collapsed Long Island duck industry in 1975—too late to be of any help to the farmers—Plum Island assistant director Dr. John Graves said to a reporter, "You see, we do have an impact."

Three infectious germs, *Bb*, West Nile virus, and duck enteritis virus— all foreign germs—have infiltrated the American landscape. All three emerged from the same geographic locus. All three occurred in the vicinity of a high-hazard, high-containment foreign germ laboratory with demonstrably faulty facilities and pitiable biological safety practices—flaws that caused proven germ outbreaks in the past, and infections among its employees. The public is asked to accept that none of these three outbreaks is connected to Plum Island.

That's what one calls blind faith.

As you read on, consider for yourself whether that assertion has merit.

PART 2

THE SAFEST LAB IN THE WORLD

Genesis

> *Plum Island will permit the Army Chemical Corps to execute required projects in connection with imported agents . . . that might become of Biological Warfare significance.*
> —DECLASSIFIED ARMY DOCUMENT (1951)

A few years before the Plum Island Animal Disease Center's dedication day in 1956, the United States launched its first biological warfare program. A glimpse into the past reveals a surprising truth: Plum Island wasn't exactly what it appeared to be to the public.

The gory details were kept secret at the time, but America's germ warfare goals—national defense—were heralded by the nation's leaders and press. A *New York Times* editorial in 1945 mused, "When the scientific story of the war is written, we have here an epic that rivals that of the atomic bomb." The paper was right. A few months after the United States demonstrated its atomic warfare prowess on Japan, it announced the development of a second weapon: killer microscopic germs. While forty-five people participated in the British biological warfare effort, the American version involved four thousand men and women. The Army ran the innocuously titled "War Research Service" (WRS) program at Fort Detrick in Frederick, Maryland. Civilian chemist George W. Merck directed the work, advised by scientists from the nation's top universities.

These men were motivated by the threats of the times. In ways, they were no different from the atomic fathers Einstein and Oppenheimer. Theodor Rosebury, a Columbia University microbiologist who worked on the WRS projects, said, "We resolved the ethical question just as other equally good men resolved the same question at Oak Ridge and Los Alamos."

The aims of the biological warfare program didn't trouble Dr. Albert Webb, a Fort Detrick scientist in the early years. "That aspect never worried me personally. People had been killing people for millennia. Whether you hit him over the head with a club, stab him with a spear, or give him a disease he might get anyway—let's not balk at that." One has to look at the whole picture and understand the enemy of the time, says Webb. "We knew other nations, Germany and Japan—*and Russia*—were working on this, and in self-defense, we had to know what the potential was. Maybe it's not a popular thought today, but I still feel that it was necessary."

Dr. Edwin Fred, Dr. Webb's ultimate boss and the patriarch of WRS, was a veterinarian, as were Merck's top aide, Colonel Arvo Thompson, Fort Detrick founder Dr. Ira Baldwin, and Dr. William Hagan (who, in addition to spawning Plum Island along with Erich Traub, helped Baldwin establish Fort Detrick). Like MDs, they were trained in infectious disease. But unlike their medical counterparts, the vets weren't ethically bound by an oath that began: "First, do no harm. . . ." Dr. Webb recalled the mind-sets of the two branches of medicine. "The MDs had this unresolved medical conflict—they weren't supposed to help kill people. Vets, I think, were much more ready than MDs," he says. "There's obviously not the same feelings about the death of their subjects. The mass killing of animals for food is an accepted part of our culture." Unlikely as it seems, the veterinarian was ideally suited for germ warfare research and development.

Before the attack on Pearl Harbor, a proud isolationism combined with the belief that biological warfare was pure fantasy left the United States woefully unprepared for—and fully exposed to—a real threat. Britain, Canada, Germany, Russia, Japan, and France had initiated germ warfare programs decades earlier. But America made up for the lost time. By war's end it built the largest and most advanced program of them all.

War Research Service launched "Project No. 1" in 1942. Dr. Hagan was chosen to take the lead. Dean of the Cornell University Veterinary College, Dr. Hagan was an expert on *Bacillus anthracis*, or anthrax, a disease of sheep and cattle. Also known as woolsorter's disease, because the germ occasionally infected people shearing wool off sheep, it was a rare human affliction—but an exceptionally lethal one. Merck thought it would make a superb bioweapon and commissioned it as WRS's first priority. Anthrax is virulent, but it carries a minimal threat of a boomerang because it is not contagious person to person.

Dr. Hagan tested many sample strains of anthrax (which WRS code-named "N") in a four-foot-tall glass apparatus called a vinegar tower.

Under the right conditions, anthrax rolls up into a ball and hibernates, or spores, and becomes resistant to threatening environmental conditions like cold temperatures. When returned to a hospitable environment, the hardy spores unfold and come back to life. Hagan found that Strain No. 99 sporulated and retained its high pathogenicity, or ability to spread disease. Dr. Hagan concentrated, purified, and dried Strain No. 99 into enough powder to make a biological bomb. In a small lab in Ithaca, New York, in 1943, Dr. Hagan created the most virulent, concentrated brand of anthrax on Earth. Anthrax became the most important biological agent developed by the American biological warfare program, and Hagan gained the dubious title of the father of weapons-grade anthrax. Late in the war, Great Britain requested samples of Hagan's anthrax, naming it "Hagan's Best."

After the war, Hagan became a driving force behind Plum Island's creation (Nazi germ warfare scientist Erich Traub would be the other). He used his clout with Congress, the Army, and the USDA to lead the charge for an island virus laboratory. He inspected Plum Island personally and lent his imprimatur to its selection. Upon its inauguration, Hagan bequeathed to the island twelve vials of "N," enough to kill about a million people, considering it takes between 4,500 and 8,000 organisms to cause an infection. To this day, Plum Island denies ever hosting anthrax or working with it, though a now-declassified catalog of deadly germs imported to Plum Island in the early 1950s clearly show that twelve vials of "N" have been kept in its freezers since the very beginning.

Presumably, Dr. Hagan believed strongly that his secret wartime research would contribute to a greater good. However, the consequences of "Hagan's Best"—particularly in light of the deadly 2001 anthrax attacks—call into question that belief.

At a full cabinet meeting at the White House on January 23, 1948, Agriculture Secretary Clinton Anderson briefed President Harry S Truman on the need for an exotic animal disease laboratory, and on the Mexican virus outbreak. Truman listened to Secretary Anderson and nodded. Two months later, the Soviet Union blockaded all roads, rivers, and rails from the American and British zones of Berlin, forcing food and supplies to be airlifted in. Then the Soviets detonated their first atomic bomb. America had a new enemy.

"Especially in view of present world unrest, and with biological warfare a distinct possibility," the USDA told Congress, an island research laboratory "would be a major asset in repelling or mitigating such danger." In April 1948, Congress passed Public Law 48-496, which established the framework for Plum Island:

> The Secretary of Agriculture is authorized to establish research labo-
> ratories . . . for research and study, in the United States or elsewhere,
> of foot-and-mouth disease and other animal diseases . . . provided
> that no live virus of foot-and-mouth disease may be introduced for
> any purpose into any part of the mainland of the United States except
> coastal islands separated therefrom by waters navigable for deep-
> water navigation and which shall not be connected with the mainland
> by any tunnel. . . .

USDA engineers and architects armed with magnifying glasses analyzed
maps of America's coastline, searching for the right place. Two weeks after
the law passed, the USDA found the perfect coastal island: Prudence
Island, off the coast of Rhode Island. The USDA had looked at many other
sites, including Fort Terry, Plum Island, a surplus island fortress of 840
acres off the east end of Long Island.

Meanwhile, the Army was separately fixing its sights on Fort Terry as a
germ warfare island laboratory. At the last minute, it abruptly canceled the
surplus sale of the island to Suffolk County and invited the USDA—which
watched in horror as wealthy Newporters, after forming the Anti-Prudence
Island Laboratory Committee, killed the aggies' plan for nearby Prudence
Island—to join them. Debating the need for Plum Island in the U.S. Senate,
New York's senators, Irving Ives and Herbert Lehman, demanded that a
provision in the law be included to protect the local New York community:

> [A]t a location to be selected by the Secretary of Agriculture after *full
> hearings* of which *reasonable public notice shall be given* to those
> who may reside within twenty-five miles from the island selected.
> [Italics added]

Senator Kenneth McKellar from Tennessee, the Appropriations Committee
chairman, expressed his understanding of the public hearings provision for
the record. "I think it is the legislative intent," he said, "that if a majority of
the people are opposed to it, then under no circumstances would the
Department [of Agriculture] establish the laboratory."

THE HEARINGS

As their "reasonable public notice" given, the USDA placed ads in the
newspapers one week before the Plum Island hearings. Despite the short
notice, 1,544 people objected through sixteen petitions, written statements,
and telegrams. Recorded opinions ran three to one against the laboratory.
An examination of the USDA's internal files reveals very few supported the
plan. The five local dairies urged "all civic-minded citizens to fight against

the menace to our community interest." The Greenport Oyster Growers' Protective Association said the pollution from its sewage would tarnish the "clarity and wholesomeness" of the oyster beds.[1] The Long Island Association, an esteemed body of business leaders, denounced the selection of *"Pest Island*, in the midst of a recreational area during a post-war suburban boom" and the manner in which the hearings were handled—"arbitrarily in conduct," with "too short a time given" for notice.

"Positively outrageous," cried one resident in the local newspaper, who added a foreboding message. "Nature has a way of doing things that scientists do not or cannot anticipate. Who knows whether some pollution of the waters may occur, or whether the disease may be carried by flies or mosquitoes." Another asked, "Why should the farmers, the fishermen, and the summer residents be placed in jeopardy for the sake of an industry unrelated to this area?" And yet another wrote, "The action of your department establishes a plague spot in our beautiful vacation land. After hoof-and-mouth disease—then what?"

Secretary of Agriculture Brannan "selected" Plum Island on July 28, 1952. "Views expressed . . . were divided concerning the location," he said, but everyone recognized the need for such a facility *somewhere*. The USDA's fuzzy logic held that 99 percent of the population fully supported the lab, since only 1 percent vocally objected. The timing and manner of the public hearings point to a far different calculation: 99 percent of the people were not at all aware of the USDA's laboratory plan or their public hearings. Those who knew of the plan vehemently opposed their government, and the government won by stifling public opinion and by keeping the true purpose for Plum Island secret.

In fact, the Army had awarded a secret construction bid to build a germ warfare lab on June 18, a full month *before* the selection hearings began. Worried locals' fears were justified: the hearings *were* rigged.

Edward L. Bernays, the father of public relations, once defined his art as the "engineering of public consent." On Plum Island, the USDA pulled off a PR masterpiece. "[I]t will become a research center known throughout the world . . . a source of income, employment, and a point of pride in the community," the USDA heralded in its hearings pamphlet. Over time, Plum Island would indeed become known, though perhaps not for the reasons they had hoped for.

---

[1] After the hearings, Doc Shahan would say the sewage "ought to enrich marine life because there's so much organic material in it." Known as Blue Point oysters (after Blue Point, Long Island), Long Island oysters had been farmed by European settlers in coastal waters since the 1600s. But by the 1970s, coastal development, pollution, and new parasites had depleted oyster populations. One foreign parasite, MSX, found its way into the Long Island Sound, and by the millennium it had wiped out 76 percent of the oysters in the Sound.

The U.S. military had a special plan for Plum Island.

In 1951, the Joint Chiefs of Staff determined that "the destruction of the enemy's food supply by the use of anti-animal Biological Warfare agents would be strategic in its effect." In a long war of attrition and "a front in Europe stabilized far to the East"—so as not to destroy the food of Allied Western Europe—"it might then be to our definite advantage to initiate a vigorous anti-crop and anti-animal campaign and weaken the Soviet will to resist and encourage defection."

The Soviets were "doing a considerable amount of work" in this area because they understood that "famine . . . provides a real threat." The time to act was now. "Immediate action should be taken," the Joint Chiefs said, "to procure an island test-site where all types of hot agents could be tested with greater freedom and an animal laboratory could be established. . . . This would also obviate the restriction imposed by law prohibiting work on certain animal diseases within the continental limits of the U.S." The military ordered Army veterinarian Colonel Donald L. Mace, Doc Shahan's partner in a 1948 virus eradication campaign in Mexico, to begin a "cooperative project" with the USDA and the Army on Plum Island, one that would be "feasible and of potential mutual benefit." The USDA would be employed as a cover to sell the idea to the New York community.

Five top secret projects were approved for the germ warfare island:

4-11-02-051	**Miscellaneous exotic diseases**
4-11-02-052	**Rift Valley fever**
4-11-02-053	**African swine fever**
4-11-02-054	**Foot-and-mouth disease**
4-11-02-055	**Rinderpest**

Army surveyors landed on a deserted Plum Island in late 1951. They set up windsocks and fans and tested wind speed. They found prevailing winds from the southwest. That was good news, because if germs from their tests escaped off island they would tend to blow east, into the vastness of the Atlantic Ocean, or south into Gardiner's Bay, and dissipate, at least in theory. The surveyors also charted a regional population map, drawing radius circles around Plum Island at ten-, twenty-five-, and fifty-mile increments. Inside those circles were the Hamptons villages to the south, the Long Island Expressway (then under construction) to the west, and coastal Connecticut to the north. These would be the fallout zones if a biological accident occurred while the wind was blowing the wrong way.

At Dr. Hagan's urging, the Army rehabilitated the old mine storage building but retained its fungible-sounding name: Building No. 257. By August 1954, over one hundred large animals—horses, cattle, sheep, and swine—milled about in outdoor pens set up in the World War I-era artillery bunkers, and scientists used one thousand mice and guinea pigs each month. The bunkers were fitted with corrals, metal railings, gratings, "dip vats," walk-through sprayers, water bins, and feeding troughs. The sight of animals idling in cavernous concrete complexes that looked like ancient Mayan Indian ruins was a glimpse into a futuristic stone age. On small creatures and in culture dishes, work with hot viruses in Building No. 257, now simply called Lab 257, was performed in "gloveboxes," designed at Fort Detrick. Each steel enclosed chamber had riveted glass windows and was fitted with thick black rubber gloves that reached deep inside for experiments.

Activating Plum Island as a full-time germ warfare island presented a number of "pressing problems," according to Army records. Foremost was the sheer difficulty of constructing the unique facility on an ocean-exposed island not connected to the mainland by bridge or tunnel. Further complicating the situation, the longshoremen ferrying the construction materials to the island went on strike, and the building laborers refused to cross their picket line. This significantly delayed the project. Recruitment of scientists from mainland Fort Detrick wasn't easy. "I went up there for a job interview," remembers microbiologist William Patrick, "and the weather was terrible—cold, foggy, and just horrible. I went over on that damn boat, got seasick, and said to myself, 'This is not for me.'" Patrick declined Colonel Mace's offer and stayed at Fort Detrick. "The weather did me in—no question about it."[2] Dr. Al Webb, who occasionally came up from Fort Detrick to visit Doc Shahan, remembers the ferry. "I remember thinking, 'This is a hell of a way to go to work!'" He saw firsthand why Patrick and others rejected invitations to Plum Island. "If the wind was blowing up the Sound, or it was drizzling or snowing, you might think it was pretty grim, too."

With all the delays, not much germ warfare research was accomplished. And the importance of that research was suddenly called into question. The Joint Chiefs found that a war with the U.S.S.R. would best be fought with conventional and nuclear means, and biological warfare against humans—not against food animals. Destroying the food supply meant having to feed millions of starving Russians after winning a war. The Army was moving in too many biological warfare directions at once, and it was time to move

[2]Patrick continued working at Fort Detrick for twenty-five years. Today he's one of the nation's top experts on germ warfare.

anti-animal and anticrop biological warfare to "other appropriate agencies of the government." The Army asked the secretary of defense to order them off Plum Island.

As Lab 257 neared completion in the spring of 1954, President Eisenhower approved an agreement between the Departments of Defense and Agriculture at a National Security Council meeting. Mace was ordered to "pursue a minimal, yet forceful, research program at Fort Terry during the Fiscal Year 1954":

> The mission of Fort Terry [Plum Island] has been changed . . . from one which encompassed studies on various exotic animal diseases to determine both their offensive and defensive potentialities as biological warfare agents to one which pertains only to the defensive aspects of foot-and-mouth and rinderpest diseases.

A declassified top-secret report stated, just before the turnover, that "even though Department of the Army plans contemplated deactivation of Fort Terry, the assistant chief chemical officer for biological warfare at Camp Detrick will retain responsibilities in the anti-animal biological warfare field." Hardly did the Army "up and leave," as the USDA presently maintains. As time wore on, it became easier for the USDA to repeat the mantra that it had nothing to do with the Army when both were on Plum Island; that there were two labs; that each agency "did its own thing;" that the USDA never so much as *looked* at the Army soldiers.

Photographs excavated from the National Archives belie those assertions.

Penned on the reverse of a curled-up black-and-white photograph is the notation LAB. BLDG. 257. FEB 2 1954 is stamped in the lower left-hand corner. Standing at a bench the size of a dining room table, under the fluorescent light fixtures in a windowless, tiled room, are four men in starched white lab coats. They are crowded around a tall, boyish-looking man who is turning on a cylindrical heater, upon which sits a large flask containing a clear liquid. The man at the end of the table is leaning over, watching in earnest, with his arm stretched out on the lab bench, clutching an unlit pipe backward. The notes on the back of the snapshot continue: "Left to Right: Lt. Col. Don L. Mace, Dr. O. N. Fellowes, Dr. J. J. Callis, Dr. H. L. Bachrach and Dr. M. S. Shahan." The first two scientists were Army, and the other three were USDA. An Army officer on Plum Island at that time remembers the scene as typical of many "demonstrations" held in Lab 257. Often, he says, Dr. Fothergill, Fort Detrick's scientific director, would fly up and "see what we were doing and how we were doing it, from technical to safety measures. Many of those exotic agents were a real tricky thing up there on Long Island."

Two more dust-caked photographs preserved in the National Archives each picture a man seated alone at a metal office desk, with a pen held in his right hand, peering into the camera lens. Behind each man's right is the same three-shelf, glass-enclosed bookcase; above it, a framed picture of bulls charging across an open swath of country. Behind each man's left shoulder is the same photograph of la Comisión Mejicana-Americana para la Erradicación de la Fiebre Aftosa (the 1948 Mexico virus campaign). On the desk is an inbox tray, a stamp, and a quill-pen cup. Colonel Donald L. Mace is in one of the photos, and the USDA's Dr. Maurice S. "Doc" Shahan is in the other. There are no other differences that distinguish the photos, shot in the exact same setting, dated the exact same day and same year. Clearly there was a USDA-Army relationship.

Retired Major Luke H. West, an Alabaman inducted at Fort McClellan in 1941, was the Fort Terry top-secret control and security officer on Plum Island. Major West devised a scheme of color-coded security cards to enter the lab compound and conducted armed patrols of the compound and island perimeters. Although only a single laboratory report describing USDA's work for the Chemical Corps is publicly available, Major West remembers "at least twenty-five to fifty" research reports coming across his desk before they were mimeographed for Colonel Mace and sent down to Fort Detrick. West wouldn't elaborate on the contents of those reports.

Did the Army's departure from Plum Island change anything? The Army may have shipped its files back to Fort Detrick, but Colonel Mace left something of infinite value behind—the germs. The "defensive" germ research performed by the men from Agriculture would be extrapolated by "standardizing" the virus into a powder and placing it in Air Force cluster bombs. It didn't matter that the USDA told the public its research was only defensive—*it was of dual use anyway*. The USDA took beneficial control over the Army freezers (the 134 strains of 14 viruses) and did whatever it pleased with them. With the arsenal left behind, research continued unchecked.

Plum Island was now far less safe. Major West and his Army soldiers no longer patrolled the island with military weaponry. As the Army went, so too went its scientific expertise, its regimen, and its deep pockets of financial resources. Now, Plum Island was a USDA site—and it would have to compete with farm-belt states and their powerful lobbies in Congress over a limited pot of agriculture funds. Unlike military installations that enjoyed the backing of enthusiastic congressmen and senators, Plum Island would have little support from elected officials, most of whom still chafed over the harsh maneuvers employed to establish the laboratory in their midst.

The USDA had control now, running not one, but *two* high-hazard biological laboratories on Plum Island. This was the same USDA whose very competence Congress previously questioned.

Only time would tell whether the veterinarians could handle such great responsibility on their own.

The first microbiology work on exotic viruses at Plum Island was performed by USDA men hired by the animal branch of the U.S. biological warfare program.

Plum Island's first lab experiment was also another first—its first lab accident.

On July 21, 1954, three weeks after the Army transferred the island to the USDA, an angry cow coughed a glob of mucus into the unprotected face of a lab worker, known as "FW."[3] The previous day, FW had injected the cow's tongue with the New Jersey strain of highly contagious vesicular stomatitis virus, a germ probably enhanced by Erich Traub during his days working with the USDA. FW went home early and climbed into bed, ill with acute flulike symptoms and chills. By 7:00 p.m., FW's temperature had spiked to 102 degrees; yellow lesions appeared on his sore throat. He plummeted into malaise and depression. The signs were clear: FW had contracted the animal virus. Hearing the news, FW's Plum Island colleagues saw a window of opportunity. They placed a call to FW's wife—who happened to be a registered nurse—in the name of science. Before FW took a heavy dose of Terramycin, prescribed by the family physician, they asked his wife to collect blood and saliva, and swab the back of his throat with a Q-tip. She complied. The samples were stowed in her refrigerator and later ferried to Plum Island.

Assistant Director Dr. Jerry Callis took FW's blood samples, spun them down in an ultracentrifuge, and stored the human serum in a flame-sealed glass ampoule for future use. Then the man's virus-rich blood cells were injected into chicken eggs. This killed the embryos, drowning them in their own blood. The new "FW Strain" was passed through four successive chicken egg embryos and then isolated again; antibiotics were found to have no medicinal effect. The first published research abstract of the *animal* disease laboratory carries an incongruous title: The isolation of virus from the blood of *man*. The report dryly notes, "The infection contracted by FW occurred . . . in a newly constructed laboratory. . . . Past experience in other laboratories has shown the dangers associated with infectious materials

---

[3]FW is the notation in the records. There is little personal information available about him. He was probably a locally hired laboratory technician.

inhaled or deposited accidentally in the eyes or nasal passages." There is no mention in the report of a safety violation or, more important, the need for revised safety procedures to prevent the human infection that sickened FW. But the USDA scientists had stumbled upon vesicular stomatitis as a promising incapacitating germ weapon. The Army was pleased with the USDA's (inadvertent) human field testing.

With a troublesome "test run" under its belt, Plum Island was ready to handle the real thing. For the first time, exotic animal germs would be unleashed on United States soil. In eerie silence, the two young scientists, Drs. Callis and Howard Bachrach, looked on with reverence as veteran USDA man Dr. George Cottral slowly and carefully unlocked the virus vault, unstrapped the box, unscrewed the canister, and carefully uncorked the ampoules of hot germs. No accidents this time. On with the work.

In their mandarin-collared white lab coats, the lanky Callis and the stumpy, bespectacled Bachrach bent over the flasks, neatly lined up on the long lab bench, stirring and shaking the concoctions, holding test tubes up to the fluorescent light, scribbling observations and mathematical equations on their pads. Any changes in color? Consistency? Evaporation? All were signs of microbiological reaction. They took samples, streaked them on a slide, peered into the microscope, and fiddled with the focus knob.

Blood serum samples of twenty-seven calves, seven cows, ten bulls, and thirty-eight steers were set out in dishes and then injected with brucellosis bacteria and foot-and-mouth disease virus to test immune actions and reactions. They had to be extra careful with the brucellosis, because it caused an ailment of fevers, sweating, weakness, headaches, malaise, anorexia, abdominal pain, constipation, rigors, enlargement of the spleen, and coughing—and lasted from four to eight weeks.[4] Wary after the FW incident, the scientists escaped harm by wiping down surfaces and glassware constantly with a generous helping of Roccal solution, a chemical that kills microbes on contact.

The experiments yielded the type of results Doc Shahan and the Army could only have dreamed of.

---

[4]In fact, this strain was a fancy of Fort Detrick. President Eisenhower had previously approved the development of "incapacitating agents," germs that severely sickened but did not kill. For that reason, brucellosis and Rift Valley fever virus, both with low mortality rates, were heavily researched at Fort Detrick.

5

The Age of Science

Germs just don't have a chance!
—SECRETARY OF AGRICULTURE EZRA TAFT BENSON (1956)

Boundless optimism. No other phrase captures the feeling of 1950s America. The United States had saved the world from totalitarianism, and its mainland shores had not suffered the physical wreckage of war. Eisenhower, the revered general who had led the Normandy invasion and crushed Hitler and Nazi Germany, was now in the White House. Thousands of returning GIs were lured by the clean air, winding roads, free-standing homes, and manicured lawns of suburban communities, typified by Levittown, built on the wide potato flats of Long Island. Automobile sales skyrocketed and families drove in new-fangled cars over freshly paved six-lane interstate roads, pit-stopping at newly constructed drive-ins and burger joints. Television, that new miracle device, entertained and informed a swelling middle class, unshackled from the poverty of the Great Depression. The future had arrived. The jet age brought people to exotic lands and beaches in hours, communication advances made it possible to telephone Europe from their homes instead of wiring telegrams, and men were being launched into space—and they came back *alive*!

Throughout the medical world, scientists were taming germ infections with revolutionary chemicals called antibiotics, developed during the war by pioneers like chemist George W. Merck. Vaccinations were becoming available, miracle injections that prevented those infections from even starting. While rocket scientists were trying to put a man on the moon, biologists and virologists were fervently exploring the eradication of disease. Polio virus, not long ago a terribly crippling disease, was being eradicated from the United States by Dr. Jonas Salk's new vaccine, and the spread of deadly tuberculosis was being controlled for the first time in history.

Unbridled economic expansion brought on a voracious American appetite: consumption of beef sharply surpassed that of pork, and the demand for all foods grew exponentially. And like the NASA scientists, the USDA also promised the people the moon—a moon not only of cheese, but of milk, vegetables, grains, and meats. The total integration of scientific knowledge would be forcefully applied to agriculture, resulting in tremendously increased productivity of crops and animal products.

Healthy animals were needed to produce wholesome food, to support America's high standard of living. The elimination of animal diseases not only meant a greater abundance of food, it prevented those diseases from jumping to man. Like Salk's success with polio, a massive outbreak of foot-and-mouth disease virus in Mexico in 1948 was triumphantly overcome, contained just miles from the U.S. border before harming America's food supply. Veteran USDA scientist Doc Shahan, assisted by Army Colonel Mace, led a ten-thousand-man campaign through the arid Mexican terrain to combat the outbreak there.

Tall, lean, and tan, Doc Shahan had thick furrowed eyebrows and a black crew cut streaked with gray. Wherever he went, he carried with him a carved wood pipe and a pocketful of fresh tobacco. His talents, according to his friends, lay in his charm. Doc's demeanor was more gentleman country doctor than egg-headed scientist, and his suave charisma proved perfect for the task. Like the Allied generals who had slain totalitarianism in Europe three years earlier, Shahan and his men of modern science vanquished the disease and saved North America. They corralled the infected animals of Mexican peasants, quarantined the beasts of burden, and then either vaccinated or destroyed them en masse. Within eighteen months they wiped the scourge clear off the continent, sparing America from outbreak and famine. But victory did not come without human cost—twenty-four Americans and hundreds of Mexicans died in the cause, in armed uprisings and brutal reprisals by Mexican villagers against the men of science.[1] Proud

---

[1] The Mexican people regarded the Americans as intruding "conquistadores." They called them *los matavacas*, the cow killers. Possibly the saddest story of the campaign is the gruesome murder of twenty-two-year-old Arizonan Robert Proctor, a technician on a six-man vaccination squad. Robert was seized by a mob of 500 people in the village of San Pedro (total population: 200). The vicious mob stabbed him mercilessly, long after he died, and gouged out his eyeballs. The thugs then buried Robert under the plaza green, and later exhumed him and reburied him atop a hill overlooking San Pedro. Fourteen villagers were later apprehended and jailed.

of their victory, Doc Shahan and Colonel Mace returned to the States, yearning for a lab where they could continue to wage war against other virulent germs.

Advanced science applied to food production would aid America's worldwide struggle of democracy over godless communism. "History has indelibly written that revolution, anarchy, and tyranny are *fellow travelers* of hunger and malnutrition," one USDA scientist said. "Our plans for the future must include an ever-abundant supply of these foods if we want our people to be *strong* and our nation to *endure*."[2] Science would make the difference in the battle between Good and Evil.

The crown jewel of this blossoming, futuristic agricultural empire would be Plum Island.

> We meet today on the site of an old fortress. This island today is again an outpost of defense, against an enemy more menacing by far than the fleets of 1898.
>
> The enemy is real and the victory we seek is a victory for every human being in every farm, village, town, and city of the Earth. Our grandparents built this country with the help of their animals. That was yesterday. Today, our farm economy pivots on animal agriculture.
>
> I firmly believe America is on the threshold of the most challenging and most prosperous decade the world has ever seen. This is the age of science and technology. The frontiers of the mind have replaced the frontiers of geography. Organized and imaginative research ... will push the scientific frontier beyond limits we scarcely dare dream today.
>
> Brains will continue to replace brawn in American agriculture and industry! Man will direct power rather than supply it! Brainpower will be more important than horsepower!

It was dedication day, and Ezra Taft Benson had the crowd on its feet. For years, Plum Island had been off limits to the curious public, under the heavily armed guard of Army soldiers and uniformed federal officers. The mysterious nature of the island lair, once a popular summer paradise, began when the U.S. War Department closed its shores and chained its piers at the end of the nineteenth century. "Tourists that float through Long Island Sound each summer," wrote the *Suffolk Sun* in 1895, "in whose fancy there seem to be enchanted islands in ideal regions, ask a thousand questions—

---

[2]Italics added.

What are they? Who dwells thereon? They gaze at it wistfully as if they would like to know more about it." Now, half a century later, feelings about Plum Island were no different. But today, on this sunny autumn morning, the gates were finally being flung open.

Visitors came from near and far to walk the plank onto the 11:00 a.m. ferry to Plum Island on September 26, 1956. The laboratories had been washed down and polished up, the weeds pulled, the signs freshly painted. The employee union readied its outdoor table with cakes, cookies, and coffee. The island smelled of freshly cut grass mingled with sea salt. The germs had been locked away in vials in the vault drawers for three weeks now.

Days before, Greenporters had noticed an uncommon buzz along Main Street—strangers darting in and out of shops, eating and drinking heartily at Claudio's and the clam bar out on the pier. Some had distinctive southern and western accents. Others spoke in foreign tongues. But all wore expressions of great anticipation. Locals opened their spare rooms, recently vacated at summer's end, to welcome them. A special ferry was arranged with the New London Freight Lines; the massive LSM transport craft that landed amphibious battalions on the shores of Okinawa during World War II a decade earlier would now land on Plum Island an excited throng of general public and VIPs. Plum Island scientists wearing patriotic red, white, and blue badges pinned to their lapels escorted local residents around the island and inside the heralded chalk-white laboratory. Members of local civic organizations came, too, from the Rotary Club, the Minnepaug Club, and the Southold Tuesday Morning Club, as did national agricultural associations and high school science teachers with their classes in tow.

Master of ceremonies was new Plum Island Director Doc Shahan— smartly dressed in a sharp two-button black suit, crisp white shirt, diagonally striped thin tie, and a handkerchief ironed into a square that peeked out of his breast pocket. Sitting in front of him, in the roped-off seating areas up front, were the VIPs: Plum Island's founding father, Dr. William A. Hagan; the island's former commanding officer, Army Colonel Donald Mace; top military brass from Fort Detrick, Maryland; and finally a slew of European research scientists. Among them was the director of West Germany's new State Research Laboratory at Tübingen, Erich Traub.

After the Pledge of Allegiance and a robust singing of the national anthem, Dr. Hagan rose to give his remarks. To most of the audience, he was known as the distinguished dean of Cornell University's veterinary school. But to a select few, he was an architect of America's biological warfare program, and the patriarch of anthrax as a weapon of war. Hagan spoke to the crowd about the lab's importance, and listeners nodded and smiled approvingly. But then he upbraided the local audience, who had initially opposed the lab and were only now beginning to accept it. Reopening old wounds, Hagan said, "Those of you who fear that germs will leak and harm your families are good but misled folk." Hagan exited the podium to tepid applause.

At first, "the neighborhood wasn't in our favor," says Diana Fish, then the tender twenty-year-old wife of scientist Dr. Ralph Fish. "They didn't want to accept us—we were ignored completely." Scientists and their families got together each week to socialize and play canasta and gin rummy. Sometimes villagers would cross the street or turn and walk in another direction when they saw Colonel Mace or Doc Shahan approaching Main Street. "We were about as welcome as a moth in the garment district," Shahan later remembered. "These men had wonderful, wonderful minds," says Diana Fish. "They carried the banner of the United States through Mexico and were rewarded with Plum Island—and the public there was not hardly ready for it."

Dedication day, however, seemed to mark a turning point in relations. One Mrs. Hallock, from one of the area's oldest and most respected families, wrote Doc Shahan that the ceremonies and public tours were

> so right—so helpful in allaying and banishing latent fears at first held by the community (I being one of those fearful!). . . . The fact that you want to think of us as friends and neighbors means a great deal. . . . We have pride in your work within—not outside of—our community.

Doc Shahan thanked the VIP scientists for attending, careful not to name Colonel Mace and the Fort Detrick scientists. "We cannot say too much for the excellent cooperation the USDA has been given by the representatives of the Department of Defense," Doc penned in his first draft, praising Mace's vision for Plum Island; but such kudos remained on the cutting room floor, and the Army germ warfare men sitting in the audience understood (and accepted) the snub. After all, their involvement had to remain top secret.

Dr. B. T. Simms spoke next, ordering Plum Island headlong into a brave new world. "We know we are facing days and weeks and years of hard work, but it will be work that we love," he exulted. "We can expect many disappointments, but we can expect them to be overshadowed by achievements."

President Eisenhower's secretary of agriculture, Ezra Taft Benson, then delivered his captivating keynote address. The former Idaho potato farmer had visited Long Island earlier that year and spoke to local farmers at Riverhead, the farming village at the head of the North and South forks, listening first to their concerns on crop supports and then their worries over Plum Island. The dapper, bespectacled Benson mounted the podium, crowned with the Department of Agriculture seal and framed by a long dais adorned in red, white, and blue bunting. With a cool breeze sweeping through the outdoor ceremony, the Mormon farmer roused the thousand-strong audience with impassioned, almost religious oratory.

Doc Shahan, sitting to the right of the podium in the first seat, beamed at the secretary and out at the crowd, applauded Benson's powerful speech as the band played a majestic fanfare. The amiable director was overjoyed that his island laboratory had arrived.

After the ceremony, Doc and his deputy, Dr. Jerry Callis, led Secretary and Mrs. Benson on a private tour down the lab corridors, pointing out laboratory rooms and animal rooms along the way. Benson gazed curiously down a chute in an animal room that led to the incinerator. Observing the tangle of decontamination machinery and pipes down the hallway that filtered the air, he exclaimed out loud, to no one in particular, "Germs just don't have a chance!" On that intoxicating day, in the hands of such capable men, it certainly seemed that way. Outside, the doctors led the secretary to the lab's cornerstone, which he dedicated retroactively. It read simply, "A.D. 1956."

The breeze had picked up, pushed by the fast-approaching Hurricane Flossy. Benson bid them farewell, tipped his white derby to the crowd, and caught the ferry back to the mainland. The other guests were also quickly shuttled off the island, and Doc Shahan moved the inaugural scientific conference to Greenport High School. Flossy wasn't the first—and wouldn't be the last—storm to batter Plum Island, which lay squarely in the path of the East Coast hurricane corridor.

Doc Shahan's Plum Island team consisted of a core of three scientists. Doc's most promising star, thirty-four-year-old biochemist Dr. Howard L. Bachrach, had been the first to isolate the polio virus, working under the Nobel laureate Dr. Wendell M. Stanley. Obtaining Bachrach for Plum Island was so imperative, Shahan wooed him east with unheard-of perquisites: the opportunity to bring in his own biochemistry-biophysics team, and the chance to design his own research wing in the new island facility. The fifty-two-year-old Nebraskan Shahan, a cowboy boot–wearing lab director, was a hard man to turn down; a reporter described him as "tweedy in dress, handsome in a wide-open Western way. . . . Doc Shahan likes people and people like him." Bachrach bit, but realized soon afterward that he got a little less than he had bargained for. "This space is, of course, much less than I have recommended as a minimum," he later lamented. "And it would help if [more] space becomes available in the basement." It never did. Shahan also drafted chief scientist Dr. Jacob Traum, who retired at the age of seventy-four from the University of California (also Bachrach's alma mater). Traum lent his reputation and immense knowledge to the fledging facility as a world-renowned expert on tuberculosis and brucellosis. But the franchise player on Doc's team was the slim, blond assistant director. At the tender age of twenty-seven, Dr. Jerry Callis was the youngest member of the Plum Island scientific staff. Callis graduated with honors from Purdue University and landed a USDA position studying animal germs overseas, where he was being groomed for the opening of the new laboratory.

In a remarkable departure from the not too distant past, local newspapers provided glowing support. The *Riverhead News-Review*, for example, commented: "Worthwhile research projects such as the Animal Disease Laboratory are in the best public interest and examples of useful expenditures of the taxpayers' money." Four years after the locals strenuously opposed the building of the laboratory, folks were now in genuine support. It didn't hurt that almost overnight, Plum Island became the largest employer in the region, with over three hundred workers making the ferry trip across Plum Gut each morning, all of them now receiving healthy federal government salaries and benefits. When the villagers' spouses, siblings, and children were hired by Plum Island, things quickly changed. "I think people are beginning to appreciate that we know how to maintain the security over any infection leaving the island," said Shahan, "and that a new industry has come to town and new money is being spent around here."

This was all a relief to Doc Shahan and Jerry Callis, who had fought hard to build the island laboratory. With the support of their mentor, Dr. Hagan, the scientists' vision was at first grandiose, perhaps too much so in Congress's eyes, as lawmakers in Washington parsed through the aggies' $30 million initial plan to build a thirty-acre behemoth. "These people came up before the committee," said the Appropriations Committee chairman, Jamie Whitten, "with the most fantastic plans for spending money and building a laboratory that you can imagine. It was entirely out of line . . . it looked like the Department of Agriculture might . . . want a Pentagon building for itself." They ended up with a more modest, $10 million complex instead.

Laboratory 101 was a pioneer effort in high-hazard biological agent containment. On that day in 1956, it appeared as a gleaming white monument. Lying on the northwest plateau of Plum Island, the lab is splayed over a ten-acre site just east of the old lighthouse. A steep cliff, towering high above huge boulders, forms a natural buttress against the churning waters at the confluence of the Long Island Sound and the Atlantic Ocean. This 164,000-square-foot T-shaped structure, with all its intricate machinery—a cluster of twisting pipes and valves and boxes and gauges taking up the entire second floor—cost $7,712,000 to build. The USDA used the remainder of the $10 million in congressional funding for the supporting cast: the guardhouse entry gate, sewage decontamination building, emergency power plant, storage buildings, fencing to corral the herds of test animals outside, and the compound fences to envelop the lab building. The two ten-foot-high barbed-wire, chain-link fences have a twenty-foot buffer zone between them; the inner fence has a concrete barrier that extends five feet above ground and four feet below. Before 101 opened, the perimeter was scorched with chemicals to defoliate the area and deter vermin from approaching.

Just west of 101, outside the compound fence, is Building No. 102, the wastewater treatment plant. Here, sewage is heat-treated to decontaminate it before it is piped into the Long Island Sound and Gardiner's Bay. Next door is Building No. 103, the emergency power plant and adjacent oil tank farm. The air that comes in and goes out, the water coming in, the sewage going out—all of it is controlled inside a master control room with large, glowing red and green buttons and throw switches mounted on the walls, and large panel grids each containing thirty-five small boxes that illuminate white with buzzer alarms.

The top of 101's T is the clean zone—the main hallway and service area, offices, the sterilizing laundry, complete with sewing and mending equipment. The wide stem of the T is the hot zone, divided lengthwise into three segments. The two outer strips house thirty-two animal isolation rooms. Each room measures ten feet by fifteen feet, and a door, painted bright red, leads through change-room air locks into "hot corridors," dimly lit by squat windows made of opaque glass bricks that refract light and cast shadows within. Together, the animal isolation rooms hold up to seventy-five head of cattle.

The middle of the stem of the T, separated from the animal wings by an outdoor moat, holds a maze of laboratory rooms, divided into four research areas. Inside is all the heavy equipment—the electron microscope, ultraviolet irradiator, ultracentrifuge, egg incubators, virus fermenters, virus freezers, wall-to-wall glove boxes. The rooms are windowless, illuminated by fluorescent lights that cast a ghastly pallor over the glazed tile walls, cabinets, and metal benches upon which tissue cultures are grown, eggs embryonated, germs pipetted (sucked by mouth into a long glass tube), cells infected, antibodies produced, and so on. The big incubators don't hatch any chicks; they grow viruses in trays of eggs and provide embryo tissues for in vitro experiments. Inside the virus freezers are germs and a stockpile of animal and human tissues grown in the incubators alongside the chicken eggs. The ultracentrifuge spins to create a great force (thousands of times the Earth's gravitational force), which separates viruses from their infected host cell and other debris.

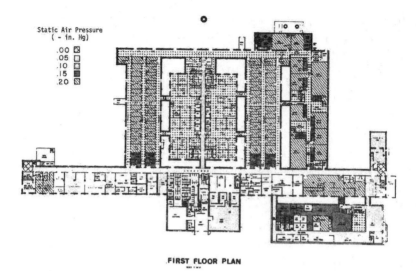

Static Air Pressure
(- in. Hg)
.00
.05
.10
.15
.20

FIRST FLOOR PLAN

At the base of the T is the incinerator charging room, where animal car-casses, organs, fluids, paper items, and other combustibles—even spent nuclear radiation—are reduced to fine ash at earth-scorching temperatures. The smoke emitting from the pile within is flushed out through a filtered stack lined with afterburner jets.

During the dedication day tours, every inch of this state-of-the-art facility was sparkling clean, but this didn't last long. After the audience left and the stage and podium were broken down, the curtain fell on Plum Island. Days later, splattered blood, animal waste, and remnants of internal organs were strewn across the freshly painted white walls and shiny floors of the animal rooms, transforming them into something more akin to tor-ture chambers than science labs. Behind one room's air-locked door, unsus-pecting cows were restrained by a brawny animal handler wearing powder blue scrubs, then injected with a menu of germs that rendered the beasts incapacitated in a matter of hours. Inside another, goats were locked in stanchions, a hairsbreadth from choking between two narrow metal bars snapped tightly into place around their necks; another room held swarms of infected soft and hard ticks gorging on the blood of two pigs; and in still another, horses were bled for serum samples.

The fauna came to Plum Island in a precise and peculiar manner. Test animals were periodically trucked in from a Virginia farm to Brookhaven National Laboratory, the federal atomic energy research facility on Long Island. A special Plum Island truck picked them up and carted them to Orient Point, where the cattle car boarded the ferry for the bumpy ride across Plum Gut. Upon arrival at Plum Island harbor, workers led the animals off the truck and through the gated animal transfer station. The animals were slipped into a "squeeze-gate," punch-tagged on the ear, and sprayed. They were led up ramps onto another truck, a permanent fixture on the island, which brought the animals to the old Army artillery bunkers where they were washed in a big vat (called a dipping), then quarantined for two weeks. Feed came in through an air lock, and a hammermill that ground up any live insects or rodents that managed to burrow inside the burlap feed sacks. No person, vehicle, or animal could cross through the harbor gatehouse without a thorough decontamination.

After an animal finished its quarantine, it was a one-way trip into the lab. There, two animal handlers readied to enter the animal's room. Adhering to strict safety procedures, they donned protective gear—two-piece black rubber slickers, boots, rubber hats, and neon orange gloves—looking something like offshore fishermen preparing to be battered by a heavy gale. When they entered the room, one went to work tying the jaw of the cow to a cleat in the wall, while the other held it steady, petting the creature's coat. Then came the injection of anesthesia. Seconds later, the men jumped out of the way as the eight-hundred-pound cow suddenly keeled over on its side. Crouched over the supine cow, one handler yanked the animal's limp tongue out of its mouth, while the other stuck a long needle into the underside, and slowly injected the liquid virus slurry du jour. Their mission accomplished, the men exited, and took turns decontaminating, stepping into a boot bath of caustic soda lye and water and washing down each other's rubber uniform with a wire brush.

Humans had their own precise methods of arrival. Entering into Lab 101 each morning, they walked down a flight of stairs and through a one-way turnstile into a change room, where they showered and changed into blue and white coveralls and white Keds. The air inside, under negative pressure, felt clammy and dry; in the most pressurized areas, it felt like being inside an airplane cabin. "We sucked on hard candy and cough drops to keep our throats moist," remembers one worker. For the burly animal handlers who moved about the lab, it was not unusual to take eight or ten showers, each supervised by guards, in a single day. "We were awfully clean when we got home at night," an employee recalls. Showers were frequent; the Plum Island records for most showers taken per shift stand at seventeen for animal handlers and twenty-three for scientists. To avoid a constant changing of clothes, workers often went about their business wearing only rubber boots, which made for quite a sight.

The only way to exit the laboratory was through the anteroom, where workers scraped under their fingernails (a perfect hiding spot for viruses), coughed, spit, and blew their noses into paper towels, and walked through a one-way metal gate that automatically triggered the spray chute of rinsing showers. Five showerheads deluged seven and a half searing gallons of water per second upon the occupant, as he washed thoroughly with hexachlorophene-infused soap. And no cutting corners, either. "Skip any part of a bath and get caught doing it," says a worker, "and you're out of a job so quick it's a pity." If a worker exited an especially "hot area," he has to take two showers with two changes of clothes in between; otherwise it was just one, but that shower had to last at least three minutes.

Even eating had rules. No going down the block for a bite to eat. Cold box lunches went into the lab through the air lock. Lab glassware, tools, and other objects went through walk-in-sized autoclaves, giant steam pressure cookers with two hot and clean doors that screwed shut and sterilized items going in, and decontaminate them going out.

Workers were required to follow other rules even after leaving the island. A sign posted at the ferry dock eliminated any misunderstandings between the director and the three hundred plus workers in his keep:

 ## ATTENTION

YOU ARE NOW LEAVING A QUARANTINE AREA

AVOID CONTACT WITH THE FOLLOWING ANIMALS

CATTLE POULTRY MICE DEER SHEEP DUCKS GERBILS HAMSTERS
GOATS PET BIRDS GUINEA PIGS RABBITS SWINE

Security was taken extremely seriously in the beginning. From the moment the Army transferred Plum Island to the USDA, the aggies disclaimed an active role in military affairs. But they set up a first-rate operation with the type of security and secrecy worthy of a military installation. Plum Island's "armored division" consisted of twenty-six trucks, four buses, two carryalls, and three jeeps, which cruised the island's perimeter on twenty-four-hour armed patrol. As employees disembarked from the ferry each morning, three uniformed guards, in dark brown shirtsleeves and black caps, examined each worker's security pass at the harbor gatehouse and checked off his identification number on a clipboard. A red pass meant the employee worked in a "hot zone"; it allowed him on the "red" bus and then gave him clearance into the laboratory building, confined strictly to the reds. A yellow pass allowed him inside the labs' outdoor compound, but not inside the labs. A brown pass permitted island access, but not near the labs. More guards patrolled the fences around the lab and admitted people into the compound at the lab gatehouse.

Obedience of biological safety and security rules was drilled into the heads of employees over a mandatory two-day orientation program. Entering a restricted area without a pass, entering an animal holding area without permission, or leaving Lab 101 or 257 without showering out were all egregious violations—even minor infractions were verboten. The degree of punishment was tied to the severity of the infraction and meted out as follows: reprimand for the first offense, ten days suspension without pay for second offense, discharge on the third offense. One worker remembers the second day of his orientation course: "After lunch, we were escorted by Pete DiBlasio [Plum Island's first chief of security], and he was held up at the door holding it for someone carrying out something, and he called out to us, 'Stop!' But we kept walking a few steps because we were in direct rays of sunlight. I took a shortcut off the path and walked on the grass, and Pete ran over to me. He said, 'Listen, I understand that this is your second day on the job. But if I ever see you walk on that grass again, I'll write you up—and after one more write-up, you're gone.'"

In the 1950s, dealing in lethal germs required loyalty oaths ("I hereby pledge full allegiance to the United States Government . . .") and State Department and FBI national security checks. Clearances at Plum Island were termed "Sensitive," meaning that most personnel had to be screened by the feds in a review that took up to three months. Sensitive positions included security (armed guards and full-time firefighters); employee-residents of the island; those with access to red areas; animal caretakers; and those with access to classified research material. The case of Isaac Gaston reflects the seriousness with which the rules were then regarded. Gaston was a truck driver who drove from building to building, carting items and people all over the island. He had been a truck driver on Plum Island since day one, but now fell under the new security rules because he made stops inside the laboratory compounds. Until the driver received his Sensitive clearance, Doc Shahan ordered his duties immediately curtailed to supply runs between nonessential buildings, far away from the restricted areas.

The rules for visitors were simple. There were no visitors. Once the germs were uncorked, no visitors, no matter how important, were admitted on Plum Island. "Nobody," said Doc Shahan to a nosy reporter, tamping fresh tobacco into his pipe for emphasis. "Nobody goes *just to see.*" He even denied access to Dr. Herald Cox from Lederle Laboratories, a dedication day VIP who had discovered the germ warfare agent Q Fever. Cox merely wanted to observe safety procedures for a lab that Lederle was building in Uruguay. Though Cox's friend Dr. Hagan called repeatedly and lobbied to allow the visit, Doc Shahan put his foot down—no visitors allowed while the lab is in operation.

The beaches and the harbor were also off-limits. Doc Shahan moved against lobstermen and fishermen setting their traps and nets in Plum Island area waters. Often fishermen, pleasure boaters, and thrill-seeking snoops had to be waved away by guards brandishing shotguns, shouting through booming bullhorns from the silky beach or from high atop the island's rocky coves. Any picnicking boaters who set foot on the island had to sign release affidavits ("I consent to any quarantine and detention imposed . . . I will avoid contact with . . . I consent in the event of emergency to be detained . . . my clothes and personal items may be held for decontamination. . . .") or they were promptly arrested, and their foodstuffs confiscated.

Once, a boater's puppy dove from the craft to chase after a piece of driftwood and swam to shore. Because the dog had set its paws on the island and might communicate disease, it was forcibly confiscated by the guards, put to sleep, and incinerated. Rules were rules, and this one was clear: "None of the animals that land on Plum Island are ever permitted to leave."

The "Nothing Leaves" policy wasn't limited to animals. To prevent contamination, only humans, their street clothes, and their jewelry could leave Plum Island. Animal remains, spent vehicles, laboratory equipment, construction debris, paints, disinfectants, chemicals, biologicals, tires, radiation, and even food were either burned or buried in multiple landfills on the island. If it fit in the incinerator, it went to the charging room; if not, into a landfill pit. Not even books from the island's medical-veterinary library could be borrowed overnight. New York Telephone placed a dedicated truck on the island, and Ma Bell's repairman, Mr. Albert A. Abersmith, kept four separate sets of tools, one for each laboratory module, and took four showers a day. On the rare occurrence a vehicle absolutely had to come to Plum Island from the mainland, its top and undercarriage were sprayed down with disinfectant, and the interior was wiped down. Then it had to drive through decontamination wheel baths and sit on the dock for two weeks minimum before it could return to the mainland with a clean bill of health. And even then, the safety office called and checked up on the vehicle after the first and second weeks on the mainland.

Some chafed at the innumerable safety rules and regulations; one called them "ludicrously careful." Doc Shahan would have none of it—if you didn't like it, then leave. "[W]e take no risks," he told a reporter. "We may be extreme, but I don't think so. It's better to be overcautious all the time than not cautious enough just once."

Speaking of the laboratory's intricate constructs, a Plum Island scientist said, "All of this planning and construction results from fear—fear which is not to be confused with cowardice—but rather the real realization of lurk-ing dangers around us." The laboratory attempted to tame those fears. "It's the most complicated in the world," said West Point Army engineer Louis Genuario, describing Lab 101's revolutionary sheet-metal air-control sys-tem he helped design. "And the largest."[3]

Jabbing his smoldering pipe at a reporter, Doc Shahan boasted, "No other laboratory in the world can match ours for all-out security." Not Eu-rope, certainly not Africa, and not even Fort Detrick, the Army's biological warfare headquarters. Shahan could make that statement with cool confi-dence: Plum Island was the latest and the greatest, an amalgamation of the best biological security techniques and state-of-the-art technology human knowledge could offer.

So when the USDA opened the Plum Island Animal Disease Center amid so much pomp and fanfare, *it really was* "The World's Safest Lab," as the USDA trumpeted in its opening-day press release.

Indeed, germs didn't have a chance.

DOC'S PRIZED POSSESSIONS
Early operations at Plum Island boomed during the prosperous 1950s. There were the early breakthroughs with brucellosis. There was the inno-vative propagation of foot-and-mouth disease virus in kidney cell cultures by Drs. Callis and Bachrach. But before any research could be done, Plum Island had to first get hold of the germs.

Doc Shahan referred to his viruses as his "prized possessions," and stored them on dry ice under lock and key. During a six-month period in 1953, Shahan and his Mexican campaign colleague, Colonel Mace, then the Army's biowarfare commander on Plum Island, secreted 131 strains of 13 different germs on the island, akin to Captain Kidd stowing his own pirate booty on adjacent Gardiner's Island some 300 years earlier. Shahan's and Mace's clandestine treasure trove was equally worthy of a white skull and crossbones emblazoned on a black flag.

Listed on the now-declassified "Inventory of Animal Viruses and Anti-sera Procured by the Cooperation Between the Chemical Corps and the U.S. Department of Agriculture and Stored at Plum Island" are the starter strains of the germs still housed on Plum Island: bluetongue, Rift Valley fever, African swine fever, fowl plague, sheep pox, Newcastle disease, goat pulmonitis, *Mycobacterium butyricum*, Teschens' disease, vesicular sto-matitis, virus diarrhea of cattle, rinderpest, foot-and-mouth disease, and

---

[3]One reporter along for the dedication day ride had this to say about Genuario's creation: "I, for one, found the 5,000-square-foot room on the second floor where the ducts, filter chambers, electrical conduits and controls come together an excellent place for inducing nightmares."

twelve ampoules (hermetically sealed, bulbous glass vessels) of a germ listed as "N."

How they got their hands on the viruses reads like a Ian Fleming novel. "It was sort of cloak-and-dagger business," recalled scientist Dr. George Cottral, Plum Island's first biological security officer. Shahan picked up the hot, live rinderpest from the Army's top-secret PROJECT 1,001 (named after the famous *Arabian Nights* tales) deep within the Kenyan jungle. He brought it with him to Great Britain's Pirbright virus laboratory near London, where he met up with Colonel Mace and Dr. Cottral. There the trio bought bovine and guinea pig tissues infected with six types of sixteen different strains of foot-and-mouth disease virus, forking over to the lab's director a U.S. Treasury check for $5,000. Federal law forbade the virus on mainland United States soil. So after overseeing its intricate packaging, Mace, Shahan, and Cottral rode back with the test tubes and ampoules across the Atlantic on a U.S. Navy freighter. They were placed carefully inside a heavy canvas bag, padlocked inside a strong wooden box, placed in a stainless steel box. The men kept it within their sight the whole trip. In the deep waters of Gardiner's Bay, just south of Plum Island, the freighter's screws came full stop and set anchor. Unloaded onto a tugboat (which looked like a dinghy next to the massive freighter) along with its three chaperones was a shiny hinged metal box—glistening in the bright sunlight, stenciled PROPERTY OF THE U.S. GOVERNMENT on all sides. The tugboat slowly chugged to Plum Island harbor, where it put its precious cargo ashore.

Soon after, other viruses, similarly sealed, made the journey to Plum Island, hailing from the far corners of the globe—North and South Rhodesia, Nanking, Tokyo, India, Thailand, Palestine, Iran, Pakistan, Turkey, Egypt, Kenya Colony, Nakuru, Kabete, South Africa, Orange Free State, Entebbe, Nigeria, Sierra Leone; from nearer reaches—West Germany, Italy, France, Great Britain, Mexico; and even from within the nation's borders—Fort Detrick and Cornell University (courtesy of Dr. Hagan). Each bug was frozen inside corked vials or ampoules, tucked in hinged metal canisters, and carefully packed in boxes bound with padlocked iron straps. The boxes were then locked in a dry ice cabinet, which itself was inside a bombproof vault. "I have trouble getting into the stuff myself," said Cottral, who held all dominion over the germs, speaking with the *New York Herald Tribune* in 1954.

That was biological security.

All research scientists share a thirst, a quest that is difficult to put into words. According to one, "The person first becomes knowledgeable in the subject, then greatly steeped in it, and finally comes to possess that which is the *feel* of the problem. So totally immersed . . . he is often in a position to render the greatest of possible scientific service, namely, the elucidation of facts previously unknown." Doc Shahan inspired his young scientists to reach for this almost spiritual place.

A year after the dedication, young Dr. Bachrach and Dr. Sidney Breese, a wizard in microscopy, announced a first: they had successfully photographed a virus using an electron microscope. Though the technology had been introduced in 1945, results were disappointing and photographic detail was difficult, if not impossible, to obtain. But recently, metallic shadowing helped add dimension and clarity to specimens, and Bachrach and Breese, after an untold number of miscues, snapped a crisp image on a plate glass negative. For the first time in the 443 years since it was first diagnosed and described by the Italian monk Hieronymus Frascastorius, one of the smallest creatures known to mankind, the foot-and-mouth disease virus, could be seen in striking detail. The micrographs revealed something never before witnessed: a spherically shaped gray ball, one-millionth of an inch in diameter and even smaller than the tiny polio virus. Other work proved that viruses were not all spherical in shape—they also came in tadpoles, rods, and cubes. Scientists could now advance with research and development at great speed.

The crown jewel was already paying dividends.

THE CHANGING OF THE GUARD

After spending a decade building Plum Island into a preeminent center of research, Doc Shahan retired in 1963. The young man who the USDA had groomed for the position from the very day he graduated from Purdue ascended to the directorship. Jerry Jackson Callis would remain at that post for the next quarter-century.

Dr. Callis told the islanders he had no intentions of toying with Doc's legacy, but instead would build upon it. His goals, he said, were threefold: develop the most successful research; maintain a high level of employee morale; and aspire to be the most respected laboratory, not only in the nation, but in the world. More than ambitious, the goals were heartfelt. His tenacious, almost childlike devotion to the island and its mission was never in doubt. As he encouraged communication between employees and management and promoted the freedom to speak one's mind, the new director promised firmness coupled with fairness. The most important resource on Plum Island is the employee, he said. He pledged to "respect the personal dignity . . . recognize length of service and work achievement . . . maintain continuous employment and realistic salaries . . . and provide work security" for all. Callis's final admonition invoked the age-old rule preached in the Gospels of Matthew and Luke—"Believe in the Golden Rule and always practice it."

Do unto others as you would have others do unto you.

ecause the "Nothing Leaves" policy discouraged most outside maintenance and support, Plum Island grew into a self-contained enterprise.
For example, the island had its own machine and metalwork shop from
which Callis and the scientists ordered "everything from roads to rabbit
restraints," says veteran Plum Island draftsman Ben Robbins. That line
became the motto of the engineering department. From completing in-
house alterations and repairs to adjusting the elaborate laboratory containment system, detailed over hundreds of blueprints, no task was too difficult
for the Plum Island workforce. By the time Callis assumed the reins, the
engineering department—along with two laboratory buildings, an armed-
guard platoon, waterworks, animal corrals, electric power plant, fire
department, sewage treatment plant, cafeteria, laundry, library, and two
marine ferryboats—had transformed the island into a realm.

Now Jerry Callis alone presided over a virtual kingdom, a fantastically
beautiful, primeval island replete with miles of white beaches and green
groves, bluffs and swamps and fields, in the cradle of the most coveted real
estate in America. All of this at his command, protected by twenty-four-
hour armed guards and 350 people manning the island each day. What's
more, no worker could pull seniority over the new director—Callis pre-
dated *every single member* of the staff. He helped design the buildings, and
he authored the innumerable safety and security rules—placing him on a
perch beyond reproach. The USDA once referred to the island as "a small
town unto itself." Others who knew better called it "Jerry's Plantation."
"They referred to it as 'Master's Island,'" remembers former Plum Island
scientist Dr. Carol House, "because he ruled it with an iron fist—like with
slaves and all."

In the eighteenth century, the locals sportively called the reclusive
Samuel Beebe, who owned Plum Island back then, "Lord of the Isles" and
"King Beebe." Now Callis had risen to Beebe's level of esteem—and never
was this more apparent as when his next-door neighbor paid a visit and
welcomed him into the clique. Remembers Callis: "Robert David Lion
[Gardiner, whose family has owned Gardiner's Island since 1638, and who
refers to himself as the Sixteenth Lord of the Manor] telephoned me and
said, 'We're both lords of adjacent islands—we must meet.'" The eccentric
codger ordered his crew to sail his yacht, adorned with a colorful flag bearing the Gardiner family coat-of-arms, due south from his manor house to
Plum Island harbor. Feasting on a light salad in the Plum Island cafeteria,
the two "Lords" hit it off immediately. Like a boy showing off his toys,
Callis toured his new friend around the island grounds. Old Gardiner
nodded and smiled and spun tales about the two island lairs from genera-

tions long ago, in an aristocratic voice not quite the Queen's English, but decidedly not American. "Lord" Gardiner then departed, thanking his companion for lunch and extending an invitation for Callis to visit his manor house. With "Lord" Callis's permission, Gardiner later sent over his niece, Alexandra Goelet, who was studying osprey while attending Yale University. In a rare dispensation, Dr. Callis allowed her and her classmates ashore to do some "birding" on Plum Island.[4] They made a special visit to the only marked grave that exists today on Plum Island, a deep pit where a Revolutionary War colonel, one Thomas "Gardner," was buried in 1786.

To his subjects, Callis was larger than life. He was greatly respected and, like many a monarch, seldom seen—spotted only occasionally in his blue seersucker suit hustling into the Lab 101 conference room. To others, Plum Island was his very own Emerald City. Callis was the wizard who, wearing a sorcerer's cap, threw the switches behind heavy curtains that shrouded the redbrick administration building perched atop the old Army parade ground, regally surveying his windswept island lair. Adding to the intrigue around the time of his coronation, Callis married Loisanne Roon, a local millionairess active in the Southold Garden Club. The couple moved into a rambling, secluded estate in Southold, tucked deep in the woods, boasting captivating views of sailboats plying Peconic Bay and the Hamptons.

Almost to a man, the workers saw him as the rare benevolent dictator who meant what he said, and practiced what he said. Stanley Mickaliger, a retired building engineer, concurs. Under Callis's reign:

> You had dedicated people. This was not just a job, it was also our home, like one big close-knit family—and we're still close after all these years and keep in touch. When I sailed on ships during the war, they always said, "It's not the ship—it's the crew that makes the ship." It was like the world of a ship over on Plum, and when we worked there, regardless of what anyone tells you or says, Callis took care of his people. Callis looked out for them.

Back then, "the place was just beautiful," remembers one worker. "There were flower beds planted everywhere—the roads were swept clean and the grass was always freshly manicured. There were fourteen or fifteen guys dedicated just to what they called 'Buildings and Grounds' who kept this

---

[4] Years later, Uncle Gardiner and Alexandra became bitterly estranged over future plans for their 365-year-old island estate. The rift between them grew so great that ninety-one-year-old Robert actually attempted to find an unrelated man named Gardiner to adopt as his sole heir, to wrest control of Gardiner's Island from his niece. He was unsuccessful.

place such that it was like being on a big estate." After all, it was the USDA's crown jewel, and diplomats and scientists from foreign lands like Sweden, Spain, Mexico, and Australia would visit often (the visitor policy had loosened a bit). The jewel required constant polishing so it could sparkle for all to see.

When Callis encouraged open communication, he meant it. Every year, he orchestrated a family picnic and awards ceremony where he grandly recognized the seemingly small contributions of workers that, together, built Plum Island into a research powerhouse. Scientists and secretaries alike were called onstage and cited for outstanding performance and dedication. Because he listened to the workforce and implemented their suggestions, new ideas came to his desk each week. Many were put into practice, and all were acknowledged with gratitude. One worker suggested installing fluorescent arrows on each dolphin marking the harbor entrance, and another thought of a temperature alarm in the Mouse House, the mice colony building. Still another proposed to install lighting and protective wire grills in front of the big oil burners in the dark basement passageways of the labs. An engineer solved the problem of ash accumulation in the incinerator room with a simple solution—install a blower that would blow the ash into the atmosphere with the smoke. He too got an award for this toxic remedy, and Plum Island burned biologically contaminated waste this way right up until the environmentally conscious 1970s.

As chief of one of the foremost scientific laboratories, Dr. Callis also served as a scientific ambassador, representing the United States in many foreign lands and helping countries stem and eradicate germ outbreaks. Like Doc Shahan, he too pushed his Plum Island scientists hard to reach new heights. And they did. VACCINE MAY END ANIMAL SICKNESS, read the front page of the *New York Times* in 1967. Foot-and-mouth disease virus, believed to be the cattle plague, the fifth of the infamous ten biblical plagues (Exodus 9:1–7)—the scourge upon Egypt's livestock—might be removed forever from the Earth. They were undoing the divine. Under Callis's direction, Plum Island developed a trivalent vaccine, capable of immunizing against the three major strains of foot-and-mouth virus, A, O, and C. Through trial and error, they grew virus in monkey kidney cell cultures they had cultivated years before, then killed the live virus with acetylethyleneimine and added two mineral oils as enhancers. Four years later, Plum Island reached another milestone, when scientists established the first rapid test, a radial diffusion test that quickly diagnosed virus infection and detected viral presence.

"At the moment," said a proud Director Callis, "we're having very good success. . . ."

Symptoms

I'm not allowed to speak. Please—please! Leave me alone!
—PLUM ISLAND EMPLOYEE TO CONGRESSIONAL INVESTIGATOR

Between 1964 and 1997, the government of Cuba accused the United States of ten biological warfare attacks after infectious disease outbreaks occurred. While none of these accusations were ever proven conclusively, one event almost certainly occurred.

On May 6, 1971, pigs in a Havana, Cuba, hog farm were diagnosed with African swine fever virus. The virus spread and some 730,000 pigs were slaughtered and set ablaze in deep trenches. The United Nations Food and Agriculture Organization called the Cuban virus outbreak the "most alarming event" of the year. Pork production, one of Cuba's few commodities, ground to a halt for months. Facing a severe food shortage, Havana residents hid their pigs with relatives in the countryside. The African swine fever virus (ASFV) is devastating to swine; close to 100 percent of infected animals die. Symptoms include acute fever, diarrhea, skin blotching, anorexia, and spontaneous abortion; the highly contagious virus is also transmitted by ticks. The virus is a hardy one—it can survive for months in meat, excretions, and secretions, and for years stored on ice. There is no vaccine and no cure.

Indigenous to East Africa, African swine fever had never before appeared in North America. Except at Plum Island. The USDA lab held no less than seven virus strains in its freezers since 1954, courtesy of the U.S. Army germ warfare program. In June 1963, Plum Island began a long-term project to "develop information on the biological and chemical properties of African swine fever virus aiming at recognition of virus strains. . . ." This research included isolating and growing various virus strains collected from around the world, running experimental vaccine trials, and testing modes of virus transmission using test pigs and different types of ticks.

President Fidel Castro charged America with waging germ warfare against Cuba. "It could have been the result of enemy activity. On various occasions, the counterrevolutionary wormpit has talked of plagues and epidemics. . . ." The "wormpit" was a euphemism for the anti-Castro Cuban exiles living in Miami. Considering the lengths taken by some Cuban-American groups in the United States—often in partnership with the U.S. Central Intelligence Agency—the accusation was not terribly far-fetched.[1] A partially declassified 1964 CIA document on Subproject 146 of MK-ULTRA (the CIA's umbrella code name for its biological warfare program) describes an unnamed plant biologist working on a "philosophy of limited anticrops warfare," including the use of cane smut against sugarcane—perhaps Cuba's most important crop—"to formulate a basic approach to an attack on [deleted material]." The redacted name of the foreign country can only be guessed.

Castro's accusation fell on deaf ears until January 1977, when Long Island's very own Woodward and Bernstein—the *Newsday* investigative duo of John Cummings and Drew Fetherston[2]—wrote an explosive story under the banner CUBAN OUTBREAK OF SWINE FEVER LINKED TO CIA. Every national newspaper carried the lead the following day. Multiple unnamed sources regaled the two reporters with a tale of intrigue and germ espionage. "With at least the tacit backing of U.S. Central Intelligence Agency officials," they wrote, "operatives linked to anti-Castro terrorists introduced African swine fever virus into Cuba in 1971." A source said he was handed the virus at Fort Gulick, a now defunct Army base in the Panama Canal Zone that hosted the Army's School of the Americas. At Fort Gulick, Green Berets trained mercenaries for jungle warfare, and staged joint Army–CIA covert operations in Latin America and the Caribbean. The virus came ashore from an unidentified vessel that landed at the Canal

---

[1] After relations between Castro and the Eisenhower administration soured, the United States broke off diplomatic relations with Cuba in January 1961. Fearing increasing Soviet influence in Cuba, a scant ninety miles south of Florida, the CIA employed many of the exiles to launch Operation PLUTO—better known as the Bay of Pigs Invasion—a coup d'état which failed miserably after a two-day struggle. In the early 1970s, around the time of the ASFV outbreak, Cuban mercenaries, working with CIA operatives Howard Hunt, G. Gordon Liddy, and others, staged numerous subversive projects that would wind up under the "Watergate" banner. In addition, it is now known that there were multiple assassination attempts on Castro, code-named Operations MONGOOSE and ALPHA-66. One plot involved an operative handing Castro a poisoned cigar; another scheme would hand him a set of diving gear impregnated with tuberculosis bacteria and a toxic fungus. Still another plan called for a chemical that would make his beard fall out. All three attempts failed.
[2] Cummings would go on to write best-selling exposés about the Mafia and President Bill Clinton.

Zone's Mindi Pier, and then it was sealed in a unmarked container at Fort Gulick.

The source said he brought the virus container to a small motorboat, which sped along the coast of Panama to Bocas del Toro near the Panama–Costa Rica border. There, the package went onto a fishing trawler. A CIA-trained source on the fishing trawler told the reporters the virus sailed north through the Caribbean Sea to U.S.-owned Navassa Island, an uninhabited spit that lies between Haiti and Jamaica. After a brief stopover, the package was put ashore one hundred miles north on the eastern shore of Cuba, near the U.S. Navy base at Guantanamo Bay in late March. There it was delivered to anti-Castro operatives. Two months later, an outbreak of African swine fever appeared for the first time in North America.

Long Island Congressman Tom Downey expressed outrage. "It is preposterous that the U.S. government tried to destroy portions of the population's food. Who is it aimed at? Is it influencing a government when you do it clandestinely?" Senators Richard Schweicker and Daniel Inouye, both conducting germ warfare investigations at the time, echoed Downey's sentiments. "He's passed the point of being able to be surprised," said a Schweicker aide of his boss. "So many of these seemingly outrageous stories came true."

Where did purified vials of African swine fever virus come from?

According to the federal government, Plum Island is the only location in the United States where African swine fever virus is permitted. No one will say on the record that virus for the Cuban mission was prepared on Plum Island and sent to Fort Gulick. However, given the frequent traffic between Plum Island and Fort Detrick, samples—with or without the USDA's knowledge of the ultimate purpose—could have been sent by courier to Fort Detrick for transshipment to Fort Gulick. Declassified documents uncovered reflect exchanges between the two labs at that time of other virulent germs, like Rift Valley fever, Venezuelan equine encephalitis, pleuropneumonia-like organisms, tuberculosis (bovine type, strain 854), and equine infectious anemia (New Hampshire virulent strain 1535).

Virologist Dr. Robert Shope (son of the late Rockefeller Institute scientist and Erich Traub's American mentor, Dr. Richard Shope), who served on Plum Island's advisory committee, is a nonbeliever. "The *Newsday* article was absurd," says Shope. "There's no grounds for it. It would have been stupid to do such a thing because of the boomerang effect [the virus backlashing on the United States]." Dr. Callis, director at the time of the outbreak, likewise denies any link between Plum Island and the Cuban outbreak. "There are a lot of rumors that the CIA planted it there—well, I'm not a CIA specialist, and I know they've done some stupid things, but I don't think they'd do it that close to the United States. Cuba likely got it from garbage they imported or from returning military staff from their African political programs in Angola."

Norman Covert, Fort Detrick's historian, shows how the CIA could easily have been involved—and unwittingly co-opted Plum Island. "There were CIA people who infiltrated the [Fort Detrick] laboratories. They did their own work, and we know now what they did with LSD and other psycho-illnesses. They had their own little cell there—they worked on their own, and I suspect that a very small circle of people knew that." This type of information isolation—informing people of project details strictly on a need-to-know basis—is the brand of secrecy that might have been used to poison Cuba's food supply with germs. Compartmentalization of each step made Plum Island an unknowing accomplice when it trafficked in viruses between Fort Detrick and elsewhere.

Efforts to explain away the outbreak as a natural occurrence do not hold up to close examination. The theory that food wastes from Spanish aircraft were fed to domestic pigs fails to address that Cuba, like the United States, had always kept their nation disease-free through strict importation quarantines. Cuban investigators claim ASFV broke out simultaneously in two distant locations; germ warfare experts say that contemporaneous sites of infection are unnatural and point to a deliberately caused outbreak. Because it is impossible to disprove, the logic of a methodical scientist dictates that a germ warfare attack cannot be ruled out. CIA assassination plots (some of which involved germs) and the Bay of Pigs invasion stand as acknowledged covert acts by the United States government to force regime change upon Cuba.

THE A-WIRE

The USDA devised a strategy to quiet the mounting concerns over Plum Island being raised by the press: it would host a national media day. Fifteen years after dedication day, Plum Island again opened its gates to reporters. Meddling local reporters with on-island deep background sources weren't telling the story the government wanted to tell; national reporters from the *New York Times,* the *Washington Post,* and the *Chicago Tribune* would. The USDA figured it could trump local reportage by spoon-feeding Plum Island positively to a captive, less informed national press.

Karl Grossman, short, rotund, and dark-bearded, was then an east end cub reporter for the daily *Long Island Press,* and managed to claw his way into the media day event. With a circulation of about 600,000, the *Press* was the seventh largest afternoon daily in the nation, serving Queens, Nassau, and Suffolk counties. The night before media day, he phoned James Reynolds, the USDA's public relations rep, for details on when to meet in the morning, and they got to chatting. The spokesman, relieved his well-orchestrated press tour was all set, casually told the reporter, "The research also involves building defenses to . . . a foreign nation utilizing biological warfare." After being egged on by Grossman ("I got him on a run," says Grossman, "and he just kept going and going"), Reynolds protested. "We're not a front for the Department of Defense," but "America must be prepared to protect its food sources. . . ." This was a far cry—wrote Grossman in a story he handed his editor before boarding the ferry to Plum Island the next morning—from the mantra that only civilian research took place on Plum. That had been the company line since dedication day, since Doc Shahan left mention of Colonel Mace and the Army on the cutting room floor.

When the press arrived, Dr. Howard Bachrach held up a tiny glass vial before the crowd of twenty-four newsmen and said its contents could infect not only all the cattle on the Earth, but all the cattle that had *ever* roamed the Earth. Billions and billions of potential infections were held between the little man's thumb and forefinger. And they had plenty more of this and other germs on hand.

Another scientist, tired of the journalists' pestering about biological warfare operations, barked at them in the hallway, "Absolutely ridiculous! Do you see any evidence of such things?" Reporters looked around the lab, eyes darting, unsure if they were missing the big white elephant.

At lunchtime in the cafeteria, Grossman placed his lunch tray down at an empty table. James Reynolds walked into the big room with his PR aides and a few of the national reporters in tow. They picked up their preordered lunches and sat down; Reynolds and his aide pointed and sneered at Karl across the room, now sinking his teeth into a soggy tuna fish sandwich.

While Reynolds had been shepherding his flock through the staged laboratory tour, extolling the virtues of Plum Island to the newsmen scribbling on little notepads, Karl Grossman's story of the night before had made the front page of the morning edition of the *Press*, and was being read by hundreds of thousands on the mainland. The PR man had been duped the night before.

That was only the beginning.

The story made the Associated Press's A-Wire, which meant that every news media outlet in the *entire country* had read an official-looking all-caps news brief off their teletype machines that said Plum Island was a biological warfare center. "These PR guys just knew—" remembers Grossman fondly. "They just knew that if something came across the AP wire, it was like it had come down from Mount Sinai." Editors around the nation were ordering their desk reporters to write stories based on the wire. With the stroke of his pen, Grossman had dashed the efforts of the USDA, singularly usurping Plum Island's canned media day. No matter what positive impressions the national reporters returned to their news desks with, they would still be colored by the contents of the authoritative AP story. Grossman wrote a second follow-up story the next day under the banner OUT OF 'ANDROMEDA STRAIN' . . . RIGHT HERE ON PLUM ISLAND, where he drew a frightening connection to Michael Crichton's 1969 novel about a deadly outbreak of viruses caused by the malfunction of a military satellite launched by the U.S. biological warfare program.

When Director Jerry Callis glanced at the cover of the *Long Island Press* on his desk, he was in utter shock. After disembarking from the 4:30 p.m. ferry off the island, he sped over to Claudio's in Greenport, hoping to find some of the national reporters to whom he could plead his case. There he came upon a young Boyce Rinsberger, then a junior *New York Times* reporter (and later an award-winning editor at the *Washington Post*, and editor of *Science* magazine). Callis shoved the cover of the *Press* in his face and gestured wildly.

"You upset?" Rinsberger said, sipping a cocktail at the bar.

"Hell, I'm very upset," Callis said.

"Well, there's not a lot you can do. You can write a letter to the editor, put it in your desk drawer, and wait two weeks. See how the other stories come out—put them in an envelope with your letter, and send it all to the *Long Island Press*. Let the editor there decide whether he has the best writer in the nation on his staff, or the worst." Callis followed Rinsberger's detailed instructions. But his letter never ran and he never received a reply from the editor.

Karl Grossman's 1971 twin stories had set the tone of dialogue and defined the standard by which the public would judge the island. The self-described "Plum Island Enemy Number One" recalls running into Callis years later on the mainland. "He screamed at me." Asked about the reporter, Callis says, "Karl Grossman? I don't know if I have a word to describe him." The reporter would continue his role as the ever-prodding thorn in Plum Island's side, and his scathing pen wasn't going away anytime soon.

CONGRESS FIRES A WARNING SHOT

The media's new interest in Plum Island piqued the interests of local officials, including freshman Congressman Thomas J. Downey. For a politician, 1974 was the year to be a Democrat. With the resignation of President Nixon and the word "Watergate" on the lips of every American, Republicans had little hope of winning elections to public office. The Democrats even won an upset in overwhelmingly Republican Suffolk County, where they had fielded Tom Downey just to have a name on the ballot. One of the youngest ever to serve in Congress, he was barely twenty-five (the minimum age set forth in the Constitution) when he took his oath on the Capitol steps. Pictured on the front page of the *New York Times* playing basket-ball with his younger brother in his parents' driveway, Downey became the poster child of the Congressional Class of 1974, the "Watergate Class."

Fresh out of college, and now a member of the House Armed Services Committee, he dove into an investigation into allegations of LSD drug test-ing and deaths at the Army's Edgewood Arsenal. "I did it simply because I wanted to get some press," he recalls. "The next day, a Capitol Hill cop comes into my office with a complete dossier including pictures and recruiting brochures and films from the arsenal, where they had him [the police officer was then an Army private] without any antidotes present, taking large doses of LSD. The materials enticed young soldiers to come to the arsenal—'Come see Washington,' they said. 'See the monument, have fun, all while helping your government.' By taking drugs!" Downey dis-covered a Long Island man had committed suicide after participating in the program. Before long, the young congressman had Army generals twice his age running for cover, forced the program to close down, and gained national attention.

Downey's next fight was led by a person even less likely than the twenty-five-year-old newly minted congressman—his twenty-seven-year-old intern.

Ron Fitzsimmons attended high school with Downey and worked on the triumphant 1974 campaign. Downey took on his old buddy, "Ronnie"—who had been horsing around on an extended-year, Vietnam-era college track at State University of New York at Stony Brook, and was unemployed during the summer of 1976—as an unpaid intern in Washing-ton. Ronnie was assigned to typing envelopes up in "The Cage," catacombs with wire-cage doors lining the attic crawl space on the top floor of the Cannon House Office Building that had storage spaces for each member. Bored, and thinking Downey was punishing him for something he did, Ronnie walked down to the office and begged chief of staff Fred Kass for something interesting to do.

Rifling through the papers on his desk, Kass uncovered a thin manila file folder and tossed it across his desk at Fitzsimmons. "Here, Ronnie. Check this out."

Inside were two pieces of paper about Plum Island with rumors about biological warfare. Though he grew up nearby, Ronnie had never heard about the island.

"See what you can find out," said Kass. "Oh, and here's the name of a *Newsday* reporter you can call."

As Fitzsimmons climbed the flights of marble steps, returning to The Cage, his thoughts raced. An odd inspiration came to him. Only a few months before he had seen *All the President's Men*, the blockbuster Watergate film with Robert Redford and Dustin Hoffman playing Woodward and Bernstein, the two *Washington Post* cub reporters who brought down the president. He remembered how they took one morsel of information and expanded on it. *It was a list—a list of political contributors,* he recalled. *And there was that scene where they went through the phone book and made all those phone calls that led to more clues. Now it all fits—things are connecting here,* he thought. A sense of purpose came over him with every step he climbed. Soon he was racing down the hallway to his cage. Ronnie harbored an envy of his childhood peer, Tom Downey, and his improbable success. Now, with Woodward and Bernstein as his self-appointed mentors, he would play investigative journalist *and* congressman. Hunched over his typewriter, he banged out a letter from Downey, asking the secretary of agriculture for a Plum Island phone book and organizational chart. He could hardly contain his excitement. The Plum Island Inquiry had begun.

Days later, the list of employees and a crude chart showed up. Grabbing the local white pages, Fitzsimmons painstakingly cross-referenced the three hundred plus names and penciled in phone numbers on the long list. He studied the chart, tracing the lines from box to box. *Now who works for who?* Then, starting with the As, he picked up the receiver and began making calls.

"Hi, I'm Ron Fitzsimmons with Congressman Tom Downey. I'm calling from Washington—we're looking into the operations of Plum Island. . . ."

"I heard lots of stuff," recalls Fitzsimmons. "But I picked up right away that—much like in the movie—people were afraid to talk. It's a small community out there—there's only that one road at the end of the North Fork. I sensed people were concerned about talking." Many hung up on him. Some agreed to talk, but not over the telephone. "In person," they said to him. "The phones might be tapped." The intern decided he must return to Long Island, convinced it was the only way to get the full story. Willing employees insisted they not meet at their homes, worried that neighbors would spy a strange car pulling up to the house. Instead, Ronnie met with them at the park, behind the school, and at their friends' homes in neighboring towns.

At the suggestion of Kass, Fitzsimmons had hooked up with *Newsday*'s Cummings and Fetherston. The team had recently uncovered a possible biological attack on Cuba, and unearthed a series of secret outdoor biological warfare tests in U.S. cities orchestrated by Fort Detrick in the 1950s. Although Fitzsimmons recalls being interested only in the "personal lives and health" of Plum Island workers, he thought the two reporters seemed more intrigued by the island's alleged connections to biological warfare. They all agreed to share sources and research, and spoke once a week thereafter.

The worker interviews, pooled together by the trio, were quite troubling. They heard that women were barred from working in Laboratory 257 for fear that they "would carry infections to their children." And men could only work on a volunteer basis on subacute sclerosing panencephalitis (SSP), a disease arising as a complication of the measles that affects children and adult males with personality changes, intellectual deterioration, periodic involuntary movements, and severe dementia. Few patients diagnosed with SSP lived past three years. Paul Rose, the Plum Island union president, told them one section of Lab 257 (he would not identify which) was closed off, and no one was permitted to enter or know what was going on inside. This didn't seem too far-fetched—a scientist had told a reporter off the record a few years before, "I knew what I and the other three people working with me were up to, but I didn't know really what the whole place was about. We kept in our own laboratory all day long . . . *there was some talk on the ferry. . . .*"

One source, taking the pseudonym "Charlie," told them there was heavy traffic between Fort Detrick and Plum Island in both directions, as well as from Egypt and Kenya. A document revealed an October 1969 shipment by military escort of "Venezuelan Equine Encephalitis (VEE) virus and antisera," to "Dr. J. J. Callis, from the Viral and Rickettsial Division, Army Biological Laboratory, Fort Detrick." VEE had been one of Fort Detrick's main germ warfare agents for decades, along with anthrax, botulism, and Rift Valley fever; human guinea pigs had been injected with VEE to develop a vaccine called TC-83, under a top-secret program called Operation WHITECOAT. "FOR RESEARCH PURPOSES," the permit announced. When the *Newsday* reporters inquired about it, Plum Island assistant director Dr. John Graves said it had been sent to prepare for the disease, which threatened to spread into the United States from Mexico.

Another document described a one-year project called "Pathogenicity and Prophylaxis of Influenza-A Viruses." Aimed at live and inactivated flu viruses and engineering man-made recombinant DNA flu virus strains, the research tested their effects on pigs and birds. Working on viruses that jumped between pigs and humans, they used pigs as virus production factories. Scientist Dr. Charles Campbell was able to combine human Hong Kong flu strain with a strain of swine flu (that in 1976 killed an Army soldier at Fort Dix, New Jersey, and touched off a national flu panic) and isolated a new, hybrid virus strain that had the characteristics of both. In this regard, Plum Island did not seem to fit its relatively innocuous title of "Animal Disease Center"—this work placed its actual research in a human realm. That year, President Ford authorized an emergency national flu vaccination campaign to combat this same "New Jersey" flu strain now on Plum Island.

Fort Detrick's animal disease chief, Dr. William Hinshaw, headed up germ warfare against enemy food. Though retired since 1966, he regularly visited Plum Island in the 1970s and served as a "consultant." Charlie also divulged that a lab chief working on a "very hot" monkey virus—possibly the forerunner of Ebola virus, the new Marburg hemorrhagic fever virus that killed German lab workers with ghastly hemorrhagic fevers at a vaccine plant in Marburg, West Germany. In *All the President's Men*, "Deep Throat" implored Woodward in a dimly lit parking garage to "follow the money." Now, the scene was re-created in the desolate field behind a school, as Plum Island sources told Fitzsimmons to "follow the viruses."

In 1958, the USDA had quietly pushed through the House of Representatives a measure to allow the transfer of viruses over the mainland to Plum Island. A decade before, when it authorized the construction of an offshore exotic virus lab, Congress specifically banned foreign viruses from the mainland, and from any island connected by bridge or tunnel. Viruses had to be unloaded from giant Navy freighters onto tugs in Gardiner's Bay that steamed their biological cargo to Plum Island harbor. Requesting the law change, USDA officials told House members, "Only four or five vials of the virus would be delivered over land twice a year" and promised to package those few shipments with the utmost care. The law change would cut costs and do away with a "an inordinately expensive and inconvenient procedure," they said. Congress agreed and granted them their wish.

The USDA was either ignorant of the level of germ traffic that would ensue or it intentionally lied to the lawmakers. By 1976, huge crates of live viruses were traveling over congested Long Island roads multiple times *each day*. Germs were shipped off Plum Island, logged only as "biologicals," with no other description. Virus and bacteria samples coming into the United States were stowed behind the pilot's seat in a commercial airliner, picked up at Building No. 80 at JFK International Airport, and shifted from one car to another in the parking lot of a department store in the Suffolk town of Sayville. Over the years, thousands of these trips were made through New York City, and over Nassau and Suffolk county roads; local health officials were unaware of potentially deadly biological voyages occurring every day right under their noses. Sometimes the shipments of live exotic virus samples were taken home with the couriers in the evening, placed in household freezers, and brought to Plum Island the next day. A USDA official later admitted to doing just that, but only "once or twice" a month. Couriers weren't instructed on what the samples contained, which could be as large as a thirty-gallon drum, or how to go about using the emergency decontamination kit in case something spilled. Often the Plum Island scientists themselves had little idea what the blood serum samples contained. Sometimes crates were locked in a freezer in the Orient Point warehouse—but the key to the freezer hung conspicuously on a wooden peg just inside the front door. There were no security guards at the Orient Point facility. They had been dispensed with long ago. And if the package arrived after the last scheduled boat left for Plum Island, it was just left outside the door to the Orient Point office, or on the side of the road, either inside or leaning against the big Plum Island mailbox. The following morning the marine crew took it to the island on the first boat out.

In June 1970, a virus courier named Alfred Von Hassel was killed in a car accident on Northville Turnpike in Riverhead, thirty miles west of Plum Island, while transporting unidentified biologicals. The container—described as an aluminum case one foot wide and one foot deep, and eighteen inches tall with latches—catapulted from Von Hassel's car on impact. It was retrieved from a field abutting the road where the vehicle had flipped over. Fortunately, Leo Golisz, an off-duty Plum Island security guard, passed by the commotion and recognized the upside-down car with its U.S. GOV'T marking. He identified himself to police and left the scene toting the shiny silver box of biologicals, which he brought to Plum Island. Other foreign shipments were less sturdy and wouldn't survive such a horrendous accident—one source recalled transporting picnic coolers and leaky cardboard containers with fluid oozing out of the bottom.

The abuses read like a checklist on how to disregard regulations and abandon all common sense. Employees told the investigators that DDT insecticide was still being sprayed liberally on Plum Island, four years after use of the toxic compound was banned by the federal government. Tons of sewage effluent flowed daily into Long Island Sound and Gardiner's Bay from the two laboratories, untested to ensure that germs had been destroyed by contamination. Radioactive materials that were incinerated required testing with Geiger counters to monitor fallout, per the Atomic Energy Commission. Nervously wringing their hands as they sat uncomfortably in their friend's house in East Marion, two employees admitted the incinerator charging room hadn't been tested for radiation exposure in *years*. Not only was security at Orient Point nil, but the guardhouses at each of the two lab compounds on the island were unmanned. The once thirty-four-man-strong security patrol had been decimated. A paltry eleven guards covered three 8-hour shifts, seven days a week, 365 days a year. "I could take you to the island on a motorboat," said a source, "get you ashore and put you in a lab—and no one would ever see us."

Then there were the illnesses.

"James Robinson," a worker in the lab glassware department, became sick in 1975 and had to retire early. The USDA offered him a 40 percent disability payment, until the union protested the settlement, stressing he had been exposed to dangerous microbes. The USDA doubled the ante to an 80 percent payment and the employee's grievance conveniently faded away. A union leader had been filing workplace grievances for years, noting that Plum Island workers occasionally contracted severe rashes. No steps were taken by management toward amelioration.

"Bruce Becker," a safety technician, also caught something. He began running a low-grade fever and became sluggish. His physician found high blood pressure and diagnosed a flulike virus, and suggested that he have his kidneys checked out. For almost a year, Becker didn't feel like himself. Then in August 1974, his condition took a turn for the worse. Any physical exertion, however minimal, caused nodules all over his skin that took weeks, sometimes months, to disappear. Plum Island's nurse, Frances DeCristofaro, arranged an appointment with another medical doctor, who diagnosed infection with an unknown virus. Becker sought a second opinion from a Dr. Georgeson in Riverhead; biopsy samples of the lumps were taken and the results came back negative. Perplexed, Georgeson suggested that Becker try the Leahy Clinic in Boston. When he did, there was still no confirmation. Samples of his kidneys and the lumps were taken at Riverhead Hospital and again at St. Charles Hospital in Port Jefferson. The Leahy Clinic doctor suggested that Becker have one of his kidneys removed, though he would not tell him why. In March 1976, the Workers' Compensation Board referred him to yet another doctor, "Tom Belford" of Greenport. After examining his patient, Dr. Belford announced, "You're either okay or you're not in the books." Belford sent for tests from the CDC in Atlanta to explore the possibility of animal diseases. Becker was given two shots, and his arm blotched and swelled in both places. When he returned in April for more tests, Dr. Belford brushed his patient away. "I'm getting too much pressure," he said. "If Georgeson wants to give you these shots, I'll give him the serum." The doctor would not elaborate.

While Woodward and Bernstein parsed through stacks of circulating cards in the ornate rotunda of the Library of Congress, Fitzsimmons and a fellow intern barged into the offices of the *Suffolk Times* and other local weeklies, spoke to the editors about Plum Island, and pored over their vertical files, reading old newspaper clippings. Then one summer day, Ronnie decided to roll the dice and personally investigate the Orient Point warehouse and ferry dock to Plum Island. Switching cars halfway to avoid detection, the intern-turned-sleuth slipped into the parking lot near the ferry launch. "I got out and went into the building," he recalls. "I saw a fridge in the corner and opened it. Inside were bottles with yellow stuff in them, labeled with Latin names. I'm not a scientist—but it looked like bacteria or something. I walked around some more." He encountered no security. Satisfied, Fitzsimmons jumped in the car and sped off.

They were probing deeper than ever before, and seemed to be getting somewhere. But by August 1976, someone pulled the plug on the investigation. One scientist who agreed to meet changed his mind, saying his superiors had advised him not to speak. Another source told Fitzsimmons, "It shouldn't surprise you, then, if I told you that Callis has told people not to talk to you." Lower-echelon lab techs and dockworkers would only mutter, "Speak to Callis," and turn away when approached. "When it became clear we were snooping around, contacts would say, 'I know who you are,' and hang up. A few who had offered information were now recanting, they were so concerned that we might publish the information, and they would be fired." One woman broke down and cried to Fitzsimmons over the phone, wailing, "I'm not allowed to speak. Please, *please*! Leave me alone!"

After compiling his extensive research, and reviewing his pre-gag-order interviews, he compared notes with the *Newsday* reporters. Fitzsimmons then typed up letters from Downey addressed to the Army and the USDA, demanding detailed responses to the results of the investigation. The Army's terse response, perhaps with a dose of condescension, was addressed to "Mr. Downey" ("It's interesting," says Tom Downey now. "Why wasn't it 'Dear Congressman'?"). The letter noted that the Army had phased out all agriculture-related activities within the last year. Seven years after President Nixon ordered Fort Detrick out of the offensive biological warfare business, the Army was just now turning over the big Detrick greenhouses used for anticrop germ warfare to the USDA. Acknowledging the agreement that likewise turned over Plum Island to the USDA, "along with defense of our livestock against biological attack," the Army also admitted the ongoing relationship between Fort Detrick and Plum Island. "Army interest . . . was one of keeping abreast of closely related efforts in microbiological research. . . . Throughout the ensuing years, there has been a cooperative effort between the Department of the Army and the Department of Agriculture in this area . . . [l]iaison was maintained at the working level. . . ." This relationship dated back to the construction of an Army germ warfare lab there, Laboratory No. 257, in 1952.

Two more letters were sent to the USDA, including a lengthy single-spaced, five-page dispatch to Dr. Callis, demanding all information on Plum Island's shadowy past and current condition.

The letters presented a thorny problem for Callis: not only was Downey a local congressman, but he also served on the Agriculture Committee. Plum Island would have to tread very carefully here. Fitzsimmons, in the name of Congressman Downey, asked Plum Island officials for names, lists, reports, correspondence, files—anything that proved an existing relationship between Plum Island and Fort Detrick. Attempting to follow leads developed by the *Newsday* reporters, the letter demanded information on a "swine flu" being sent to Fort Detrick, the Panama Canal Zone, and any other installation. In all, fifty-six specific and open-ended requests were made of Plum Island, many of which contained multiple subrequests.

The letters closed, "Due to the nature of the information requested above and the importance of this subject, I ask that you act at greatest dispatch. Your prompt attention to this matter is greatly appreciated." Truth was, there really was no rush. The summer was winding to a close, and intern Ronnie Fitzsimmons had to go back to college.

Awaiting the USDA's reply at the end of August, Ronnie typed up a "summary" for his boss. The three-inch-thick report bowled over the congressman and chief of staff Kass. "Tom was flabbergasted," says Fitzsimmons, "I wasn't getting paid—I did it on my own—and I guess I just ran with it." They each took a piece of the tome and fanned through its legal-sized pages. They scanned the interview transcripts and flipped through the "Items" section, which included a possible link between Plum Island and an outbreak of swine flu in Cuba. It also documented hearsay of a conversation hinting "President Ford knew of some kind of leak of shell-fish toxin and swine fever within the past several years."[3] They read the one-page "Possible Stories," that spoke of the island's "shabby operation," "off-limits" laboratory, and unwarranted gag order ("Why can't a congressman's office get straight answers from government employees?").[4] Downey was visibly impressed and clearly surprised.

Then the USDA replied to Downey, answering the five-page inquisition in the most general of terms. For the first time, however, it acknowledged research on behalf of the Army in the early 1950s on vesicular stomatitis, brucellosis, and Newcastle disease, an exotic viral illness that affects the nervous and respiratory systems of birds. Laboratory infections of the latter disease have been known to occur in people, and it causes the death of nearly 100 percent of flocks that contract it. The USDA continued working with the Army, maintaining a liaison at Fort Detrick. In 1962, said the letter, the USDA provided farm plots in Oklahoma and North Dakota for the Army's open-air trial on cereal rust of wheat, one of Detrick's major

---

[3]The year before, CIA Director William Colby told the Senate's Church Committee, then looking into intelligence activities, that the CIA had a secret cache of ultradeadly shellfish toxin stored in a Fort Detrick vault, even though President Nixon had ordered the destruction of such weapons back in 1969.

[4]The final part of the otherwise sobering report reflected a vindictiveness on the part of its author, perhaps brought on by the gag order Fitzsimmons faced when trying to interview sources: "For *The National Enquirer*, Dr. Callis is reportedly homosexual."

anticrop germ agents. While the Army told Downey the relationship was at a "working level," the USDA stated it was "executively directed." It was probably both.

Throughout the investigation, Dr. Callis repeatedly offered Fitzsimmons and his boss an opportunity to visit Plum Island to obtain information firsthand. Downey hardly knew the extent of Fitzsimmons's efforts, and the intern balked. He would say almost three decades later that an official visit would have amounted to "total bullshit. . . . We felt they would show us a laboratory building we wouldn't understand or know anything about." Instead of attending a dog-and-pony show, he preferred to stand on his interviews with current and former employees, who he figured knew the real Plum Island.

To Fitzsimmons's chagrin, the story that reporters Cummings and Fetherston published in *Newsday* that fall noted the links between Plum Island and Fort Detrick, but didn't address employee safety and health, issues he believed were vital. Titled "THE PLUM ISLAND LAB: FOR MANKIND OR AGAINST IT?", the article did not mention employees like Bruce Becker with his mysterious illness brought on by the lab. It failed to mention the lone medical official stationed on the island, a part-time nurse who confessed she did little more than dispense aspirin and Ace bandages. Nor did it address the assistant director who placed a package of live virus on the seat next to him on a trans-Atlantic commercial airline flight. Fitzsimmons had shared his painstakingly detailed notes with Cummings and Fetherston, and little of it made print. "I was frustrated," he says, "hoping [the *Newsday* investigators] would report about the people and how they were getting sick from exposure to these deplorable conditions." Plum Island management branded the article as a distortion of facts written to sell newspapers, brazenly denying any germ warfare research. "This center has not been, nor is it now, engaged in biological warfare," said Director Callis.

Making matters worse for the probing intern, the *Suffolk Times* rushed to the aid of Plum Island and publicly whipsawed the "college students" and the "irresponsible" actions of *Newsday*. "The desire to uncover another Watergate scandal runs strong in the hearts of all investigative reporters," the paper acknowledged, but tarnishing reputations and suggesting the island was a cover-up for biological warfare was appalling. Fitzsimmons had acted recklessly. And noting that the Marxist-Communist publication *Daily Worker* charged Plum Island with germ warfare back in 1952—sending comrades to fight the laboratory during the public hearings—the editorial mused it was "somewhat surprising to find *Newsday* following the [Communist] party line" today. The *Suffolk Times* had done everything but brand Ronnie Fitzsimmons a hippie Communist and *News-*

day an organ of the Soviet Politburo. The editorial marked the apex of the local community's support of Plum Island, support that would wane soon enough.

"I felt they did *something* on Plum Island—even though they always denied it. The very nature of the secrecy of the place gave fuel to the rampant speculation that existed," says Downey, today one of Washington's most powerful lobbyists, still youthful three decades later at age fifty-four.

"Because they wouldn't tell you anything, everybody believed the worst," Downey says. "And who would believe *them*? This was a period of time when lying was part of the operation. People wouldn't tell you the truth—they didn't tell the soldiers the truth [at Edgewood Arsenal] and they're giving them these drugs." After the LSD scandal, "It would not have been a long shot to think that the Army was testing biological agents at Plum Island or working with the USDA—it would have been an easy connection to make. I never believed that—with all due respect to the Army, not that they are inherently dishonest people—but I never believed they told the whole truth. We just assumed they were not going to tell the truth about things, until you beat them to death with it. And they didn't."

"It was a rough time, a time very, very different from today," Downey continues. "The 1970s was the end of an era for them, too. You had a post-Vietnam period, and the military was up against it. And to [be browbeaten by] some freshman member of the House Armed Services Committee? They weren't going to pay attention to me, they were just hoping I went away."

Which he did. Or rather, his intern did. A *Newsday* editorial called Downey's Plum Island inquiry "not enough." Busy with other high-profile matters, Tom Downey launched no formal investigation. And Ronnie Fitzsimmons went back to Stony Brook to finish up college.[5]

Years after the Downey–Plum Island inquiry, Fred Kass confessed to Ronnie that they never expected anything to come of his work.

It was, after all, just another summer intern project.

---

[5]Later, he would find the recognition in the political arena he so desperately sought, when he became the first man ever hired by the National Abortion and Reproductive Rights Action League, better known as NARAL. Today, he lives with his wife and two children in the Washington, D.C., area, and is the national spokesperson for an association of abortion clinics.

7

"The Disastrous Incident"

> *A safety-designed building should be constructed in such a manner that it is virtually fireproof and so that each floor and room can be considered relatively waterproof, airtight, and insect and vermin proof. Thus, only the controlled entrances and intakes and exits and exhausts would need continual surveillance for trouble on the outside. . . .*
> —PLUM ISLAND DIRECTOR JERRY J. CALLIS

September 15, 1978, 7:00 a.m.—*This is going to be a rough day,* Billy thought, holding his stomach and trying to gain some composure as the ferry bumped and tilted through the choppy waters of Plum Gut. Thursday nights were always fun nights to put back a few with the guys at happy hour, their practice run for the weekend. It was Indian summer, the best time of the year on the North Fork. The weather was warm, the humidity finally beginning to break and, thank heavens, those damn city summeristas disappeared on Labor Day. Every Memorial Day weekend an exodus begins from New York City. America's rich and famous (and those aspiring to the same) disperse into the two forks of eastern Long Island to escape the city's blistering heat and cramped quarters. The population of a quaint little enclave like Southold or Southampton swells to innumerable tallies. For those without a private helicopter the traffic becomes murderous; and the local, modest-living populace waits out the summer in agony.

Just eight more hours left in the week, but for now, Billy Doroski would have to deal with the consequences of the booze. The early boat to Plum Island was typically a unique but enjoyable commute—like an oversized carpool where a shared and sheltered camaraderie reigned. The regulars were always there to goof around with on the way over, and most of the stuffed-shirt scientists usually caught the later 8:00 a.m. boat. No fun this morning, though. A tall coffee nursed the pressing pain at the back of Billy's neck, but didn't do a thing for his ailing stomach, which churned with each methodical thud of the boat over the Gut's strong currents. A laboratory technician, Billy was responsible for all the preparatory work for the scientists—inoculate the animals with viruses, draw blood serum samples, culture viruses in dishes, and prepare sample slides. In reality, Billy handled the lion's share of the labor so the doctor could stroll in, take a look at his work, and postulate.

Leaving the ferry, he boarded the old school bus stenciled LAB 257, and slowly eased himself down into the small seat. He exhaled. *Yeah, this is going to be some day—some day. Now why the hell did you have to go and do that to yourself?* He slumped over and rested his palm squarely on his forehead as the bus careered south along the narrow path, past the marshy lake, before screeching to a halt a few minutes later at 257. Billy waited patiently for his turn to stand and exit, then filed into the compound through the guardhouse door at the outer fence along with twenty-five or so other techs and building engineers, before passing through the gray air-lock entrance into the animal corral chute. Today, he would be assisting Dr. Ahmed H. Dardiri, the chief of Lab 257, who was conducting the annual foreign animal disease school, teaching doctors and veterinarians from around the country how to recognize the symptoms of exotic virus and bacteria infections. That morning, the doctor planned to demonstrate the effects of bovine herpes mammillitis virus, and asked Billy to set things up. Or, in other words, to inject the animals with the virus.

Billy entered the change room, closed the door behind him, switched from his street clothes into his government-issued pale blue lab smock, and, big virus syringe in hand, opened the door to the animal cubicle. He gazed at the animals through his tired, bloodshot eyes. Then he walked closer and squinted at them.

Oh shit—they already got it?!

Billy was startled by what he found. Two steers were stumbling around in the yellow-tiled cellblock, drooling profusely and foaming at the mouth, muttering low groans. *This can't be,* he thought, *there's no way.* Maybe he was in the wrong lab. Maybe the pounding headache was mixing him all up. *No,* he thought, *this is the lab room I'm supposed to be in—I know it is! They have to be clean, these animals, the handlers just brought these two in here from the Old Cow Barn yesterday.* The aftermath of last night's festivities was impairing his judgment. *Well, there's still a little time left—let's try this over again.* Perplexed, Billy rubbed his tired eyes, left the cubicle, closed the air-lock door behind, took a decontamination shower in the change room, and regrouped for a moment. The symptoms he saw bothered Billy, and with good reason: he had once contracted an exotic animal disease after accidentally injecting his cuticle with live viruses. He sat down, scanned the newspaper, and sipped some more coffee—the brief rest would sober him. Then he slowly got up and went back in.

Same thing. He looked at the weeping cattle, checked his clipboard, then looked at the animal again, shaking his head. Something was terribly wrong. He showered out and tried a third time, but same thing again.

This isn't me anymore—I've got some really sick animals in here!

Billy phoned Dr. Dardiri, who had just arrived on the later boat, and the doctor rushed up to the animal containment room. Opening each animal's jaw and inspecting their hooves, the doctor observed signs of disease in the mouth and feet and confirmed his tech's dire assessment. Dardiri called Plum Island Director Jerry Callis at 9:30 a.m. and informed him of what they discovered, noting that the animals came from the Old Cow Barn, officially known as Building 62, one of the three outside animal holding pens. Dr. Callis instructed him to swab some samples and start a laboratory diagnosis immediately.

School would be cancelled that day.

With the receiver still in his hand, Dr. Callis toggled the hook and rang Dr. Louis Jennings, the chief of animal supply, ordering him to Building 62 to see if the animals there were showing any unusual symptoms. Racing into the pens, Jennings grabbed hold of the nearest steer. It looked pretty bad—sullen-faced, the poor beast was drooling and foaming at the mouth. Around 11:00 a.m., Jennings phoned Callis and confirmed their worst fears—the very first animal he examined was ailing from . . . well, from something. He scraped samples of the steer's tongue tissue and gave them to a safety officer, who rushed them over to Dardiri in Lab 257.

Dr. Dardiri had long experience in diagnostic methods—he and his staff had designed scores of them for countless germs. Science, by its nature, is inexact, a constantly evolving body of knowledge among learned men and women. Identifying viruses is no different, as a microbiologist must put an unknown sample through the step-by-step process of reasoned elimination. So first, Dardiri looked at the virus's history. In this case, this step was unusually easy, because it came out of a lab upstairs, rather than from a test tube flown in from the jungles of the Congo. So the historical deductive reasoning was a snap: clear symptoms of foot-and-mouth disease (FMD) virus meant he could proceed straight to the reagents. Normally, he'd have to grow a sample to provide enough material to set up various test cultures and find the culprit. Since there are seven strains and multiple subtypes of the virus, a bunch of reagents were brought out of the freezer. A reagent is a specific substance that reacts only to the disease substance (called the antigen) it binds it to. For example, reagents can be antibody proteins that a test mouse produces to fight a certain strain of a virus. Fluorescent antibody stains light up on a slide when they come in contact with the viruses, which are invisible to the eye (except with the help of an electron microscope). However, one can witness the known, visible reagent substance *reacting* to the invisible virus, which can often be identified by the ruptured cells, where they have attacked. Measuring the holes created by the dead cells indicates the potency of the virus. By manufacturing reagents using live viruses on different test animals, the Plum Island staff had assembled a veritable library of most known germs.

Pulling out the test kits from the walk-in freezer, Dr. Dardiri first tested the samples Billy extracted from the animals upstairs against each strain of FMD and a match was made: Type O-1. Then the moment of truth. The samples from Animal Supply Building No. 62 also reacted to Type O-1 virus.

Oh God—an outbreak!

Viruses had escaped from at least one of the two labs. Foot-and-mouth disease virus had broken out. Who knew what other germs had escaped and where they had gone? A rattled Dr. Dardiri left the laboratory room and typed a terse message addressed to J. J. Callis, Director: "Summary: Foot-and-mouth disease positive."

> **cull:** 1. to pick, to gather . . . 3. to select and separate out as inferior or worthless.
> —*Webster's New International Dictionary*

Armies across Europe erected giant funeral pyres and set carcasses of cattle, pigs, and sheep ablaze. Open pits of fire dotted the countryside. They crackled, hissed, and popped with burning fat and flesh. Clouds of thick black smoke billowed high into the sky and traveled for miles. In the end, fifteen million animals were slaughtered over a span of four months, reduced to tons of black-gray ash and smoke. The governments called it "culling."

A plague had ravaged much of Europe and parts of the Middle East. Though it sounds like a historical account from the Middle Ages, this culling took place in February 2001. For some, the grisly images of the mass graves and carnage were reminiscent of Nazi concentration camps and Stalin's Great Purge.

The culling was a response to foot-and-mouth disease virus, the Plum Island Animal Disease Center's cause célèbre.

Foot-and-mouth disease has been among us for many centuries, and maybe far longer. Some believe it was the fifth of the ten biblical plagues, the disease with which God smote Egyptian livestock. The journal of Hieronymus Fracastorius, a fifteenth-century Italian monk from Verona, noted an epidemic of cattle herds, an account that closely mirrored the disease's symptoms. It mostly attacks cloven-hoofed animals—cattle, pigs, and sheep. Large blisters form on the animal's mouth and feet, and grow up to two inches in diameter. Healthy cells underneath liquefy, and pus swells inside viral blisters until they burst, leaving behind painful ulcers. The pain of the infection on the foot renders the animal lame and immobilized. Blisters on the mouth and tongue make it impossible to eat. Compounding the problem, opportunistic bacteria enter the raw, exposed areas and cause infections ranging from the shedding of hooves and tongue to sudden death. While the virus itself rarely kills the victim, complications from it can be fatal. For many who have seen the effects of foot-and-mouth disease, it is extremely saddening to observe a stricken animal, standing helpless as its ailing body gradually wastes away.

To this day, there is no cure.

Were it not for the founding fathers of Plum Island, foot-and-mouth disease never would have been wiped out of North America in a 1948 eradication campaign.

And, but for the existence of Plum Island, it would not have returned to the continent in 1978.

While Dr. Dardiri was confirming his worst suspicions, Dr. Callis had gathered the lab chiefs and management staff to his office in Building 54, perched on a bluff overlooking Gardiner's Bay and the faint shoreline of Montauk Point in the distance. He briefed the group and dusted off a copy of the 1969 emergency operations plan, the most recent one they had. Though the plan called for a secret code to be used in the event of an outbreak inside U.S. borders to avoid public alarm—"Rumen" for rinderpest virus, "Nada" for foot-and-mouth disease virus—no code names were necessary internally.

The plan called for Armageddon—destroy every living thing on the island except humans. Somebody provided a head count of all fauna: ninety-four cattle, eighty-seven pigs, sixty-six lambs, twenty-eight rabbits, twenty-seven chickens, thirteen goats, six horses, and two ducks, and colonies of mice and guinea pigs. The test animal population on this isle, combined with the flocks of wild birds, deer, insects, and other small critters, was larger than Noah's Ark.

"We can't dispose of the carcasses by burial as the plan calls for—it would

take far too long. The risk of epidemic increases every moment infected ani-
mals remain alive. I'm going to need everyone's support here."

When Dardiri called and confirmed a virus outbreak, the group unani-
mously agreed to burn the place up and scurried off to tell their subordinates. To ease their minds, the line workers were told the escaped virus
rarely affected humans, though many of them knew about Billy Doroski's
infection, when he accidentally injected himself with a healthy dose of the
virus.[1] Billy stuck himself with a needle when inoculating guinea pigs, and
soon after, his finger swelled and he contracted flulike symptoms. Plum
Island scientists confined him in quarantine, inside the containment laboratory, for a full week. When they tested his blood, they found antibodies and
confirmed he had indeed been infected with foot-and-mouth disease virus.
They freed him only after the virus had run its course. After being released
from biological quarantine, he went home and wore a rubber glove on the
infected hand for five days, and then returned to work.

All workers were directed to cease all work and await further instructions. The first order of business was to decontaminate and evacuate everyone on the island, except for the outbreak control team. Then every single
animal would be hauled into the lab and, after blood and tissue samples
were taken, incinerated. Meanwhile, everything on the island would be
sprayed down and chemically scorched. Washington would have to be notified and, only if absolutely necessary, the public at large. Any chance there
was to keep this under wraps, however, was negated by the presence of
sixty-five private contractor workers then building a massive extension
onto Lab 101. Carpenters had been hacking and sawing away at the building for two years now, and there wasn't a soul in Callis's office that afternoon who didn't suspect they were the cause of this mess. The foreman was
called over to be briefed on measures his men would be expected to follow.

Dr. Callis's phone rang. It was Dardiri again. The director's expression
turned grim. More bad news. The people aboard the ferry that had been
launched at noon for Orient Point were now potential traveling mobile disease vectors. Alarmed, they radioed the ferry captain just before the boat
docked at its destination. Now the mainland knew something had gone
wrong. A crowd of people, who had gathered to meet their family members
after a half-day shift, watched curiously as the boat approached and then
mysteriously circled around and sped back to Plum Island. Dr. Callis

---

[1] The first human case of foot-and-mouth disease virus was reported in the hamlet of Yetlington, England, in
1966. There, farmhand Robert Brewis contracted large blisters on his hands and inside his mouth. He also
suffered grogginess and other flulike conditions. Attacked by a swarm of reporters, he managed a joke for
the English tabloids. The doctors said he would recover and would not have to follow the fate of lower animals who caught the virus. "My only worry was that I might have to be destroyed," Brewis said of his infection.

ordered the emergency plan into immediate action. From this point forward, Plum Island would be under a full lockdown. No one was permitted to leave. The director now faced the most painful task of his career—issuing a press release announcing the island's very first virus outbreak.

Everybody got a call to go in and see if their animals were sick," remembers Dr. Carol House, who was working in Lab 257 at the time. "Then we were told to sit tight." Researching in the windowless, fluorescent-lit containment lab rooms on a typical day was difficult enough. "You don't know what time it is, you lose track of things and become very focused. Some people developed light deprivation and seasonal affective disorder," recalls a Plum Island researcher. Once the air-lock doors began to hiss and inflate behind them, researchers hunched over their lab benches, amid the low hum of the air blowers, burying themselves in their work with no connection to the outside world. At Plum Island, isolation and compartmentalization became a way of life.

Now, the walls on the "inside" were closing in. Those with telephones contacted their spouses (which took infinite attempts because the switchboard was jammed) to tell them what had happened and that they wouldn't be home for dinner, but they were okay. Dr. House's lab partner that day was a visiting scientist in from Paris. She planned to train him on the latest diagnostic techniques, but all research work was called off and the two sat around twiddling their thumbs. "It was kind of funny, because he thought we did this every day, and my French wasn't good enough to tell him this day was very different."

Hours later, the 390 workers finally permitted to leave Plum Island—but they had to be decontaminated first. Crews went into each laboratory change room, bagged and labeled everyone's street clothes, and took them away. In place, they laid out one-size-fits-all sterilized white coveralls and white slip-on tennis shoes. Personal belongings, like watches, jewelry, pocketbooks, and wallets, were removed from the change-room lockers. Car keys and eyeglasses would be permitted—after all, people had to drive their cars home—but only after they were carefully dipped in an acetic acid bath. Idling in the cramped containment rooms, pacing endlessly back and forth, began to drive people batty. Finally, minutes before 8:00 p.m., they were released. Entering the change rooms, the staff members showered out without towels and put coveralls on their wet bodies. They were going home.

Exiting the laboratories en masse, the workers looked like a horde of invading aliens in baggy white spacesuits. The 325 staff and the 65 construction workers evacuated in a quiet, orderly fashion. By then they were too tired to say anything. The sharp odor of burning hay was in the air as they filed onto a bus, which splashed through a makeshift decontamination lagoon before pulling up to the decontaminated ferry. Other than the thirty-five-man cleanup crew, everyone was off the island by 8:45 p.m. Frances Demorest, Plum Island's assistant librarian, recalls getting home at midnight and greeting her husband. "I woke Harrison up, and he just looked at me and let out a great laugh. My hair was hanging and I was wearing size forty white coveralls and plastic sneakers. Oh, what a mess!"

The first load of Hereford steers left the animal pen, Building 62, at 9:00 p.m. and boarded the cattle truck bound for the east animal wing of Lab 101. Animal handlers hauled the goats, sheep, horses, and pigs from their stalls into the west animal wing, then onto the loading platform. As they emptied the pens, the areas were sprayed down with a lye solution to kill live germs that escaped. Within two hours, all of the animals moved out of Building 62. The men showered, changed clothes, and went on to the larger task of removing the animals held in Building 21 and in the old Army bunker, Battery Steele. That took until daybreak on Saturday. Downstairs, in the rear of the building, workers cranked up the oil-fired incinerator.

Up the ramp, at the head of the necropsy room's "disassembly line," stood veterinarian Donald Morgan, dressed in a white smock and skullcap, who closely inspected each cow marching toward him in a jagged line. He drew a blood sample and, donning rubber gloves and grabbing a scalpel from the knives and chisels laid out along the cutting table, clipped off some pieces of fleshy body tissue.

Two animal handlers then alternated kills with overdoses of anesthetics or a special gun that fired a lethal bolt of compressed air. Each animal was then hacked up with power saws and knives and disposed of. Big chunks of flesh were sent down the chute into the incinerator. The necropsy room was a muddle, strewn with blood and pieces of flesh. Inside, a cacophony of animal moans, groans, and squeals filled the air, along with saws buzzing and chains clanking, and the echoes of parts thumping down the chute. Animals were euthanized outside of the laboratory, too; there was simply not enough kill space inside Labs 257 and 101 to do all the killing. Those carcasses were carted into Lab 257 and sent up the cargo elevator to Dr. Morgan on the second floor.

Each time the room filled up, the phone rang in the incinerator charging room. *"Okay, ready? Here it comes!"* One floor above, two animal handlers chained up the hindlegs of each carcass and swung it over to the chute built into the floor. One lifted the steer's forelegs up over the lip of the chute while the other pushed it over. Then they lowered the chain over the winch and let it slide.

The incinerator area was considered the "hottest" area on the island, for the high levels of contamination and for the searing heat emitting from the furnace. Regular incinerator duty was grueling, exhausting work, the worst job on the whole line. Plant management crews took turns doing it and they loathed it. There were some interesting perquisites, however. For starters, there was a wide assortment of pornography piled high in the corner of the charging room. "You never had to worry about any women being in that room," says one worker. "And if a woman or a supervisor came in, you could always quickly toss the porn into the fire." The other perk strains credulity, had not multiple sources confirmed it. "On kill day, the guys upstairs would carve up steaks sometimes, put them in plastic bags, and toss them down the chute to us." From sirloins, to chicken breast cutlets, to pork chops—not USDA Grade A, but USDA Grade V (for virus riddled)—they were all broiled up in the incinerator wing kitchen. "Everyone, to a man, would deny it," says another employee. "But we did it—we ate the meat."

During the outbreak, it was without a doubt the most dangerous place to be, putting staff members at great risk. Down the chute came animal parts and cell cultures, teeming with germs, mixing together, floating around the charging room area, and then entering the fire pit. Normal laboratory operations called for "kill day" once every other week to dispose of test animals. This would be "kill weekend." Down the stainless steel chute came bloody animal parts of all shapes and sizes—legs, midsections, entrails—pouring into the wagon cart bin in the charging room. It was a macabre sight. Cattle, sheep, and horse heads rolled down the slide after their detached bodies.

Once the cart was full, the weary incinerator crew, dressed in bright red coats and skullcaps (to indicate the severity of contamination), chained it to a slaughterhouse hoist and ran the cart along a track to the other side of the charging room. There they opened the two black cast-iron sliding doors and were immediately blasted with heat from 2,000 degrees of oil-stoked fire raging seven feet below. "The smell and the heat were unbelievable," said one worker. "The closest thing to hell you've ever seen or felt." Each man grabbed a side of the cart and tilted it, dumping the cargo in. Flames immediately shot up from the nine-by-twelve-foot lake of fire. Burning fat and flesh popped and hissed. Sliding the hatch back into place, the charging room cooled a bit—down to 120 degrees Fahrenheit—and they watched the pyre grow higher and higher through the tiny thick glass porthole. When the flames retreated, workers dumped in more animal refuse, 1,500 pounds every hour, cartload after cartload. The roast lasted into the night and through the next day, for forty-eight consecutive hours, a period that far exceeded the design capacity specifications of the twenty-six-year-old oven. Some said the bricks ran so hot the eighty-four-foot-tall stack glowed red in the night sky like a molten beacon. Others recalled that the concrete wall of the incinerator cracked that weekend and had to be repaired. That Sunday evening, the crew pressure-washed the heaps of guts, excrement, and blood caked all over the charging room. One employee who was there describes the odor that came forth that night. "Imagine a roast beef left on high heat for eight hours, then left out rotting for eight more—it was a butcher shop gone wrong on the rainiest day."

The laboratories would need disinfecting with caustic soda lye, para-formaldehyde, and acetic acid. There was plenty of paraformaldehyde on hand, but only small amounts of the two other chemicals. Two tons of lye arrived within twenty-four hours on a barge. The safety officer ordered an employee to the mainland to buy up every last bottle of vinegar on the shelves in area supermarkets. Crews garbed in thick black rubber suits and fishing hats mopped the laboratory floors with acetic acid solution (using the vinegar), then scrubbed the yellow tile walls and ceilings. Over the next three days, the three outdoor animal holding pens were sprayed down with lye, as were the roads and walkways, the exteriors of all cattle trucks and cars (the insides were wiped down with the vinegar), and the administration buildings. Emergency crews wore protective suits to avoid chemical burns. Truckloads of manure were sprayed with lye and buried in a six-foot-deep trench covered with topsoil, which was also sprayed. Clothing was gassed with paraformaldehyde, and bales of hay, straw bedding, and animal feed burned in a colossal bonfire.

"The whole island was lyed, everything burned and died," says a decon team member. "Grass, trees, animals—all dead. The whole island was covered with ash. It looked like snow—all white." To complete the cleanup, Plum Island remained shut down on Monday. As the spray on the roadway was drying, the first ferry with returning employees pulled into Plum Island harbor at 8:00 a.m. Tuesday morning.

It took the incinerator over a week to cool down. Then workers carefully opened the crematory, crawled inside and shoveled the tons of ash and bone that had accumulated. Wearing flimsy paper masks and boots, they packed the ash into drums, topped them off with water and lye, and let them sit in the air lock for a day. The contaminated ash mixture was then emptied from the drums on the island grounds, and piled into large mounds.

fter the lab chiefs left his office, Jerry Callis slumped into his office chair. How could this have happened? It seemed impossible, unthinkable even, and now he had to tell the whole world. This was the worst day of his scientific career. Well, it was a Friday. By the time they cobbled together a press release and distributed it late in the day, it would get buried in Saturday's newspapers. Weekends were the best times to leak bad news; people paid more attention to recreation than hard news. And the significance could be dampened further with the right word crafting.

"Foot-and-mouth disease has been diagnosed in cattle in a pre-experimental animal holding facility at the Plum Island Animal Disease Center," began the press release. The first line of the release was word-smithing at its best, patently unintelligible to people unacquainted with the science. But lines like "Authorities do not consider this a threat . . ." and "This is the first infection to occur outside the high-security laboratory . . ." and "It rarely affects humans. However, the disease could cause economic disaster . . ." made the reality slowly begin to register with weekend desk editors. Weekend or not, this was a major story to be covered.

In 1971, the USDA had total confidence its exotic virus laboratory, boasting,

> Plum Island is considered the safest in the world on virus diseases. As proof of this statement, there has never been a disease outbreak among the susceptible animals maintained outside the laboratory on the island since it was established.

Now, sorting through the hazily written press release, the message became clear: "The World's Safest Lab" had failed.

By Sunday, *Newsday* had no less than four reporters on the case, phoning all over Plum Island and hunting the mainland for leads. The paper ran three successive stories that first week with increasingly ominous banners: U.S. TRACKING PLUM I. VISITORS, SOURCE OF PLUM ISLAND GERM LEAK SOUGHT, and EXPERTS PROBE ANIMAL DISEASE.

When the fourth *Newsday* story was published in as many days, the sputtering Plum Island spin machine kicked into gear. Laboratory chief Dr. Charles Campbell told a reporter, "I believe all the people here feel we've contained the disease." Applying blame, he postulated that violations of the biological safety procedures by construction workers caused the outbreak. Rumors that migratory fowl like ducks, Canadian geese, osprey, and seagulls were dying were flatly denied, and no one came forward with evidence of any birds killed. Was there a possibility of sabotage by a visitor? A trespasser? A disgruntled worker? "One or two people have mentioned that possibility," said Campbell. "It's conceivable, but we feel it's highly unlikely." What began to alarm people most wasn't the fear of being attacked by foot-and-mouth disease virus, but by other germs from Plum Island. Was this an isolated incident?

THE VIRUS HUNTERS

One federal outfit that took the outbreak seriously was the Emergency Disease Organization (EDO), based out of Washington, D.C. By close of business Friday, the USDA swung EDO into action. The next day, Dr. Stanley Newcomb and his eight-person team descended upon Riverhead, Long Island, and set up a crisis command center at a local motel. The team conferred on the epidemiology of the virus—the how and where of an epidemic. If a germ outbreak's evidence reveals how it occurred, then scientists can trace its path and guess where it might lead.

The team studied what little evidence was available. Less than a full day after discovery, it was too early to determine the specific cause of the outbreak. To the epidemiologists, speculating was wasting valuable time. Instead, Dr. Newcombe and the team shifted from determining the *what* to examining the *how*. They ventured three guesses. The virus could have "(1) escaped from the high security research building on Plum Island to the quarantine area; or (2) been introduced onto Plum Island from elsewhere, off the island," and in either case, "(3) been carried from Plum Island to susceptible animals on Long Island and elsewhere."

Looking through the records of the two laboratory buildings, they first determined that Type-O virus was being worked on. Lab 257 hadn't worked on it for months, but three of Lab 101's four lab modules were studying it. So the virus likely originated on the island, and Lab 101 was likely the culprit. Then they addressed what could easily transfer virus to the mainland: humans, deer, and air. Like mosquitoes carry West Nile virus and ticks carry Lyme disease, people can carry millions of germs on their skin, in their hair, even inside their mouths. When a reporter asked Dr. Newcombe about deer that swam from Plum Island to Long Island, he replied that herds of deer could be infected, but the urban setting of Queens (which is really western Long Island) formed a natural barrier where there were few, if any, susceptible animals. Still, that left all of Long Island in the virus's path. There was also the question of airborne transmission. Some argued the virus lasted only a few hours in the open air. But Dr. Callis believed the virus could travel long distances by "hitchhiking on air particles."[2]

---

[2]Later, the "Yellow Winds" of China would be blamed for an outbreak of virus that traveled clear across the Yellow Sea to Japan in early 2000.

First, the team put the Plum Island employees aside. They were low risk because they followed systematic decontamination procedures every day and knew what to regularly avoid on the mainland. Then, from the visitor logs and contractor's employee manifest, they compiled a list of non-employees who'd had contact with Plum Island in the last sixty days. All were potential vectors. Of the 103 people they pinpointed, all resided within the New York City metropolitan area; other than the Frenchman in Dr. Carol House's 257 lab room, there had been no visiting foreign scientists or students.

The immediate priority was to get to the seventy-seven construction workers. They posed the highest risk because they didn't know a thing about infectious disease. Their actions over the next few days could spread the virus far and wide, resulting in monumental disaster. Each team member worked the phones from the command center asking questions. "Have you been near or at any animal farms? Stockyards? Packinghouses? Sale barns?" None of the urban hard-hats had. Some of the questions even prompted chuckles and snorts. "Do you keep any cows, goats, or pigs at home?" Other questions were more alarming. "Have you sent or received any packages or had any contact with people from a foreign country?" And some hit closer to home. "Have you been to any pet stores that had birds? How about any state fairs? Have you been to the Bronx Zoo? Any other zoos or amusement parks with those drive-through animal kingdoms? Have you traveled out of the New York metropolitan region? Where exactly did you go and what did you do there?" The replies brought good news. There had been a few family excursions within the suburbs and one visit to a pet store, but no potentially dangerous biological contacts. Spot checks at their homes confirmed they owned no livestock animals.

Scientists traveled in pairs to each known animal farm in the region: seven dairy herds totaling nine hundred milk cows. Standing at a healthy distance, the medical officers observed the animals for signs of disease, then sprayed themselves down with disinfectant, changed clothes, and proceeded to the next farm. There was real concern about the Long Island Game Farm, a three-hundred-acre zoo in nearby Manorville. This popular children's zoo made an ideal breeding ground because it was home to many susceptible animals, including ring-tailed lemurs, buffalo, squirrel monkeys, cows, cougars, horses, giraffe, sheep, zebra, and ostriches. If the virus existed on the mainland, then it probably would have turned up here. Sighs of relief overcame the medical inspection team when the game farm received a clean bill of health.

Within six days, EDO completed its germ hunt. Dr. Newcombe filed his report with the best analysis he could render. Most scientists refuse to render absolute conclusions (the possibilities, after all, fuel the discipline of science). Dr. Newcombe was no different. "Transmission to susceptible animals off of Plum Island is improbable, although not impossible . . . it would appear equally improbable that the outbreak on Plum Island resulted from virus introduced from off the island." He told a reporter, "Our faces are red, but we don't think it got off the island." This disease is not normally contracted by humans, he said. "That's not true of all animal diseases, but it is true of this one." The virus was one of over twenty-five pathogens regularly studied on Plum Island. Why was it the only one being traced by the virus hunters?

Back in Lab 257, Dr. Donald Morgan was working feverishly with his team testing the two hundred plus samples they had taken from the destroyed animals. When a virus is discovered, the strain isolated is often named after the location where it was found. Morgan named the virus "P.I.S.S.," for "Plum Island Sub Strain." The USDA chiefs in Washington went livid. A foot-and-mouth virus outbreak in the United States would prompt a worldwide ban on American meat imports, tank the agriculture sector, and wreck the economy. They ordered Plum Island to cease using "P.I.S.S." or the word "outbreak." Henceforth, the gregarious Morgan and his colleagues were only to use the phrase "The Incident." The USDA refused to admit its folly to the world. Dr. Morgan paid no mind and fla-grantly violated the edict, dubbing it alternatively "The Outbreak" or his favorite, "The *Disastrous* Incident." Morgan's eight-year-old daughter Margaret evidently shared his sense of humor. Upon his return home after kill weekend, she greeted him at the door wearing a T-shirt he had given her. It read "Prevent Foot-and-Mouth Disease—Vaccinate Twice a Year." For the first time in days, a weary Morgan cracked a smile.

To this day, the official line from Washington, blindly echoed in papers from the *Washington Post* to the *New York Times* to the *Wall Street Jour-nal*, is that the last outbreak of foot-and-mouth disease virus in the United States occurred in 1929. However, it is clear that the last outbreak in the United States occurred on September 15, 1978—on Plum Island.

September 1978 was to be the month of the grand opening of Dr. Cal-lis's newly renovated, state-of-the-art facility. Instead, it marked a low point of Plum Island and its director's storied career. By this time, the con-tractor had only finished "a rather small portion of the whole thing," said Callis. Barely halfway completed, the skeletal addition—amid all the dirt, piled-up materials, scaffolds, and equipment crowding the whitewashed Lab 101—looked and felt more like a war zone than a construction zone. Worse yet, it would soon fall under federal investigation for virus leaks and defective construction.

The contractors were working to extend Plum Island's scant laboratory space. In 1952, the USDA hadn't been able to build its desired facility, as Congress approved only one-third of their original request. They settled on just two buildings, and each lab wing could host one germ at a time due to the possibility of cross-contamination. Plum Island simply had too little lab space to study too many exotic foreign diseases emerging on the world scene.

The new state-of-the-art 70,000-square-foot addition to Lab 101 would include a third animal holding wing and associated research and autopsy laboratories; a vaccine plant capable of making 30 million doses a year (8,220 each day) and a refrigerated warehouse for virus storage; and a stylish entrance foyer in place of the menacing perimeter gatehouse. The additional space would bring the facility up to the size the USDA originally envisioned, enabling more research and more staff (adding 50 employees to the existing 330). More research meant more chances for scientific breakthroughs.

Or so they thought.

Over the next couple of weeks, Dr. Callis turned up the rhetoric on the contractors. "We feel [the outbreak is] connected to construction or the changes in operating procedures that had to be made to accommodate the construction. We let people inside the compound wall [whom] we normally wouldn't have let in." Some were confused by his reasoning. Admitting construction workers inside the fenced compound could not, on its own, cause an outbreak, since all experiments were conducted inside the laboratory. Unless, of course, an agent escaped from the building into the outdoor compound. Or unless there was outdoor testing on Plum Island.

INVESTIGATION

Putting aside the serum samples from the animals in Building 21 and the Battery Steele holding pens, Dr. Morgan focused his investigation on the samples from Building 62. Of the animals killed in the outbreak, those in the middle pens, Nos. 5 and 6, had high antibody counts, while those in the side and end pens were low, indicating that the virus may have been walked in by an animal handler (lurking on his hands, or his sleeve, possibly) or brought in with the airflow current (if it had leaked from the laboratory construction site). The antibody counts indicated two to three cycles of disease. The virus had been multiplying and spreading for an entire week.

Ten days after the outbreak, Merlon Wiggin wrote a five-page memorandum titled "The Committee Dilemma." As chief of engineering and plant management, Wiggin was in charge of every function on Plum Island other than research: marine transportation, wastewater treatment, water supply, electricity, the emergency power plant, motor pool, metal shop, fire department, grounds crew, laundry, boiler plants, cafeteria, general maintenance, and biological containment. Second only to Director Callis, Wiggin oversaw nonresearch functions on the island and had twice as many people under his command. If the buck stopped with anyone other than Callis, it stopped with Merlon Wiggin. Hailing from a small town in Maine, Wiggin married his wife, Isabelle, just six months after he met her on a bus ride to nearby Portland. After working at Air Force bases, the thirty-year-old engineer with tousled brown hair applied for a position at Plum Island. Doc Shahan hired him while the facility was in its infancy, providing more than enough intricate mechanical projects and glitches to satisfy the young engineer.

Most disturbing in Wiggin's memo was his casual admission about the most dangerous part of the facility:

> *The Incinerator Area.* Tests of this area show that hundreds of cubic feet of air a minute from a highly potential contaminated area have been escaping.

"Another area that comes to mind would be the exhaust air from the laboratory." Wiggin suggested testing the 107 filter units on the roof that trapped pathogens as air exited the building. "For nearly ten years, I have recommended that these filters be tested after each change (we have the equipment on hand to do it) and have also recommended that their efficiency be improved by replacement." He did not mention who exactly was preventing Wiggin, the chief of all plant management, from ordering a test inspection and replacement of the filters. Regardless of what an inspection of the air filters might yield, "It seems quite obvious the virus was transported by humans to the animals in Building 62," he wrote, which "seem[s] to rule out transmission by aerosol all the way from [Lab] 101." If indeed that was correct, it meant the biological safety measures were disregarded by an animal handler—a scientific staffer—and therefore not an employee in Wiggin's bailiwick.

Oddly, Wiggin ruled out the heavy construction going on on all four sides of Lab 101, particularly in the area of the incinerator. "We have carefully checked and tested other areas, such as the Contractor Sites, and have found nothing." This too, was in Wiggin's self-interest. He was the federal government's contracting officer and project liaison to Joseph Morton Company, the contractor. "I think we need to realize at least for the present our laboratory buildings are not perfect secure envelopes," he concluded. Despite that, shutting down was not in Plum Island's best interests. "I see little to gain and much to lose (monies, work, training, improvements, repairs, etc.) by delaying the return to normal operations. . . ."

Earlier in the day, the Lab 101 foreman, Truman Cook, and his men were busy spraying down the incinerator room with lye, even though he had been ordered the week before not to decontaminate or change any filters in the incinerator room until further notice; the investigation committee planned to inspect the area. After Wiggin learned Cook was deconning the area and attempting to change the old air filters, he told Cook to stop immediately.

The next morning, when investigators entered the incinerator charging room, they noted intense heat emanating off the sliding hatch doors and the smell of charred, rotten meat. Cook and his maintenance crew were nowhere to be found. All had called in sick that day, something that had never occurred before. Safety officer Dr. Jerry Walker—once described by a reporter as "thin, of average height, sort of Southern-square looking . . . reserved to the extreme . . . in all, decidedly odd"—was chair of the investigation committee. He led an examination of the crime scene. The air intake had been blocked, and the supply intake ducts to the incinerator were jammed wide open. Outside light shined into and air flowed out of the room, a chamber that was supposed to be sealed to the outside world. A committee member pointed out lead tape applied across the latch of the emergency door, which now hung slightly ajar. Dr. Walker ordered the openings blocked at once. The filter housings had been sprayed with lye by Cook's crew, so they couldn't be tested. But then they spotted an unthinkable sight. "Several filter units had media improperly installed with gaps up to three-quarters of an inch," the committee reported. That meant that at any one given time, no less than 750,000 viruses could be exiting the building, marching out side by side. When Wiggin turned on the air pressure system, the supply fan started before the exhaust fan. For the air pressure to contain germs properly within the room, it had to start in reverse order. Stunningly, the electricians had wired the safety interlock backward.

Dr. Walker paged through the incinerator log and spoke with the employee who kept it. He shook his head as he read the same static pressure number recorded for each and every kill day. Either the gauge did not function, or the worker dummied the logs. Standing over the shoulder of the log keeper, listening to him try to explain the log entries, Walker realized the numbers "had little meaning to him." Concealing disgust, he flatly reported, "It is apparent that a knowledge of pressures is essential by someone during incineration with responsibility to maintain control of the pressure required."

During the following week, Merlon Wiggin and his assistant climbed up onto the Lab 101 roof and inspected the 107 air-filter units. On the roof stood a mechanical garden of metal stacks that sprouted out of the black tar roof—some short and some long, some wider and taller than the men—all anchored with wire stays. As a nippy October wind swept off the Sound and crept up their backs, they heard the collective whirr of the air filter fans.

If the incinerator was in bad shape, the condition of the roof was even worse. The rubber gaskets designed to seal the gaps between the air filters and roof masonry were brittle, cracked, and leaky. Some gaskets were altogether missing. The filter housings were installed in such a way as to leave openings "allowing the complete passage of air without the benefit of filtration." It was as if someone with no training ripped an old one out and slapped a new one in, giving the task zero thought. The ductwork had holes punched into it. Replacement filters were much thinner in size than those used in years past. Wiggin's roof tour was disturbing, and its ramifications were nothing short of terrifying.

The inside review was just as scandalous. The air pressure logbook for experiment rooms during four weeks prior to the outbreak wasn't even dummied like the incinerator logbook—the pages were blank. Wiggin found many lab rooms way out of proper air balance and some were positive to the outside, meaning that germs were being circulated through the entire laboratory building. Bad air, "hot" with viruses, was being forced to exit somewhere. It had a choice of possibilities—through heavy paper and charcoal filters designed to catch germs or through the easier path, via gaping holes in the roof. To top it all off, he found no air filter maintenance. The rubber gaskets hadn't been changed in nearly thirty years (when they were first installed in the early 1950s), rendering each air filter between the defective gaskets virtually useless. And they weren't the only filters in question; some of the sewage vent filters, which strained biologicals out of animal wastes, were in poor condition, and a few were even missing. The appalling condition of the laboratory shocked Wiggin.

It became painfully clear why Wiggin's foreman, Truman Cook, and his crew were feverishly tinkering with repairs the day before, and why they called in sick today, the day the committee began its investigation.

Try as he may, there was no way the chief engineer could bring himself to certify the blatantly porous building as safe. "Recommend that Lab 101 *not be considered* as a safe facility in which to do work on exotic disease agents," Merlon Wiggin told the investigation committee, "until corrective action is accomplished." That included smoke-testing each filter after the repairs to see if any air continued to escape.

Digesting Wiggin's disquieting findings, Director Callis and safety officer Dr. Walker grew increasingly uncomfortable. Research experiments were backing up and a special project with the Rift Valley fever virus project was far too important to be delayed. Unwilling to accept the dire assess-

ment, they ordered Wiggin to begin smoke tests and continued virus production and animal experiments under a pack of faulty air filters. "No new
activities are expected to be taken on over the next several weeks . . ." wrote
Dr. Callis in a letter to Washington, ". . . beyond those commitments
already made." In other words, on with the work. Setting off smoke bombs
in the incinerator and lab rooms, Dr. Walker and Mr. Wiggin watched harmless white wisps of smoke waft through cracks, gaps, and holes, visible
stand-ins for deadly germs hundreds of times smaller that were escaping.
Not one of the laboratory's 107 air filters was salvageable. All of them had
to be scrapped and new mountings installed.

But well before the air filters were fixed, Plum Island scientists continued their infectious virus research, in a porous lab facility they knew was
inadequate to contain the germs safely.

The committee rendered its final report to Dr. Callis on January 9. Of
Plum Island's three lines of defense—the containment laboratories, restric-
tions on personnel and material, and the island location—the first two had
failed. Their best guess as to what happened went like this: one of the animal handlers walked into the Building 62 holding pens to feed the animals
as he routinely did. While they ate, the handler cooed at the beasts and petted them lovingly, as they were prone to do, and unwittingly fed them help-
ings of viruses along with their meals.

But that didn't explain how one of the animal handlers, restricted by
the rules from entering the laboratory building, became a carrier of the
virus. Investigators turned their attention to the incinerator room and the
107 air filters on the lab roof. Dr. Walker's impressions of the incinerator
logbook were true. For years, no one set the internal air pressure equal to
the atmosphere, as required by the rules during a burn. Because of this negligence, "the entire area had been pressurized to the outside atmosphere,"
which meant the exhaust was literally blowing contamination out of every
crack and crevice into the sky.

The other theories, not fully reviewed by the investigators, involved
the construction work. During heavy rains, water seeped under a plywood
barrier into the incinerator room and then retreated back out. "They were
digging the hole for the flue and chimney stack and we had torrential rain
for a week," remembers one worker. Dirt near the plywood was excavated
and hauled to a site three hundred yards from the Building 62 pens. The
committee omitted another occurrence that may have played a role. "Supposedly it was caused by a leakage of sewage," says Ben Robbins, a retired
Plum Island engineer. A local subcontractor ruptured an underground pipe
that carried contaminated wastewater from Lab 101 to Building 102, the
decontamination plant. Perhaps infected sewage seeped into the dirt that
was later moved next to Building 62.

The investigators interviewed workers and explored related worries. They learned from their support staff that emergency backup power generators were "old and in critical need of updating." They listened to "much concern" among the ranks for the deterioration in the island's biological safety measures. Frequent travel-related absences of safety officer Dr. Jerry Walker were cited. The committee told Dr. Callis there existed a "general breakdown of security procedures that [once] served psychologically as constant reminders of the need to practice safety."

How and why the facilities were in such wretched shape were questions "which have to be answered," they said. "Certainly, however, the increase in scientific work . . . age of the facilities . . . reduction in staff . . . restricted budgets . . . [and] inflation" had something to do with it. Lack of sufficient plant management and safety personnel was a factor. "[W]hether there are in fact too few or those available are not being properly utilized is a question to be answered."

"It is unlikely that the exact route of infection . . . will ever be known," the committee concluded. Among the men who signed the report were the two most responsible for the disaster: safety officer Walker and chief engineer Wiggin.

It was time to run for cover.

D r. Callis notified Washington that swift measures were taken to ameliorate the deplorable conditions. Building 62 was closed because the contractor stored much of its materials in the vicinity. Air filters were replaced except those over Lab C, which were expected to be completed by February. Two new safety positions were added to assist Dr. Walker in carrying out daily air filtration checks. He assured the laboratory brass that Lab 257 checked out fine and was ready for the Rift Valley fever virus research program. "We are anxious to complete these remaining operations . . . so that we can all get back to more productive endeavors."

Dr. Callis formed two more committees, an ad hoc operations review committee to evaluate inadequacies in material and staffing, and a safety committee to conduct the witch hunt. For reasons unknown, the latter included Dr. Walker as one of the members—the "safety officer" minding the store during the outbreak. On February 14, 1979, the committees sent a secret memorandum to Dr. Callis that placed the outbreak squarely upon the shoulders of Merlon Wiggin. "[M]ismanagement of Engineering and Plant Management operations," began the valentine, "resulted in circumstances which led to the escape of the virus from [Lab] 101 and certainly violated the spirit and intent of the Agent Safety Program at Plum Island." Work orders for new items and change orders with the contractor took precedence over routine laboratory maintenance. An "appalling lack of effective communication" between Wiggin and the lab foreman, Truman Cook, existed, they said. Furthermore, lab foremen had no appreciation for containment, "specifically, air-handling equipment and its function in the building." Wiping the egg off committee member Dr. Walker's face, the memo noted that no one had notified the safety office about any of these conditions. This, they concluded, was the sole responsibility of chief of engineering and plant management Merlon Wiggin, and Lab 101 foreman Truman Cook.

Dr. Callis chose this time to abscond to Florida on a winter vacation, leaving the most onerous task for his assistant director, Dr. John Graves. Within days of Callis's departure, Dr. Graves ordered Wiggin and Cook suspended immediately without pay. Dr. Graves told the press that Wiggin and Cook were being held responsible for not maintaining the air-filtering system. No action was taken against Dr. Walker or against Dr. Jennings, who was in charge of the animal supply pens. One scientist says Jennings regularly shuttled between the lab and animal supply without following proper showering-out procedures. "They didn't want to go after Jennings," says Merlon Wiggin, mentioning his senior scientist, tenured status. Shortly after the outbreak, Dr. Jennings quietly retired from Plum Island.

The next morning, Graves and security guard Ed Wolf physically blocked Wiggin from boarding the ferry to Plum Island, ordering him to remain in the Orient Point dock warehouse office.

One unnamed official called the two fingered men "scapegoats . . . [We] could all be blamed for the outbreak. The buck stops at the top." Wiggin and Cook filed grievances with the USDA. Then they took their case to the people. "There are other people who have just as much responsibility as we had. It was a kangaroo court," Truman Cook told *Newsday*. "When the right time comes, I'm sure that I will get a chance to say who's responsible," said Merlon Wiggin. Back from his sunny Florida sojourn, Dr. Callis criticized their "poor supervisory and management practices . . . poor safety practices . . . [and] lack of communication."

After three weeks, Wiggin was readmitted to Plum Island, but barred from entering his office. Dr. Callis then called a rare all-hands meeting inside the old Army chapel to announce that Lab 101 was returning to full operations. The ad hoc committee, he said, had completed its report on recommendations that were now being implemented. At that moment, Wiggin realized that in addition to everything else, he had also been frozen out of the ad hoc committee, to which Callis had appointed him a member. Wiggin went home and stewed over the weekend. *I am not the only person responsible for this!* He decided to give the director a piece of his mind.

"Perhaps I was not notified as to when the committee was to meet because I ask questions the Chairman [Assistant Director Dr. Graves] and the Safety Officer [Dr. Walker] find embarrassing," Wiggin told Dr. Callis in a letter he handed him on Monday. Safety officer Dr. Walker appeared not to know whether the filters were tested to catch germs as small as 0.3 microns or 3 microns wide. "This to some may be not important, but it is off by a factor of 10." To those who knew, it meant the difference between no viruses escaping the building and thousands escaping every second. Most alarming was Wiggin's charge that Drs. Walker and Graves brushed off seriously flawed test results of the new Lab 101 air filters. "[Dr. Walker] was frequently getting test results of 100% and that was an impossibility using deep bed media [air filters] which only test 85 to 90%. . . ." Before placing the lab back on-line at full capacity, Wiggin implored Dr. Callis, "I wondered if you were concerned about the following:

> Vent filters—I understand they have all yet to be checked and the leaking ones changed.
> Sanitary drainage system—We know from the recent findings that we had some badly corroded piping.
> Material support—A lack of warehouse parts and supplies was a deterrent to the level of maintenance in this [laboratory] building.

Referring to the draft report Callis circulated, Wiggin questioned why it stated unequivocally that Plum Island could return to full operations. Callis replied, *"We'll discuss it later, Merlon."* But Dr. Callis paid little mind. To him, Merlon Wiggin was a disgruntled former engineering chief now trying to make hay. Lab 101 was back in business. On with the show.

Meanwhile, the Plum Island blue-collar workforce seethed over Truman Cook's suspension. They knew that Cook was management's fall guy. "Truman was a good guy, and everybody liked him," remembers one employee, "or was related to him." His family had been in the Southold area since the 1600s. The Cook family was a proud, honest family, and Truman was no different. No way these out-of-town scientists were going to turn a local good ol' boy into a poster boy of shame. More than 150 employees—most of the support staff—signed a petition that protested Cook's suspension and vehemently demanded his reinstatement. Fearing open mutiny on the island (and loss of community support off the island), management buckled and restored Cook to his position. "I'm very relieved," he told a reporter who broke the news to him. "I think that's about all I'd better say."

A second petition drive began, this time to reinstate Wiggin as chief engineer, but management nipped the nascent movement in the bud, order-

ing it withdrawn from circulation. Unlike Cook's petition, this one specifically stated the safety office was to blame for the outbreak. The union representing the Plum Island workforce wrote Dr. Callis asking why the safety
office was not held at least partially responsible for the outbreak. There was
no reply. An investigation committee member then came forward, anonymously, to a local newspaper, and condemned the singling out of Merlon
Wiggin. "You can't really put the blame on one person. To say one or two
people are responsible is ridiculous. In fact, our report strongly implies
involvement by the safety office." Against Dr. Callis's scientific appointee,
Dr. Walker, Wiggin lost the wrestling match. Scientists, after all, were the
reason Plum Island existed.

One former employee dispenses with the finger pointing and recrimi-
nations, and gives due credit to the policy of keeping animals outdoors on
the Plum Island—animals that acted as dutiful sentinels. "Thank God we
had those animals in the Old Cow Barn [Building 62]," says an employee.
They were, after all, the canaries in the mine. Without them, the outbreak
wouldn't have been discovered until it was too late.

On the same day that the suspension ax came down on Merlon Wiggin
and Truman Cook, the USDA's Office of the Inspector General
announced a federal investigation into the Lab 101 rehabilitation project.
"We're concerned with the quality of the construction. Some things that
happened . . . just don't add up," said a spokesman.

Since the groundbreaking two years earlier, the bulk of Merlon Wiggin's
time was spent tending to an unusually large number of change orders,
which were revisions to the design specifications. Over 240 change orders
were issued. Original plans called for masonry, but concrete had to be used
for proper biocontainment. Metal air-lock doors were coated with the wrong
epoxy and severely blistered; it was so bad that 320 doors had to be removed
and replaced. The roof was faulty and moisture seeped into the lining. The
contractor, Joseph Morton Company, estimated an additional $5 million in
cost overruns, 50 percent above the original budgeted cost of the project.

As Plum Island's house of cards came falling down, Wiggin fired off a
lengthy missive detailing scores of design flaws and contractor blunders to
the USDA's head contracting officer in Washington. It addressed the faulty
construction project he said he had labored to rectify, and addressed what
he saw coming:

> [The rehabilitated laboratory building] will not be an agent safe,
> energy efficient, bio-containment facility that can be utilized without
> repair or changes.

Within the next few months, I expect there will be an attempt to discredit me and my reputation. There is possible evidence of criminal neglect in relation to the recent outbreak at Plum Island.

To preserve the record and protect himself, he sent copies to a friend in Texas for safekeeping.

The USDA owed Joseph Morton Company over $1.5 million for all the change orders. "If we are paid for the work we've already done," Morton told the department, "we can complete the renovation in four months." A dark cloud continued to swirl over Merlon Wiggin. "He's not free of responsibility on the design errors," a source told the *Suffolk Times*. "Some of them might have been caught by him ahead of time." Were his hands unclean because of the virus outbreak, the construction debacle, or now both?

On March 5, 1979, a letter of default went out from USDA headquarters addressed to Joseph Morton Company, canceling its $10.7 million rehabilitation contract. Just the week before, Morton had wired Washington, demanding back payment by the end of the week of at least $500,000 to continue. Morton laid off all of the construction workers except a few supervisors, and expected to cancel the contract themselves.

Then things got even uglier. April showers rained down on the construction site. No one in management thought to tell any of the 150 plus physical plant employees to secure the partially completed construction from the elements. Corrosive saltwater seeped into seams and in between beams, rusting and decaying the steel framing and construction materials. In June the USDA advertised for proposals from contractors to complete the project, and by the middle of July, it had fourteen bids. But the USDA sat on those bids.

Dr. Callis informed Wiggin that his suspension would continue. The Office of the Inspector General was now focusing on allegations that Wiggin received kickbacks from the contractor on the construction project. And so, too, was a federal grand jury. The grand jury issued subpoenas to Joseph Battaglia, president of Joseph Morton Company, and Samuel Semble, a Morton executive. The subpoenas called for documents, books, and records. The two men testified to the grand jury but provided no records. The records had been shredded.

Wiggin decided to swing back at Plum Island to get some leverage before the federal grand jury closed in on him, so he filed another employee grievance with the USDA. In a slap at Callis, USDA headquarters in Washington sided with Wiggin in a letter they wrote him: "We do not concur with [Plum Island's] conclusion that the outbreak was a direct result of performance deficiencies on your part since the file contains inconclusive and contradictory information." Washington admitted Wiggin alone had been hung for the collective negligence of the whole gang.

Wiggin and his wife then filed a $1 million defamation lawsuit in New York State Supreme Court, charging slander and defamation of character. The complaint laid out his side of the outbreak and stressed that for years he warned higher-ups that the buildings had cracks and leaks, and told them—Drs. Callis, Graves, and Walker—that the Lab 101 incinerator was ineffective in containing viruses. Plum Island administration officials "conspired to destroy [Wiggin's] reputation by not only falsely imputing blame to him for the outbreak, but in addition, to defame him of committing felonious crimes," including allegedly accepting $500,000 in bribe money by the Joseph Morton Company.

Dr. Graves refuted the charges by noting that Wiggin never sounded any specific alarms. "Any problems were in his sphere to remedy," he said. "He was not a third party standing on the sidelines—he was the one to do it."

With the tide turning in Wiggin's favor, his comeback came to a screeching halt.

MERLON WIGGIN CALLS IT QUITS AT PLUM ISLAND, announced the *Suffolk Times*. "I don't feel there's a future here," the forty-eight-year-old engineer said, announcing his departure at the end of August after eighteen years on Plum Island; twenty-five years of federal service now made him eligible for early retirement. Ten years ago, he had been tapped by NASA to design a containment lab at the Houston Space Flight Center to hold moon rocks. The Australians wanted him to help set up their own version of Plum Island. Now he was leaving under a dark cloud.

"They offered me an early out," says Wiggin. "Financially, it was almost too good an offer to refuse. But at first I didn't like the idea." He initially wanted to return and set the record straight and clean things up. USDA officials in Washington, who had absolved him of any personal wrongdoing, told him he wouldn't be returning to the most pleasant work environment, and that he should strongly consider retirement from federal service. He says he had also heard that there would soon be a move to privatize the Plum Island workforce, and he felt that privatization wouldn't go over well. Though Wiggin claimed he rode off into the sunset smiling, others said there was more to it. One administrative source says he retired "after we told him he would lose his retirement benefits if he was indicted by the grand jury." It seems Plum Island traded his freedom for his silence, a silence concerning problems so embarrassing, so potentially dangerous that, if disclosed, would have put Plum Island out of business.

---

hey say time heals all wounds. Merlon Wiggin proves that axiom with his reminiscences. "I think it was the nicest job I ever had," says Wiggin. "I enjoyed the ability we had to make anything and do everything. We had a great group of people—real pleasures, a real family. It was good working for Drs. Shahan and Callis." Let bygones be bygones, he says. Life after Plum Island has treated him well. He embarked on a private consulting career and worked on biocontainment projects around the globe. He participated in community efforts as president of the North Fork Environmental Council, founded Suffolk for Safe Energy, and now leads the effort to save the crumbling Plum Island Lighthouse. Today, the seventy-three-year-old engineer refuses to surrender to autumnal drift, still working as national expert on the mechanical aspects of biocontainment.

After Wiggin left Plum Island, the Office of the Inspector General dropped their kickback inquiry, and Wiggin dropped his defamation lawsuit and cooperated with the OIG investigation. Participating in a sting operation, he met with executives from the Joseph Morton Company on several occasions at a local restaurant, wearing a concealed wire recording device. "When the man over at the bar took out his cigar case, that was my cue to turn on the wire," Wiggin recalls with a chuckle. After the meal G-men would follow his car, recording more revealing information that would be later put before a federal grand jury. The federal grand jury indicted Morton company president Joseph Battaglia and Samuel Semble—but not chief engineer Merlon Wiggin—on counts of fraud and obstruction of justice. Both Battaglia and Semble were convicted at trial and sentenced to prison for crimes of fraud and conspiracy, having bilked taxpayers out of millions. The Joseph Morton Company then filed for bankruptcy in March 1980 at the same time that the USDA was trying to recover the money it paid the company for its sub-par construction work. Hearing of the bankruptcy, Deputy Agriculture Secretary James Williams said, "Everybody thought it was going to be worked out. . . . I'm mad as seven hundred dollars." Williams should have been about $1,999,300 angrier, because the USDA would be out $2 million in taxpayers' money.

"INCREDIBLE BUNGLING"

The Office of the Inspector General issued a damning 190-page report condemning the USDA's poor management of the Plum Island rehabilitation:

> [T]he project, which originally was to have cost $11,000,000, is now estimated to cost approximately $25,000,000 to complete.
>
> The contracting officer . . . performed only a superficial credit

check. Had a sufficient check been performed, the information would have raised serious questions about [Joseph Morton Company's] ability to perform. . . .

[M]anagement knew that the contractor was not: correcting major construction deficiencies; protecting equipment and facilities from deterioration due to the weather; paying subcontractors; and paying workmen the prevailing wage rate.

In addition, they found that the USDA paid Morton 80 percent of the cost when only 58 percent had been completed, and withheld not one dollar for defective workmanship. They found leaky roofs, beams practically rusted through, misaligned foundations, and stainless steel pitted from a leaking wall. They came across an entire incinerator unit ruined, left uncovered in a ditch and half submerged in murky rainwater. In short, they found a total mess. While the Inspector General's Office called for disciplinary action, they didn't name any names.

When the USDA came to Capitol Hill asking for another $13.4 million (a far cry from the $1.8 million Dr. Callis told the press he needed at the start) to complete the construction project, the Senate had this to say:

The history of the use of the $10,000,000 [appropriated in 1976 by Congress for the Plum Island rehabilitation] including its management . . . is a unique "horror story" in contemporary contracting. Although the Committee has, through its normal oversight mechanisms, attempted to keep apprised of the situation, it has been hampered by disingenuous statements of Federal officials, incompetence, and near-incredible bungling.

The Committee believes that given the present situation, with the contractor having been defaulted by the Government and under indictment in a U.S. court, with facilities only partially completed and unusable, with a severe report by the Inspector General, and with a history of the Department's unwillingness or inability to keep the Congress properly informed, firm action must be taken immediately. Our preference would be that all responsible parties be disciplined prior to providing any additional appropriations.

While Congress agreed to appropriate another $10 million to complete the project, the money would be available only after the secretary of agriculture demonstrated to the committee that he was taking "severe disciplinary measures" and legal action to recover funds from the surety company. The USDA promised to act, but apart from Merlon Wiggin, no one else was "severely disciplined."

The money from Washington didn't arrive. First, Dr. Callis announced the construction would be completed, finally, by spring 1981. Then, Dr. Graves estimated, June 1982. Then they simply said "no comment." Congress showed its displeasure with the USDA's negligence by impounding the funds.

Armed with fraud and obstruction of justice convictions, the U.S. Attorney General filed a $14 million lawsuit against Joseph Morton Company's insurer to seek payment on a performance bond. In April 1987, the U.S. Court of Appeals for the Second Circuit affirmed a lower federal court's jury award of $10,631,326 to the USDA—$17 million with judgment interest applied—the largest federal jury award in the history of New York State at the time. Rehabilitation of the old laboratory facilities began again, this time with new contractors and under new Plum Island management.

Plum Island's two-year laboratory rehabilitation project took nineteen years to complete. It was finished in 1995.

Rift Valley Fever

> *In and out of rivers, streams of death in life, whose banks*
> *were rotting into mud, whose waters, thickened with slime,*
> *invaded the contorted mangroves, that seemed to writhe at us*
> *in the extremity of an impotent despair. It was like a weary*
> *pilgrimage amongst hints for nightmares. . . . I raised my head.*
> *The offing was barred by a black bank of clouds, and the*
> *tranquil waterway leading to the uttermost ends of the earth*
> *flowed somber under an overcast sky—seemed to lead into the*
> *heart of an immense darkness. . . .*
> —JOSEPH CONRAD, *HEART OF DARKNESS*

Contrary to what the public was led to believe, some animals on Plum Island were kept alive after the virus outbreak. Though the emergency plan that called for the sacrifice of every animal to prevent disease transmission, the animals earmarked for the January 1979 Rift Valley fever project were spared. "The only animals kept alive were the mouse and guinea pig colonies," said Dr. Graves at the time. Dr. Douglas Gregg, a Plum Island scientist working at the time of the outbreak, remembers it differently.

"The Army was there testing sheep," says Gregg. "They had just been vaccinated outside, in the middle of the island, to save space. They were the only animals that weren't destroyed. Those animals were not showing any signs of disease. They were saved, brought into the laboratory, and held there. And they continued experimenting with them there in the lab, which they would have done anyway."

It is hard to fathom that, after the hours toiled during "kill weekend," the virus hunters' search for possible mainland spread, and the vigorous decontamination efforts of both laboratory buildings and the 840-acre island grounds, a single animal capable of spreading exotic viruses would have been kept alive. But sixty sheep were. And they vaccinated the sheep outdoors with Rift Valley fever vaccine—inactive viruses that could revert to a virulent state and turn the test herd into virus incubators. This violated the cardinal rule established by Plum Island's first director, Doc Shahan: no experiments shall take place outside the containment laboratory. "No virus studies are conducted except in the enclosed laboratory buildings, and no contact is possible with other livestock or with birds and insects." Scientists were taking an immense risk to free up laboratory space. After the outbreak, Dr. Callis asked a panel of national biosafety experts a question: could animals vaccinated against foot-and-mouth disease virus (far less dangerous to humans than Rift Valley fever virus) be held outside, to save space? The experts said that animals could be injected with vaccines outside, but only if they were genetically engineered vaccines of U.S. origin, essentially synthetics that didn't contain actual virus. This reduced to nil the risk of reverting to virulence and transmitting disease. But that wasn't the case with Rift Valley fever virus vaccine, which was authorized to be tested outdoors on Plum Island.

The jagged Great Rift Valley runs along the eastern spine of Africa, the result of a series of faults that first appeared when Africa separated from the Arabian peninsula thirty million years ago. The Rift snakes some four thousand miles, from Zambia in southern Africa up through Mozambique, Tanzania, Kenya, Uganda, and Ethiopia, and under the Red Sea to the Jordan River valley. In Olduvai Gorge, deep within the lower reaches of the Rift Valley, shadowed by towering volcanoes such as Kilimanjaro and Mount Kenya, scientists unearthed the oldest fossils of the human species *Homo erectus*, dating over 500,000 years. From this cradle of human existence has spawned some of the deadliest natural threats to human life known to mankind.

On a farm near Lake Naivasha, heavy rains in July 1930 were followed by 3,600 lambs and 1,200 ewes falling violently ill and dying within days. Perplexed by the sudden tragedy, local veterinarians and scientists rushed to the scene to examine the remains and take blood and tissue samples for further study. Back at the lab, four staff members came into direct contact with the tainted blood and contracted an acute febrile (fever producing) disease that resembled sandfly fever. It turned out that every farmhand in the area had been suffering a similar ailment. To prove conclusively the mysterious culprit was infectious to humans, the scientists inoculated a man with a sick lamb's blood, then used his serum to successfully pass disease on to healthy sheep. Further tests revealed that the offending agent was a virus. More lab workers contracted the volatile disease, named the Rift Valley fever virus, working on it back in England. One year later, the disease struck again while sixty miles from the original outbreak. Both cases were matched to the same germ. The new germ wasn't particularly deadly, but it did cause severe sickness, and could cause blindness.

During World War II, Plum Island founder Dr. Hagan realized the terrific potential of Rift Valley fever as a germ warfare weapon. He put the new virus at the top of the list of agents the Axis powers might use against the Allies, along with anthrax, yellow fever, and botulism. Hagan deemed it "particularly dangerous, since the virus was hardy" and "difficult to diagnose." In addition to causing near-fatal 107-degree fevers and severe flulike symptoms, it ruptured blood vessels and caused the eyes to hemorrhage profusely. The bleeding blocked the optical nerve and often caused permanent blindness.

In a top-secret wartime report to George Merck, Hagan's colleague Dr. Thomas Rivers noted that the virus would make a fine weapon. "Rift Valley fever in humans does not kill people as a rule, yet a widespread epidemic would be demoralizing." Rivers, a head scientist at Rockefeller (and the boss of Dr. Richard Shope), also suggested use of a biological cocktail—similar to a designer germ weapon the Nazis were designing—using mosquitoes. "[I]t would be interesting to find out whether the same mosquitoes could be simultaneously infected with [Rift Valley fever, yellow fever, and dengue fever] viruses and be capable of producing concurrent epidemics in a susceptible population. . . ."

After the war, the Army biowarriors put Dr. Rivers's plans into practice. Pentagon scientists briefed President Dwight D. Eisenhower on using Rift Valley fever as a nonlethal biological weapon that would only "incapacitate" the enemy, rather than kill him. Used correctly, it could deter and demoralize the enemy and, at the same time, spare buildings and infrastructure from incendiary bombs. The president approved funding in this new area of weaponry, calling it "a splendid idea." Research on incapacitating germ agents began.

Starting experiments at its new Plum Island laboratory, the USDA asked a South African scientist for instructions for handling Rift Valley fever. He warned, "Never have we handled a virus which spreads so quickly, easily, and rapidly as virulent as Rift Valley fever virus. . . . It is now a special instruction here that this virus is handled in complete isolation."

NAMRU-3

In October 1977, James Meegan, commander of United States Naval Medical Research Unit No. 3 (NAMRU-3), noticed thousands of Egyptians in scattered parts of the country falling mysteriously ill. Some 200,000 people, young and old, flooded into hospitals with eye infections, fevers, and myalgia. In the towns hardest hit, two-thirds of the populations contracted disease. Many contracted encephalitis (an acute inflammation of the brain) and hemorrhagic fevers (uncontrolled internal bleeding). Hundreds were dying. An epidemic had erupted, coinciding with noticeable large swarms of mosquitoes. Doctors feared it might be the Marburg virus discovered in Germany in the late 1960s, the older brother of the Ebola virus.

NAMRU-3 collected samples in one of the high-infection locations and shipped the materials to Yale University's Arbovirus Research Unit (YARU). Dr. Robert Shope, head of YARU, found that Egyptian samples reacted to antibodies of Rift Valley fever, an arbovirus, a family of viruses that are communicated by bloodsucking insects. The diagnosis was surprising, since previous experience had shown Rift Valley fever to be rarely fatal. Major C. J. Peters—the intrepid virus hunter from Fort Detrick featured in Richard Preston's harrowing chronicle of the Ebola virus, *The Hot Zone*—called it a "unique situation. We didn't know of any other virus that caused such a wide range of illnesses. . . . They were like totally separate diseases." Shope was extremely uncomfortable with the thought of a now-lethal virus strain in his Connecticut laboratories. "Rift Valley fever was not supposed to be worked with in the United States—it's a foreign animal disease," he recalls. "As soon as we found out what it was, we wrapped up work, disinfected the place, and sent the materials out to Plum Island." The new Rift Valley fever virus was carefully shipped across the Long Island Sound to Plum Island and deposited in the virus library alongside five other earlier, less deadly strains the Army had procured for the USDA:

STRAIN	VIRUS	CHEMICAL CORPS. ACCESSION NO.	ORIGIN
Smithburn, Neurotropic	Rift Valley	118	Entebbe, 1951
Smithburn, Attenuated egg passage	Rift Valley	119	South Africa, 1951
84674 Ovine, Virulent	Rift Valley	120	Orange Free State, 1951
Van Wyk egg adapted	Rift Valley	200A	South Africa, 1952
Unknown strain	Rift Valley	127	Kabete, 1947

OPERATION WHITECOAT

The commanding officer of Fort Detrick, Colonel Richard Barquist, asked Major C. J. Peters to courier vaccine to NAMRU-3 in Cairo, test it on the U.S. troops there, and trace the epidemiology of this strange new virus. The seasoned virus hunter couldn't resist the challenge; a decade later, this same gutsy, aviator sunglasses–wearing scientist swaggered into a Reston, Virginia, monkey house and swiftly put down an outbreak of the feared Ebola virus. Peters brought with him "Rift Valley Fever Vaccine, Formalin-Inactivated, Tissue Culture Origin, NBDR-103, Lot 6," an experimental vaccine designed by the Army during the early days of Fort Detrick. Listed in "Final Development" stage in the now-declassified 1977 congressional report, *U.S. Army Activity in the U.S. Biological Warfare Programs*, it hadn't gone through extensive testing and it wasn't licensed safe for use by the U.S. Food and Drug Administration (FDA).

The only testing of this experimental vaccine occurred between 1958 and 1974 in the Army's top-secret human guinea pig operation, code-named Operation WHITECOAT. WHITECOAT was a milestone on two counts. It was the first time the Army reached out to a religious organization to form a singular military unit, and the first time it drafted humans for biological warfare experiments. At the urging of Theodore R. Flaiz, the physician who oversaw the worldwide health outreach of the Seventh-day Adventist Church, 2,300 Adventists—nonsmokers and teetotalers all—signed consent forms and volunteered to be deliberately infected, injected, and exposed to concentrated virus and bacterial mists, genetically altered germs, and investigational therapeutic treatment. Conscientious objectors to the draft by virtue of their faith, their church endorsed this "service to the nation" on the grounds that it saved lives in conventional battle.

In this case, Whitecoats (named for their standard issue white lab coats) were injected and then challenged—like a mouse or sheep might be—with live Rift Valley fever virus. In strict isolation, they were closely monitored, poked, and prodded by scientists in white-hooded biospacesuits.[1] Conducted in 1958, Operation WHITECOAT Project No. 58-2 tested three men with Rift Valley fever virus who spent seventeen days in the hospital deliberately infected with the disease. Another three were tested in 1969 as part of Project No. 69-9. In Project No. 69-10, twenty "volunteers" gave their arms to needles and suffered through a nine-day hospital stay. Eight of the men became so ill they had to be put on convalescent leave to recover. Further details of these human vaccine tests remain classified.

---

[1]Interestingly, one of the scientists prodding the WHITECOAT men was a young Dr. Robert Shope, who had joined Fort Detrick's Project WHITECOAT team in 1955 after finishing his schooling at Yale.

The Army knew full-scale human trials could not be conducted at home without causing public uproar, but they could make use of a disease outbreak in Egypt and obtain useful data through "peaceful" medical intervention. "Rule One of biological warfare development," said Peters in his book *Virus Hunter*, "was that you don't weaponize an agent until you have a vaccine or other treatment for it. Otherwise, if the wind blows the wrong way or something else goes wrong, you're in a whole bunch of trouble." Egypt's misfortune gave the United States the Rift Valley fever vaccine trial it wanted.

Peters went to Egypt and tested the experimental vaccine at NAMRU-3 on young naval men and women who weren't quite volunteers. Major Peters simply said to the troops, "We think you ought to have it." Naturally, they all responded exactly as they were trained to: *"Yes, sir! Thank you, sir."* Some three hundred Egyptians working for NAMRU, happy with their steady paychecks, joined in the chorus. It may have worked, since none contracted the fever.

Endemic in sub-Saharan Africa, Rift Valley fever somehow migrated to the Nile delta on the Mediterranean Sea, and somehow morphed into a hemorrhagic killer. While there were plenty of potential natural vectors, including mosquitoes along Lake Nasser, most disturbing were reports the germ was introduced intentionally. "There were a lot of scary possibilities which couldn't be definitively proved or disproved," Dr. Peters said. "[U]nless you caught some guy up on a hill with a blower and gas mask generating aerosols from a can, you couldn't be sure it came from man or nature." During this period, the Soviet Union and Egypt enjoyed a robust economic and military relationship that included furnishing Egypt with all kinds of weaponry. Further raising the specter of a man-made biological event, be it accident or otherwise, it is now generally understood—due in part to ex–biological warfare scientists who defected to the West—that the Soviets conducted a immense germ warfare program at this time, which included creating genetically engineered organisms (such as a strain of antibiotic-resistant anthrax) designed for their ability to kill.

The new, deadly version of the Rift Valley fever virus is again on the move. In September 2000, rampant outbreaks occurred on the Arabian Peninsula in Saudi Arabia and Yemen—with the virus poised to extend into Europe. The results of the initial Egyptian outbreak were devastating: 82 million in livestock infected, 200,000 humans infected, 18,000 of those human cases clinically confirmed, along with over 700 deaths. A survey of blood serum taken prior to 1977 proved the virus was not present in Egypt before the epidemic.

Today, Egypt and two other Arab nations—Syria and Libya—refuse to join the 1993 Chemical Weapons Convention, which 165 other nations have agreed to. Presumably another nonmember, the newly constituted

Iraq, will join. In a 1994 Congressional Research Service report on biological warfare programs, Egypt was listed in the "suspected program" column. Egypt's reputed biological warfare germs include plague and anthrax bacteria, an ailment called "West Nile" fever virus, and a germ that infected 200,000 of its own people in 1977—the Rift Valley fever virus.

Like most other nations in the region, both friendly and unfriendly to the United States, there has never been an independent inspection of Egypt's biological warfare capabilities.

IN OUR BACKYARD

While Egypt was the scene of the deadly outbreak, the American research on Rift Valley fever took place on Plum Island. Dr. Gerry Eddy, a veterinarian and head of virology at Fort Detrick, obtained USDA permission to work on the virus in 1977. Though even the milder *Entebbe* strain from the 1950s was labeled number six on the list of most dangerous viruses by the secretary of agriculture, the Army's Fort Detrick and the USDA's Plum Island did cooperative research on the killer strain, *Zagazig 501*, named after one of the hardest hit villages in Egypt.

Until now, Plum Island had worked on five animal diseases fatal to people. This would be their sixth zoonotic—a germ capable of both animal and human infection. However, unlike previous studies, Rift Valley fever came on the heels of a confirmed human epidemic, so Colonel Eddy sent along the experimental WHITECOAT vaccine that would ostensibly protect the staff from exposure. The collaboration proved to the Army's germ warriors they didn't need to run Plum Island themselves, not when the USDA gladly aided when necessary. The vets were there for the Army during World War II and they were there with them in the cold war. The key would be to see that minimal public interest over a "famine disease" in Egypt faded from memory.

But somebody talked. Somebody read the internal memo explaining the proposed research, and Peters's account of people in desert streets suffering from massive hemorrhaging, blindness, and encephalitic brain seizures, with herds of cattle and droves of sheep carcasses strewn across the countryside. Safety Officer Dr. Walker told over one hundred workers in contact with Lab 257—lab techs, janitors, electricians, boiler operators, clothes launderers, engineers—they had to sign health waivers and be injected with a vaccine that wasn't FDA approved. Though it was "voluntary," there was really no turning it down, not if the workers wanted to keep their current jobs. "You either took that shot or you had to go—or push a lawn mower around the island for the rest of your days," says a former worker. A few brave souls wrote the words "under protest" beneath their signatures before being pricked with the syringe. Talk spread along the island paths, in the cafeteria, and on the ferry. People began to worry.

At the 1978 annual summer retreat on Plum Island, Drs. Callis and Walker took turns briefing the Plum Island outside advisory board on their plans for Rift Valley fever—not only with the mild *Entebbe* strain, but on the dangerous *Zagazig 501* strain. Plum Island wanted to begin right away.

Dr. William Scherer headed up Plum Island's board of consultants, outside scientific experts who met once a year on the island to offer collegial advice and encouragement—"a dog and pony show," as one member put it. As chairman of Cornell University's microbiology department and a member of the National Research Council, Dr. Scherer was the type of scientific heavyweight Plum Island needed. He understood the meaning of viral reservoirs, hosts, and vectors like few others.

"Jerry," Dr. Bill Scherer began, brows furrowed. *"You know there are some important things to consider with Rift Valley. Look at this place,"* he said, extending his arm in a wide gesture. *"You have swarms of insects flying around on this island. There's a lot of water and marshy land here."* Scherer pointed out it would be much safer to wait until after mosquito season, and study the virus in the dead of winter. He offered to send entomologists, insect experts, to catalogue Plum Island's pest population and determine the precise level of protection required.

Callis listened quietly and politely nodded. The research was scheduled to start in early October, past the normal insect season, but hardly in the middle of winter. As for the insect study, there just wasn't enough time this year—they were too busy readying the virus experiments. Maybe next year. The consultants' prescient advice unfortunately was just that: advice. It wasn't binding. While the advisory group was "always concerned something would get loose from the laboratory," recalls member Dr. Robert Shope, its counsel remained "mostly reactive."

Soon thereafter, the Suffolk County Department of Health got a call from an anonymous Plum Island employee, *Newsday* got a hot tip from an anonymous source in the Department of Health, and a story ran on September 9, 1978. "You can't keep anything secret today," Dr. Callis complained grudgingly. After receiving word, Suffolk County's health commissioner, Dr. David Harris, placed a concerned phone call to Plum Island. Harris wanted to know what the Rift Valley fever was and how it related to some illness going on in Egypt. Dr. Walker assured Harris that the disease caused only mild flu symptoms in people, and in Egypt some "complications" had resulted in a few deaths. Mosquitoes apparently had something to do with the disease. Walker said the virus study would begin the following month and wrap up by the end of March, well before the 1979 mosquito season. No mention was made of other insects known to transmit Rift Valley fever, such as flies and ticks.

As for the workers, Walker assured, everyone on Plum Island would be vaccinated, the virus be kept in the high-containment lab, and employees wear face respirators. If by some remote chance anyone should become infected, they will be airlifted by military plane to "The Slammer," Fort Detrick's human isolation ward. Employees suspected of infection would be locked down in an insect-proof building on Plum Island until a firm diagnosis. There was, of course, the possibility the work may go over schedule, Walker said. Harris graciously agreed to "institute extensive mosquito control measures" by spraying chemicals with crop-dusting airplanes should the research continue past the first of May.

Plum Island refused to acknowledge they had been given up by frightened employees. "As a professional courtesy," Walker said, "if some new disease comes on [Plum Island] that is infectious to both animals and man, of course we would notify the health department." But contacting local officials was far from standard policy for the federal enclave. In fact, Plum Island officials never before informed the county about its zoonotic disease research—on multiple strains of at least five different germs—or about what germs were stored in its cavernous freezers.

In a feature editorial on September 11, 1978, four days before the Plum Island virus outbreak, *Newsday* declared, "We consider [the new openness] a welcome innovation," but "[u]nder no conditions do we think the lab should continue its Rift Valley fever research into a Long Island mosquito season." Falsely soothing the public, *Newsday* told its readers "[T]he disease itself is not fatal, but it's often accompanied by complications. . . ." In fact, the seven hundred people who died in Egypt didn't fall prey to complications. In each case, it was the Rift Valley fever virus disease that either putrefied the victims' organs or swelled their brains into lethal shock. Worse, reporting only fifty deaths in Egypt caused by the virus, the *New York Times* was off by a factor of fourteen.

Then the Plum Island virus outbreak occurred. Callis and Walker realized there was no way they could clean up the island in two weeks' time, vaccinate one hundred employees, and begin the Rift Valley fever project with Fort Detrick. "We really can't think about it until we get the island back together," said Walker. But, he added, "We intend to go on with the project" as soon as test animals arrived. Apparently, many of those animals were already there.

One man who refused to be played by Plum Island or misled by media reports was Suffolk county executive John V. N. Klein. Hailed as an "honest country gentlemen," the young public servant had a promising political future, and the charge of a county of over 1.3 million people. A week after the outbreak, Klein resolved to take matters into his own hands. Going over Callis's head with a two-page appeal to Secretary of Agriculture Bob Bergland, he spoke of a "grave concern" and "apprehension" over "the potential spread of disease once exposure has taken place." Exercising strict protocol, Secretary Bergland didn't respond to the county executive's dispatch, but instead referred it to a USDA research office underling. The acting associate deputy director of the Federal Research, Science, and Education Administration, United States Department of Agriculture, T. B. Kinney, a name of unknown gender, wrote a three-page letter accompanied by glossy Plum Island brochures, all extolling the virtues of animal research, while addressing none of Klein's worries. His letter made no mention of Fort Detrick's, Colonel Eddy's, or Major Peters's involvement, or that the study was in reality Army biological warfare research.

John Klein nearly hit the roof as he read Kinney's letter. The feds were stonewalling him. Plum Island and the USDA had no regard for state or local government, or for the public. The facts were almost too fantastic to be true: a biological research laboratory flatly pronounced it was going to fiddle with a deadly virus on a ranch full of animals, days after germs were proven and acknowledged to have escaped their lab! This time, Klein would not mince words. The experiments raised "a basic and profound issue with respect to the relationship between the U.S. Department of Agriculture and the people of Suffolk County," he wrote Secretary Bergland. "I cannot any longer permit this administration of county government and the people it represents to remain unaware of the total spectrum of activities carried on on Plum Island. . . ." Klein requested a personal meeting with Bergland.

"The People," Klein exclaimed, "can no longer exist 'in the dark' on this issue and must have the right of being fully informed with the opportunity to object . . . to any activity deemed inconsistent with the interests of the people of Suffolk. . . ." Without an opportunity to be heard, to be recognized, to be adequately represented with such base interests at stake, Klein all but promised that a Boston Tea Party would ensue in the local waters off Plum Island. Indeed, the public began to foment, pelting Plum Island with scores of phone calls and letters objecting to the precarious research; and the editorial boards of newspapers weighed in as well. Local groups organized town meetings, hailed Klein's decisive actions, and blasted the feds' proposed research. "How tragic would be the irony," proclaimed one resident, "if an agency established to control and find cures for diseases caused instead their proliferation?"

There was no reply from Secretary Bergland. Incensed, Klein felt that come hell or high water, Plum Island would not start this project without his direct input, whether it meant a federal lawsuit or a local picket demonstration. He fired a final warning shot by way of telegram:

TELEX CABLE U S
SECRETARY OF AGRICULTURE BERGLAND

REPEAT REQUEST OF MY LETTER OCTOBER 26 FOR PROMPT MEETING WITH YOU
RELATIVE TO PLUM ISLAND LAB EXPERIMENTATION STOP IT IS ESSENTIAL THAT DIRECT
AND PERMANENT MECHANISM OF COMMUNICATION BETWEEN USDA AND SUFFOLK
COUNTY RESPECTING PLUM ISLAND BE ESTABLISHED IMMEDIATELY STOP THE PEOPLE
OF SUFFOLK ARE ENTITLED TO NO LESS THAN FULL KNOWLEDGE OF THE NATURE AND
SCOPE OF EXPERIMENTATION WITHIN SUFFOLK COUNTY AND WITHIN ONE AND ONE
QUARTER MILES OF INTERSTATE FERRY TO CONNECTICUT STOP PLEASE RESPOND
PROMPTLY WITH DATE FOR MEETING WITH YOU STOP

JOHN V N KLEIN

Days after the telegram, Klein received a Washington-ordered phone call from Dr. Jerry J. Callis, director of the Plum Island Animal Disease Center.

"*Come on over,*" said Dr. Callis in a low Georgia drawl. "*Come see how safe we are over here.*"

BEYOND THE GATES

Minutes before 7:00 a.m., Klein's entourage drove through the barbed-wire checkpoint at the Orient Point lot to board the ferry. It was the dreariest of Mondays—gray skies, rainy, and bone-chilling cold. Klein's group included township supervisors, the county health commissioner, the county attorney, and ten reporters, including newsmen from the *New York Times* and *Newsday*. In the guard building, they took turns reading and signing the visitor affidavit required to board the boat. With their signatures, each acknowledged they might be detained in the event of emergency, and promised to obey the safety officer and not come into contact with animals after leaving Plum Island. Two burly, gray-haired security officers called the group to attention. "There are no cameras, no tape recorders taken to the island," warned the older of the two. "I know you can have these little things in your pocket . . . we'll make sure on the boat you won't have one on you."

The journey took about thirty minutes through the bumpy chop of Plum Gut. Awaiting them at the harbor guardhouse was the muttonchop sideburned Callis, in a dapper checked suit and dark tie punctuated with shimmering lodestars. He greeted the landing party warmly, then shepherded them onto the bus, which roared away toward the old Army parade ground. On the way through the woods, past the marshy wetlands, Callis lectured the group on the island's three-hundred-year history. With impeccable timing, he brought his story to a close as the bus pulled alongside the double doors of the old white Army chapel. The heavenly spire and the pews and crucifix had been removed long ago; the smell of frankincense and myrrh had given way to a deep musty odor. Inside the brown-paneled chapel were the laboratory chiefs, assistant director Graves and safety officer Walker. Shades were pulled down over the long rectangular windows, and a staffer dimmed the lights and turned on the projector so Callis could begin the presentation.

Klein and his team shifted uncomfortably in their folding chairs in the dark as Callis narrated a colorful slide show. As Callis clicked away, he informed the group about the current breadth of America's livestock ("136 million cattle, 25 million swine, 25 million sheep, 25 million goats, 8 million horses"); the history of foot-and-mouth virus outbreaks in North America—two-thirds of it anyway ("Mexico outbreak 1948, Canada 1952"); the USDA's careful, thoughtful selection of Plum Island in 1952 ("a coastal island separated by deep navigable waters"); and so on.

"I want to talk about biological warfare," said Callis. Everyone perked up. A dozing reporter in the back row stirred. "We don't do any of it on Plum Island." He clicked the slide button again, and large black-and-white headlines emblazoned across the chapel walls:

U.S. TESTS GERM WARFARE OFF L.I.

THE VIRUS HUNTERS OF FORBIDDEN ISLAND

ANIMAL DISEASE DETECTIVES WORK ON ISLAND OF NO RETURN

PLUM ISLAND FOR MANKIND, OR AGAINST IT?

"I solicit your assistance in dispelling these very unfortunate errors," appealed an unhappy Callis. "It's not good for the United States to have these perpetuated." Finally, facing the spin control head-on, the dam broke. A reporter asked about Karl Grossman's AP wire story seven years before in which Plum Island admitted a hand in biological warfare.

"I'm not knowledgeable about defense," said Callis, mindful that the Rift Valley fever project—the reason for his slide show—was a joint affair with the Army's Fort Detrick biological warfare facility, a fact unknown to his audience. Klein stood up. He'd had enough.

"Dr. Callis, please explain this Rift Valley fever program to us. Can we put into place a program to know what you are working with that poses a risk to humans and discuss your safety measures? We cannot continue to learn of your work by accident," Klein scolded.

Rift Valley fever was spreading through Africa into the Sinai Peninsula, Callis replied, focusing his answer not on safety, but on an urgent need for research. "Last week we heard it was in Nigeria." One look at the expressions on the local officials' faces told their response.

Nigeria? Who cares? they thought. *Nigeria is the other side of the world! In Africa! What about our people—the people RIGHT HERE?* Not acknowledging the nonverbal clues, Callis continued on, noting that Plum Island had a distinguished advisory board. He even extended an olive branch: one county health official would be permitted to attend their annual meeting on Plum Island. This, thought Callis, was a generous offer from an island unaccustomed to sharing its activities with anyone. He clicked quickly through several more slides until he reached one of selected zoonotic diseases. Some of these, he casually mentioned, had been investigated on Plum Island in the past. Though Callis attempted to gloss over this, Klein and the reporters sharply interrupted.

"Which germs did you work on? Which ones affect humans? What types of illnesses? Are any of them fatal?" Five of the diseases—virulent influenza, Coxsackie B-5 virus, louping ill, Venezuelan equine encephalitis, and Rift Valley fever—were lethal to humans. Five more—Newcastle disease, vesicular stomatitis virus, contagious ecthyma, foot-and-mouth disease, and Nairobi sheep disease—caused human illnesses. No mention was made of the other viruses and bacteria studied or the scope of the vast Plum Island germ library.

"Has anyone been infected in the lab?" Six had taken ill from accidental exposures, Callis said. Three people had contracted Newcastle disease, two caught vesicular stomatitis, and one caught foot-and-mouth disease. There were no fatalities to date.

Discussion turned to the outbreak. Sounding all too rehearsed, Plum Island officials assured the group that every effort was being made to test systems and ensure virus agent containment.

"We've had twenty-four years of safe work," Callis said to his tormentors. "Let's not get it out of proportion now."

After the slide show, the scientists led the tour to the ominous-looking Lab 257, before directing them into the perimeter guardhouse. Each guest forked over his personal items—watches, wallets, wedding rings, coins, paper notes, pencils—which were placed in small gray envelopes personalized with their initials.

"You have to go into the lab like Mother Nature brought you into this world," Callis drily quipped to his tour group. Everyone was ordered to strip in the change room and don white coveralls, white socks, and white sneakers. Then, with the door behind them closed and air-sealed, the door at the opposite side of the room was depressurized. It hissed loudly, then opened into a maze of hallways, dimly illuminated with fluorescent light. After observing farm animals in various stages of different infections, the group stripped naked and waited single file to take decontamination showers before leaving the lab, following the same strict procedures required of workers in the early days. As they changed back into their street clothes,

Walker peeled two Band-Aids off a reporter's fingers, informing him that the crevices harbored viruses.

Dr. Callis saw the group of fifteen men to the harbor and graciously thanked them for their interest. Did the press learn anything new in this, the second glimpse of Plum Island? "The whole thing was pretty redundant," says one attendee, comparing the tour with the earlier 1971 press junket. Another observed "a serious lack of mission at the island. . . . One would expect to find a sense of pride and accomplishment among the workers there. Based on our conversations, such is not the case."

After the grand tour, Klein reflected on the experience. "Now it turns out, there's a whole laundry list of such diseases they've been researching." Had Klein known the true condition of the facilities Plum Island was concealing; or that the insect study recommended by the advisory board was snubbed; or that the study was a joint affair with the Army; or the deadly effects of human exposure to the Rift Valley fever virus; or the true number of casualties that the new *Zagazig 501* strain wrought upon Egypt—things might have been different. Instead, without precaution the work went full steam ahead. Of Klein and his visit, Dr. Callis says twenty years later, "I didn't give a damn." The federal government wasn't going to be told what it could and could not do by some local politicians.

Try as he might, county executive John V. N. Klein didn't slice through the veil of secrecy over Plum Island. A noble failure, it had little if any impact on the island's activities. After the 1978 virus outbreak and the flap over Rift Valley fever virus, the island dropped the "World's Safest Lab" moniker. In fact, it dropped public contact altogether. The gates would not reopen until another Plum Island biological debacle fifteen years later.

THE RIFT VALLEY FEVER EXPERIMENT
Unlike bacteria, which can be seen easily under a traditional microscope, viruses are too small to be detected. To measure the concentration of viruses, scientists use the plaque-reduction method, whereby a visibly clear area in a culture dish, called a plaque, is left behind after the virus destroys the healthy culture cells. By measuring the plaque—essentially the damage the invisible viruses have done—you can get an idea of the titer, or how virulent and concentrated the virus is. Viruses are measured in plaque-forming units, or PFUs. In the Plum Island study of the deadly strain of the Rift Valley fever virus, the lab techs took a starter culture of African green monkey kidney cells out of the freezer and propagated them in culture plates using a growth medium of amino acids and heated calf fetus serum. To brew large quantities of virus, they fed *Zagazig 501*-plus on the monkey

cells they cultured inside sterilized fermenters, not unlike the ones seen at a microbrew pub. *Zagazig 501* was the Rift Valley fever strain pulled from a fatal hemorrhagic human case during the Egyptian outbreak. Peters added some kick at Fort Detrick—the plus—when he passed the virus through two sets of lung cell cultures taken from rhesus monkey fetuses. It was potent as hell. From the fermenter, they decanted the slurry into hundreds of 1-milliliter ampoules and stored them in nitrogen-cooled freezers at minus 70 degrees Celsius. They thawed each ampoule immediately before pacing briskly through the "hot corridor" to the animal room to inject a test animal.

Rift Valley fever virus affects its victim much like the well-known Ebola virus. Ten of the sixty sheep were used as controls, meaning they were not afforded the luxury of a vaccine. Twelve hours after the unlucky ones are exposed to the virus, viremia—the onslaught of a virus infection within the body—begins. Their temperatures rise sharply and their hearts beat rapidly in an attempt to circulate an immune response to parts of the body suffering from localized attack. But their immune systems cannot counter the multiplying germ. The blood circulates viruses to all quarters of the body, where they attach to the surface of organ and membrane cells, then infiltrate those cells. Within the cytoplasm, the virus hijacks the cell's machinery and replicates itself exponentially. Inside of one hour the cells lyse, or burst, unleashing thousands of new baby viruses upon neighboring healthy cells, where they attach and repeat the vicious cycle ad infinitum. The exploded cells cause edema, or large buildups of fluid in the connective tissues throughout the body. The sheep spasm and lose their gait, leaning up against the cell-block wall of the windowless room to stay upright. A pregnant ewe spontaneously aborts her young onto the floor in a grotesque scene. The sheep vomit blood and discharge red-tinged mucus from their noses. The dizziness in their heads makes them slide, one by one, down the wall and onto the concrete floor.

Within forty-eight hours, eight will be dead. Samples of each animal's blood-riddled excreta are taken to see if the virus passed through. The two that make it to recovery are then killed, along with the fifty sheep in the test flock. Each is lifted onto a cart by an animal handler and wheeled into the necropsy room. On kill day, veterinarians necropsy (animal autopsy) each of the animals under bright examination lights, snap photos of the diseased internal organs for later analysis, clip tissue samples from here and there, and send the carcasses down the chute to the incinerator charging room.

The initial experiment was a success. The human vaccine produced immunity in all of the vaccinated sheep and warded off the virus.

With the April 1 safety deadline met, the Plum Island scientists decided to push their luck. They revaccinated employees and continued the groundbreaking research, extending the study through the end of April and into the start of the mosquito season. Supplanting the ubiquitous foot-and-mouth disease virus work, Rift Valley fever fast became the island's germ of choice. Eight Shetland ponies injected with Rift Valley fever were observed daily and their rectal temperatures recorded for forty days to see if they were susceptible hosts or carriers. Necropsied on day forty-one, the ponies showed no internal signs of disease or clinical signs. However, the vets found that the Arabian horses maintained low levels of viremia, and, as such, were potential reservoirs to infect biting mosquitoes.

Next, they brought in another sixty-three sheep for a third Rift Valley fever project. This study revealed that infected sheep could pass USDA food inspections for human consumption. Drs. Dardiri and Walker observed the virus lingering in spleen tissues, but reasoned that "the spleen is not normally eaten." The scientists also announced in a press release that they produced "antigens and reagents" for future use. Put more plainly, they were brewing *Zagazig 501* virus in gross quantities on Plum Island.

As the studies progressed, tensions eased between the Plum Island and Suffolk County officials. Together they drew up emergency plans for hurricanes, civil disobedience (like violent animal rights demonstrations), fires, sabotage, radiological incidents, and unauthorized removal of an exotic pathogen. But the détente was short-lived.

Dr. Martin Mayer, the county's disease control chief, got a call from an investigative reporter over the weekend of April 29. Did he know Plum Island extended research until May 8, eight days into the mosquito season and six weeks after they promised to conclude? "No, I'm completely unaware," Mayer said. He called the island first thing Monday morning, and Walker promised him they would be finished—honestly, this time—by week's end. "A couple of extra days aren't going to hurt," conceded Mayer, calling it a "minor" deviation. Yet mosquitoes in the region were hatching and beginning to fly, according to the county's mosquito control chief, who told *Newsday* the pests would not be swarming for another week or two over "dry areas" like Plum Island. This "dry area" boasted three freshwater swamp beds covering 120 acres of an 843-acre island, and over twenty different species of mosquitoes called Plum Island home. Dr. Callis said the project ran late because researchers were in the middle of their studies.

While Plum Island faced three ongoing investigations—one for a virus outbreak, one for a construction fiasco, and one for criminal fraud—scientists uncorked vials of deadly *Zagazig 501* (and its progeny) and infected cattle, horses, sheep, pigs, and even seals. Few of the air filter, incinerator, and negative airflow repairs were complete.

The labs were vulnerable, and the USDA was compounding the risk.

A TROUBLING SCENARIO

Discussing Rift Valley fever—over a decade before the West Nile virus became part of the American lexicon—Dr. C.J. Peters and Plum Island director Dr. Roger G. Breeze wrote, "[T]he most probable routes of introduction of Rift Valley fever into the United States are via a viremic person who will be bitten by a mosquito or an infected mosquito aboard a plane. The infected mosquito will then infect a susceptible animal in which the virus will be amplified for the infection of more mosquitoes."

But what if the virus performed an end run, not from the nation's international airports, but from within, set free from a poorly run top-secret facility where it was kept in large quantities? What might have happened? And how?

It's a Thursday in early May, just a few days before the Rift Valley fever virus experiments concluded:

As Long Islanders open windows, inviting the crisp spring air to fill their homes, and venture outdoors to garden, mosquito larvae begin to stir, hatching from their eggs. Soon the winged creatures take flight and begin to feed. On Plum Island, scientists there are wrapping up a large-scale study of Rift Valley fever. Seven young Plum Island Culex pipiens, Aedes vexans, and Aedes atropalpus mosquitoes fly into Lab 101 through a three-quarter-inch gap between an old air filter and the roof. Two of them buzz their way into an animal experimentation room. Two more find the "hot zone" corridor. One flies into the clean corridor. The remaining two meet their demise in an exhaust fan.

The first two head straight for two test sheep dying of Rift Valley fever virus (the Zagazig 501 strain), and take a blood meal. Viruses disseminate into the insects' midgut and move to their tiny salivary glands. Two days later, an animal handler, Jeff, opens the air-lock door to check on the sheep. As he opens the door, the two mosquitoes in the corridor rush into the animal room. While he draws blood from one of the ailing sheep, one stings him on the arm while the other bites the sheep. Busy holding the frightened animal steady, he doesn't feel the pinprick. The mosquito takes an ample blood feast, leaving behind virus in the unknowing man. As Jeff opens the door to leave, the mosquitoes rush out. In the "hot corridor," one of the scientists on her way out of containment brushes by the animal handler. "Have a good night, Jeff," she says to him. "You too, Jane." One mosquito targets her and bites her on the neck. She brushes her collar, thinking the irritation is from the lab-coat tag rubbing against her skin.

That evening, Jane meets her family at Claudio's in Greenport, their favorite seafood restaurant. It's balmy outside, and the wait is long but worth the wait. While the couple chit-chat outside and ask the kids about school, a few mosquitoes zone in on them. "Damn gnats," Jane's husband complains, swatting away a small swarm. "We're getting eaten alive by bugs! Guess summer's officially here, huh?"

Back in the lab, the four mosquitoes buzzing around in Lab 101 die. A sole survivor makes its way outside through a hole in a rubber gasket attached to the room's air filter. It flies with purpose to the outside animal pen for a new meal of sheep blood.

Separately, one of the doses of experimental vaccine used to vaccinate the sheep outdoors is contaminated with live virus. The dose was injected into a ewe the day before. The ewes aren't scheduled for laboratory testing until after the weekend.

Two fuses are lit.

Earlier that afternoon, Jeff finishes tending to the ill animals in the test rooms. He spends a few extra moments comforting them with pats and coos. Then he showers out of containment and takes the ferry home to Long Island. The next morning, he wakes up and shakes off a little nausea. "Maybe there was something wrong with that fish Bobby caught and cooked up for dinner," he thinks. Convinced the feeling will pass soon enough, he readies for work and drives to the Orient Point ferry dock. While waiting to board the boat, a mosquito flies over and bites him on the leg. Preoccupied with his queasiness, he doesn't notice and climbs aboard the Plum Island ferry. Jeff shakes off his sluggishness, finishes work, and heads home for the weekend. He figures he must have mild food poisoning, and he'll beat it by Monday for sure.

Over the weekend, virus replication rages inside the two sheep—the one bitten by the mosquito and the other with the contaminated vaccine. They're both viral amplifiers, virus-manufacturing facilities literally spewing infectious particles. Swarms of mosquitoes from the marshes and landfills are in the animal pens now, biting the two sheep and the other animals.

A feeding frenzy begins.

Hovering near the island's harbor, a swarm takes a ride to Orient Point on the afternoon ferry and continues the journey on Long Island. Some catch the cross-sound ferry to New London, Connecticut. Others bite Canadian geese and gulls flying south to perch on Hamptons beaches. As the birds head toward Connecticut, they are bitten again by more hungry mosquitoes. People are migrating, too. America's well-to-do are traveling to Long Island's East End to spend their weekend in share-houses, bungalows, and waterfront country estates. Locals are hiking, cycling, swimming, and fishing in the environs.

The insects spread far and wide and bite more "amplifiers"—pets, live-stock, and people. The mosquito that bit Jeff at the Orient Point dock splatters on the windshield of a car. But the one that nabbed Jane outside of Claudio's bites another patron the same night, and then a golden retriever.

The cycle continues.

On Friday afternoon, the Rift Valley fever virus travels to Jeff's lymph nodes, which engorge with virus. Through the lymph and the bloodstream, the virus surges through his body all day Saturday as he pops open a Bud and watches the NBA playoffs. He's sluggish all day. The day lulls slowly into night. On Sunday morning, instead of feeling better, things take a turn for the worse. Jeff's temperature spikes to 105 degrees. The nausea, chills, headache, and achy-weak feeling intensify. He becomes dizzy in the afternoon, doubles over in the kitchen, and vomits. This is where the flulike symptoms end and Rift Valley fever begins. Jeff, who's never been afraid of heights, has vertigo just standing up. His eyes are suddenly sensitive to light, so he draws shut the blinds. His stomach feels strangely full, though he hasn't been able to hold anything down since breakfast Friday. The fullness turns to pain, and his eyes hurt more and more. He realizes this is far more serious than mild food poisoning or some twenty-four-hour bug. He hates doctors, but now he relents. Summoning the strength to drive to the hospital in Greenport, Jeff checks into the emergency room.

Unlike other hemorrhagic fevers like the Ebola virus, Rift Valley fever has a special affinity for the liver, where it concentrates necrosis, the destruction of cells. Each virus is 100 nanometers long (500,000 could fit on the head of a sewing pin) and has an envelope that binds to the wall of a liver cell and enters it. Inside Jeff's liver cell, the envelope opens and releases a second shell, which in turn releases three strands of nucleic acid that unfurl and replicate inside the cell, until the pressure of the baby viruses against the cell wall forces it to burst. Dead tissue is mottled with yellow blotches and blood. His lungs and spleen are attacked, too. The gallbladder swells to four times its usual size, congested with fluid and blood.

Jeff staggers dizzily into the emergency ward and he's admitted immediately. The doctor on duty pulls the curtain closed and helps his patient sit up. Contact with the virus is easy enough—through the skin via the mouth and nose, inhaled directly into lungs, or by the bite of a mosquito. Weak and semidelirious, Jeff coughs without covering his mouth and sends virus particles through the air. The doctor's next breath draws them into his own lungs.

The virus breaks down the integrity of Jeff's smaller blood vessels, and when they give way there is noticeable bleeding. Jeff has his first bloody nose, which lasts for an hour even though the nurse applies pressure and tilts *his head. While the nurse controls the nosebleed, her skin comes into contact with his blood. Vessels in his small and large intestines begin to collapse. That evening, he defecates diarrhea and dark blood. Blood shows up in his urine. Hemorrhaging in his eye covers parts of the retina and the optical nerve so that Jeff goes blind in his right eye and can see only shadows through his left. Hematomas appear on his torso, broken blood vessels near the skin. Capillaries near the skin collapse and cause red dots on his arms and legs.*

The medical staff is perplexed. The ER doctor calls in the head of infectious disease to have a look. Their patient is slipping out of consciousness; he can no longer speak. Before they can do anything meaningful, they need to know what they're dealing with. Jeff's blood test results are in, and alarmingly, the results indicate nothing—or nothing that the hospital has the capability to test for. The two MDs argue about the situation as they frantically flip through their Merck Manuals. *Could this be a viral hemorrhagic fever? Could it be Ebola virus? With no established course of treatment, there's little the doctors can do. They place an emergency call to the county health department and warn them of their preliminary diagnosis. Despite futile efforts, within twelve hours Jeff suffers near total vascular collapse. He goes into a state of shock and loses pulse. The medical team rushes to his side and tries desperately to revive him.*

It's too late. Jeff's gone.

Next morning, the doctors and nurse contract flulike symptoms themselves. If they are bitten by an insect out in the parking lot or in their backyard, or if they breathe on another family member or use the telephone, the virus will gain more momentum.

As Jeff lay on his deathbed, Jane began to suffer a severe headache. A migraine, she figures. But on Sunday night, she suffers a seizure. Panic-stricken, her husband rushes her to the hospital, the same one Jeff went to. In her case, the Rift Valley fever virus chooses to cross the blood barrier and enter neurons, and it attacks her brain. Edema, cell fluid from the broken-up neurons, swells and inflames her healthy brain tissues. Hemorrhages occur. The emergency room doctor on the second shift doesn't connect Jane's symptoms with the patient who died earlier. That was an isolated case with no confirmed diagnosis. Instead, this doctor guesses Jane's ailment is a case of encephalitis—like West Nile virus—and administers Acyclovir intravenously (as prescribed in her Merck Manual) *to help ease the brain inflammation. It works, but moments later she lapses into a life-threatening coma. Jane manages to pull through, "lucky" to have contracted the encephalitis brand of Rift Valley infection, instead of the hemorrhagic fever that stole Jeff's last breath hours before.*

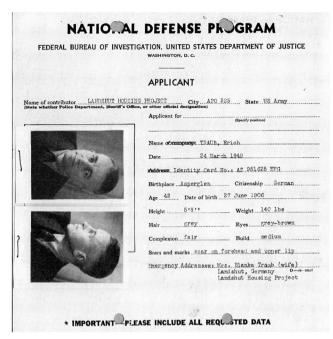

NATIONAL DEFENSE PROGRAM

FEDERAL BUREAU OF INVESTIGATION, UNITED STATES DEPARTMENT OF JUSTICE
WASHINGTON, D. C.

APPLICANT

Name of contributor ___LANDSHUT HOUSING PROJECT___ City __APO 225__ State __US Army__
(State whether Police Department, Sheriff's Office, or other official designation)

Applicant for _____
                              (Specify position)

Name of occupancy __TRAUB, Erich__

Date _____24 March 1949__

Address __Identity Card No.: AZ 951625 EPG__

Birthplace __Aspergian__   Citizenship __German__

Age __42__ Date of birth __27 June 1906__

Height __5'8"__   Weight __140 lbs__

Hair __grey__   Eyes __grey-brown__

Complexion __fair__   Build __medium__

Scars and marks __scar on forehead and upper lip__

Emergency Addresses: __Mrs. Blanka Traub (wife)__
                              Landshut, Germany
                              Landshut Housing Project

* IMPORTANT—PLEASE INCLUDE ALL REQUESTED DATA

Page one of Erich Traub's Operation PAPERCLIP application. The Nazi germ-warfare scientist was smuggled into the United States in 1949 and worked with the CIA, Army, Navy, and USDA. He was a founding father of Plum Island. *(National Archives and Records Administration)*

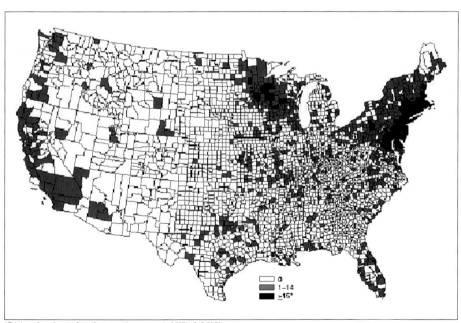

* Total number of cases from these counties represented 90% of all 2000 cases.

Number of cases of Lyme disease, by county, United States, 2000. *(Centers for Disease Control)*

U.S. Fish and Wildlife Service illustration depicting the Atlantic Flyway. Countless birds traffic between Plum Island and Old Lyme, Connecticut, the epicenter of the initial Lyme disease outbreak. *(U.S. Fish and Wildlife Service)*

Two views of the Lone Star tick, named after the unique spot on its back. Thousands and thousands of Lone Star ticks were bred on Plum Island and impregnated with exotic germs during the years of the initial Lyme disease outbreak. *(State of Michigan, Centers for Disease Control)*

Female, male, and nymph deer ticks, and larvae, placed along a metric ruler. Note the deer tick nymph is about a millimeter in width, and the larvae even smaller. *(Centers for Disease Control)*

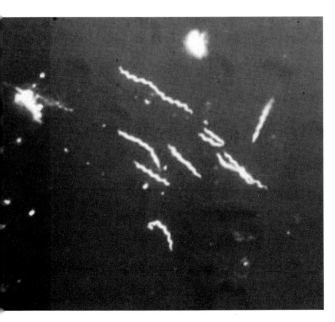

Borrelia Burgdorferi, or *Bb*, the bacteria that causes Lyme disease, magnified four hundred times under a microscope. *Bb* was discovered a few years after the Old Lyme, Connecticut, outbreak by Dr. Wally Burgdorfer. *(Centers for Disease Control)*

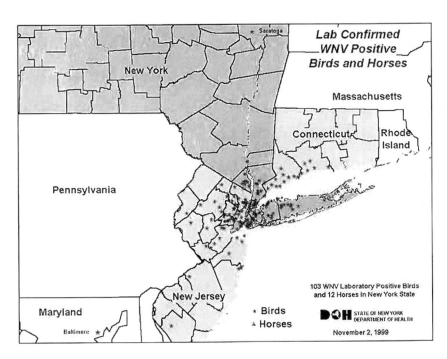

West Nile: regional map depicting horses and birds infected by West Nile virus. Note how the infected horses are concentrated in the shadow of Plum Island. *(New York State Department of Health)*

L AB 2 5 7

The biological warfare brass on Plum Island in 1954. (top row, left to right: Dr. Ralph Fish, Dr. Oliver Fellows, Lt. Bethea, Lt. Skanchy. Bottom row, left to right: Lt. Col. Harry Cottingham, Col. Donald L. Mace, Lt. Hunt. Not pictured: Lt. Luke West). Colonel Mace was the commanding officer of Plum Island when it was an Army germ warfare center. *(Plum Island government archive photo)*

Colonel Donald L. Mace, biological warfare commanding officer of Plum Island, February 1954. *(Plum Island government archive photo)*

Maurice S. "Doc" Shahan, first USDA director of Plum Island, sitting in the exact same chair and office as Colonel Mace, also February 1954. The USDA maintained for decades that there was no interaction between it and the Army germ warriors. *(Plum Island government archive photo)*

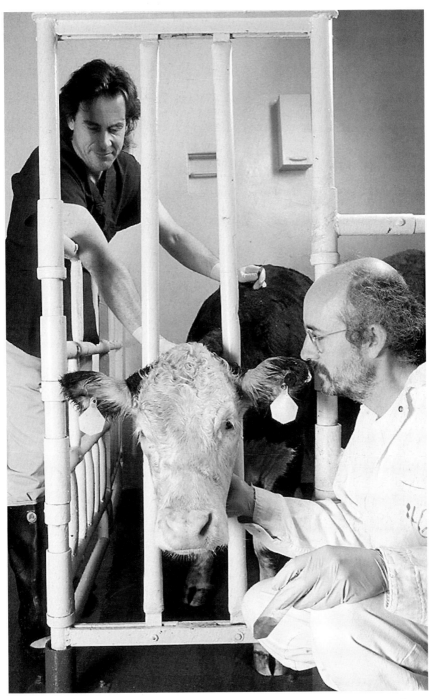

An animal gets its neck sandwiched by a squeeze gate used to hold test subjects in place for a foreign germ experiment. *(USDA photo)*

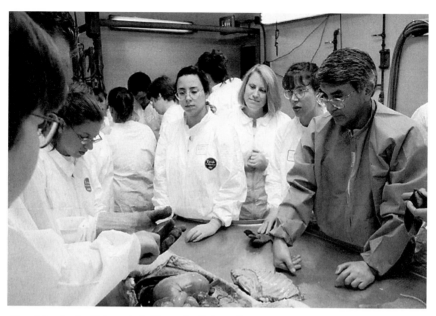

Plum Island scientists and visiting scientists gather around for an animal autopsy of a diseased carcass, called a *necropsy. (USDA photo)*

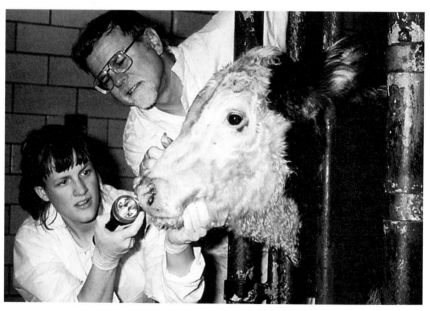

An examination into exotic animal germs by a Plum Island scientist wielding, of all things, a flashlight. *(USDA photo)*

Some of the 107 biological air filters atop Lab 101. Although they were designed to keep germ-ridden air from escaping into the atmosphere, many of the filters had inch-wide gaps between them and the roof. *(© by Edward Gajdel)*

A view through the airlock door into the secret world of the Plum Island biological containment laboratory. *(© by Edward Gajdel)*

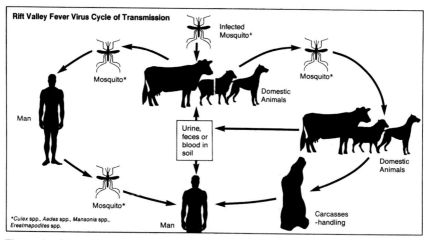

Rift Valley Fever Virus Cycle of Transmission

Infected Mosquito*

Mosquito*

Mosquito*

Man

Domestic Animals

Urine, feces or blood in soil

Domestic Animals

Mosquito*

Carcasses -handling

Man

*Culex spp., Aedes spp., Mansonia spp., Ereatmapodites spp.

The cycle of transmission, from insect to animal to man, of the feared Rift Valley fever virus, a viral hemorrhagic fever closely related to the Ebola virus. Rift Valley fever is transmitted between animals and humans by mosquitoes. *(USDA)*

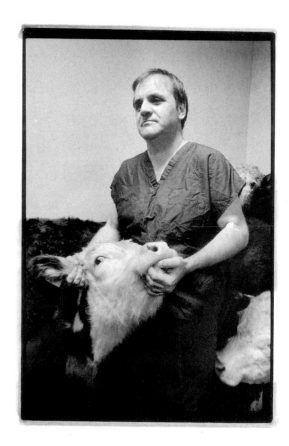

Former Plum Island director Dr. Roger G. Breeze inspects a cow for signs of disease. *(© by Edward Gajdel)*

These photographs show
the type of garbage
dumped in multiple places
on Plum Island, an exam-
ple of the island's
"Nothing Leaves" policy
gone wrong.
(confidential source)

Frances Demorest,
the assistant
librarian and one
of the long-
standing veterans
of Plum Island,
shown working in
the old library.
Demorest would
pay a terrible
price for speaking
out about the con-
ditions at Plum
Island. *(Plum
Island government
archive photo)*

Hurricane Bob, readying to smash into Plum Island—and Lab 257—dead-on.
A biological meltdown would occur inside Lab 257 that night. *(NASA)*

Lab 257 as it looks today, an archaic relic of the shadowy biological past of Plum Island. *(Sandra Lauterbach)*

Plum Island support workers on strike. The strike began in August 2002 and was still ongoing in October 2003. *(Suffolk Times and photographer Judy Ahrens)*

U.S. Senator Hillary Rodham Clinton speaks to a standing-room-only crowd of worried residents at a community meeting in February 2003 to discuss the Plum Island strike. *(Confidential source)*

Another view of Lab 257. *(James Muscarella)*

An aerial view of Lab 101. The bottom of the photo shows just how accessible to intruders the outdoor compound has become. *(USDA photo)*

Plum Island in the distance, from Orient Point. Note the cross-sound ferry to Connecticut at right, carrying hundreds of passengers and vehicles. *(David Tonsmeire)*

Plum Gut and Plum Island Harbor, another unguarded entry point into Plum Island. The Connecticut shoreline lies in the distance. (*USDA photo*)

Plum Island looking west. Orient Point, Long Island, is less than two miles away. The mainland lies in the distance. *(USDA photo)*

A cartoonist pokes fun at the strange and secretive nature of Plum Island. (Suffolk Times *and cartoonist Rob White)*

I'M SORRY, JACOBS, BUT THE **CONFIDENTIALITY** OF OUR WORK HERE ON **PLUM ISLAND** MAKES IT NECESSARY FOR ME TO INSIST THAT YOU REMAIN ON THE ISLAND FOR THE **REST** OF YOUR **DAYS!**

Meantime, Laura, the other patron at Claudio's bitten while in line, feels nausea as she readies to leave her summer cottage—where she's been tilling the garden and picking out new kitchen cabinets—for work Monday morning. "Maybe I'm pregnant," Laura hopes to herself. They've been try-ing for a baby for months now. With an extra spring in her step, she walks to the train station to catch the train to Manhattan. "I'll be fine," she thinks, smiling. "If it's what I think it is, a little morning queasiness is nothing." At work that day, she slices her finger on the paper cutter; a co-worker rushes to her aid. Now he has come in contact with the virus running through Laura's veins. He lives across the Hudson River, in Metuchen, New Jersey.

That same morning, Sunny, the golden retriever nabbed by the same mosquito as Laura, collapses on the living room floor. The owner of the spry three-year-old knows immediately that something is wrong. He carries Sunny outside, lays her gently on the bed of his Jeep, and rushes her to the vet. The vet takes a look at her and shakes his head in doubt. He instructs the lab tech to draw a blood sample. "The test should tell us something," he says in a comforting voice. But by the next day, Sunny succumbs to an unknown. The lab tech develops a bad cough and has sharp pains in her stomach that evening. She calls in sick the next day. She is one of hundreds already in direct contact with an infected friend, relative, or co-worker.

And this is just the beginning.

AFTERTHOUGHTS

Plum Island scientists voice their *own* fears, in a piece submitted to the New York Academy of Sciences, only a few years after their experiments:

> Introduction of Rift Valley fever virus into northern North America in the spring, when mosquito activity is on the rise, would pose a greater threat than an introduction in the late fall, when a frost would kill the mosquito population and potentially end the outbreak. Numerous North American mosquito species are competent labora-tory vectors of Rift Valley fever virus. The development of an epi-zootic [disease in large numbers of animals] / epidemic also requires the prevalence of amplifying hosts, such as cattle, sheep, and goats [and people], with levels of viremia high enough to infect vectors.

A self-indictment. The record shows Plum Island was well aware of the dangers posed by its research in a dilapidated laboratory facility.

The virus outbreak forced Plum Island to renew its commitment to biological safety—but this transformation took years to effect. Brushing off a reporter's question about the virus outbreak that *truly* occurred and the decision to continue research, one of the lab chiefs replied matter-of-factly,

"Rift Valley fever is transmitted only by mosquito. For our security to break down, a mosquito would have to get into the lab, bite an infected animal, get out through our intricate security system, and bite a susceptible man or animal. We think that possibility is extremely unlikely."

The documented holes in the roof of the "intricate security system" of the USDA-described "World's Safest Lab" shifts this glib assessment from "extremely unlikely" to frighteningly possible. It is easy to see how a second outbreak could have occurred on Plum Island, one of unimaginable ramifications.

In the meantime, the deadly Rift Valley fever virus research charged ahead.

Fortunately, they got away with it.

Crossing the Rubicon

> *A potential for agent escape is still present. . . . It is time to*
> *reconsider Plum Island Animal Disease Center's future.*
> —NATIONAL ACADEMY OF SCIENCES (1983)

To Dr. Jerry Callis's disappointment, the virus outbreak and the Rift Valley fever debacle overshadowed Plum Island's positive contributions, including an achievement that would change the future of science. While attention focused elsewhere, Callis's chief scientist, the unassuming Dr. Howard Bachrach, was quietly achieving scientific miracles deep inside Laboratory 101.

Working for Nobel laureate Wendell Meredith Stanley after the war, Dr. Bachrach was the first to isolate the polio virus. He was also the first to photograph polio with the lab's primitive electron microscope—though Stanley, according to scientific traditions, received credit for both as lead researcher. These accomplishments led to the development of Jonas Salk's polio vaccine and the end of the virus as a menace. Soon after, the promising young biochemist accepted Doc Shahan's generous offer to come to Plum Island. The parallel paths of Jerry Callis and Howard Bachrach represent two threads running through Plum Island's history. Both Callis, the youngest of ten siblings on a rural Georgia farm, and Bachrach, whose father owned a boys' haberdashery in Minnesota, were drafted by the USDA to learn hands-on science in Europe while America prepared to research exotic animal germs for the first time. While the outgoing Dr. Callis was a tall, officious administrator type, one who relished giving authoritative speeches about scientific research, the bespectacled Bachrach was a gnome-sized scientist who preferred tinkering on the workbench in the windowless lab fortress. When Callis was elevated to director in the early 1960s, Bachrach was named chief scientist. "Callis was very supportive of Howard," says Dr. Robert Shope.

Dr. Bachrach had been using chemicals in the mid-1970s to fractionate viruses, or break them up into component protein pieces. He theorized that one or more of a virus's proteins might provide enough of a "signature" to fool an immune system into creating a response, triggering immune memory cells and thus providing protection from being subsequently infected by the real germ. Vaccines, traditionally killed or weakened viruses that provided immunity, always left open the danger of reverting to virulence and causing disease to the host. But if an inanimate viral subunit vaccine could be designed, not only would it never revert to a virulence, but it would eliminate vaccine contaminants that sometimes caused harmful side effects. He discovered that the foot-and-mouth disease virus's capsid, or shell, was made up of four distinct proteins that latched on to healthy cells to infiltrate and infect them. After many years of trial and error, in 1975 Bachrach found his subunit protein: VP3. When he injected this third capsid protein into pigs, it protected them from a later challenge with the virus.

This was a bittersweet discovery, however, because extracting and producing VP3 in large quantities was too complicated to be practical.

That is, until genetic engineering came into the picture.

Scientists figured out how to manipulate DNA, molecule chains coded in specific patterns that instruct cells how to build proteins. By snipping portions of DNA that made insulin and splicing those portions into cells, they caused the cells to manufacture large amounts of synthetic insulin. Insulin, traditionally extracted from animals, was at last widely available to diabetics, helping them lead normal everyday lives. Scientists called the technique gene splicing or recombinant DNA, and it promised to advance microbiology to infinite heights. In the world of disease prevention it held even greater promise. This new technology, thought Bachrach, could produce VP3 en masse! Dr. Callis agreed with the chief scientist. He presented Bachrach's proposed work to a special committee at the National Institutes of Health to obtain their approval. The NIH liked the idea of using an animal virus as opposed to a human virus—the stakes were lower, and there was less of a chance that an accident or an unexpected new resultant germ would attack human populations.

In 1981, aided by three assistants on loan from the brand-new San Francisco–based biotechnology company Genentech, Bachrach entered the air lock, donned his white lab coat, and went to work in Lab 101.[1] The infective part of the foot-and-mouth disease virus (called the virion) consists of eight thousand nucleotide building blocks, housed inside the capsid. These blocks code the information for the virus's proteins. Dr. Bachrach knew that VP3 was made up of protein numbers 8 through 211. Using special enzymes, he pasted the DNA building blocks that coded for the VP3 proteins into a plasmid. A plasmid is a small ring of DNA that acts as a transporter, or magic carpet. The plasmid was then spliced into an *E. coli* bacterium, a benign microbe found in the human digestive tract. Lodged inside the bacterium, the plasmid directed the *E. coli* to assemble VP3 protein molecules; it essentially taught a living germ how to cobble together a foreign protein. When the *E. coli* replicated, the DNA instructions to make VP3 also replicated with it. Bachrach grew the modified bacteria—now an entirely new life form—in large kettles, and the newfangled bacteria grew easily in unlimited amounts as unmodified *E. coli* normally would. Each new *E. coli* produced an astonishing two million VP3 proteins.

Seven weeks into the project, they extracted the synthesized VP3 from the vats of slurry. Healthy cows and pigs were vaccinated with it, then held in animal rooms with disease-carrying swine infected with a highly virulent virus strain. Of the thirty-three animal species that can transmit foot-and-mouth disease, pigs are the worst offenders, spewing virus everywhere in their vicinity. On the tenth day of exposure, the scientists walked into the animal rooms and cheered—not one of the vaccinated animals had caught the disease. Bachrach and his team were thrilled with the results of the experiment. They created the first genetically engineered vaccine in history. And they did it on Plum Island in 1981.

"We believe this to be the first production through gene splicing of an effective vaccine against any disease in animals or humans," Secretary of Agriculture John R. Block proudly announced. With Bachrach's ingenious recombinant DNA vaccine, disease control would be safer and far less expensive. Genetically engineered vaccines for this and other viruses could be produced anywhere in complete safety, because modified *E. coli* with VP3, and not the offending virus itself, would be produced. There was *zero* chance of VP3 becoming virulent on its own. And unlike traditional vaccines, genetically engineered vaccines didn't require refrigeration, no small benefit given the prevalence of disease in the developing world.

True to character, for all the newswire stories and national press attention the revolutionary achievement received (including front-page billing in

[1]Genentech was founded by scientist Herbert Boyer, the co-inventor of recombinant DNA in 1973. The company spawned the biotech business, an industry valued at over $430 billion today.

the *New York Times*), the timid Bachrach gave but a single quote, letting Dr. Callis handle the publicity. Eliminating the threat of virus "is our goal and is what we expect to do," the diffident biochemist told the *Washington Post*. That was the extent of his public comment.

At a February 1985 ceremony at the White House, President Ronald W. Reagan presented Bachrach with the National Medal of Science, the United States' highest honor for scientific achievement. Welcoming him into the elite order, President Reagan spoke of his "pioneering research in molecular virology" and his "role in using gene splicing to produce the first effective protein vaccine." Equally impressive was his induction into the National Academy of Sciences, a select fraternity of the nation's brightest scientific minds. Dr. Fred Brown, a current Plum Island scientist and the world's foremost authority on foot-and-mouth disease virus, calls Bachrach "one of the foundation stones of Plum Island." Dr. Howard Bachrach had come a long way from Faribault, Minnesota. He had become a scientific legend.

Obtaining the achievement of a lifetime at age sixty-one, Dr. Bachrach decided to retire in 1983, after placing the capstone on a thirty-year Plum Island career. Employees gaze upon the reticent scientist—who returns on the ferry now and then to tinker with viruses—with measures of admiration and awe. "He is kind of an icon," says a former administrator. After retiring, he turned to his three other passions: golfing in Florida, gardening in the backyard of his Southold home, and playing shutterbug, just like the old days in the 1950s when he snapped the first-ever glimpses of strange viruses.

The breakthrough did not merely revolutionize the future of an animal disease, it sounded the starting gun of the biotechnology race. Before long, genetically engineered hepatitis B and rabies vaccines were developed. The innovative work on Plum Island changed the face of biological science forever. It promised to resuscitate Plum Island's reputation. But with Bachrach's departure, one of Plum Island's two "foundation stones" was cast into the water. The absence was deeply felt. Without the anchor of the no-nonsense, determined chief scientist, the place shifted further downward.

HAZARDS

While officials on Plum Island thought the virus outbreak was a trifling event the public ought to forget and move on, Washington took the blunder far more seriously. A safety review board was appointed, made up of biological safety officials from USDA headquarters, the National Institutes of Health, the Centers for Disease Control, and Fort Detrick. These men were the top germ safety experts in the nation, and what they had to say in their *fourth* annual review of Plum Island since the outbreak speaks volumes.

"We believe there is a potentially dangerous situation and that without an immediate massive effort to correct deficiencies, a severe accident could result. . . . [L]ack of preventative maintenance, [and] pressures by management to expedite programs have resulted in compromising safety." After they described the failure—the "breaks"—of three of the four degrees of containment during the virus outbreak (the fourth degree was the island itself), they noted two more incidents of this nature. "In August 1980, a break was due to the use of facilities for purposes for which they were not designed, and in January 1981, a break was due to the absence of first-degree containment and lack of maintenance of second-degree containment. We find both unacceptable." In the first incident, Plum Island experimented with animals, not in the specially filtered and drained animal isolation rooms, but in corridor hallways never designed to hold infected animals. The second infraction showed smoke tests revealing gaping holes in the animal test rooms. "These are signs of unsafe facilities," the safety review board said. "[B]reaches from known deficiencies are never acceptable."

One passage written by the experts shows just how dangerous Plum Island was at the same time it mass-produced deadly Rift Valley fever viruses:

> It is distressing to the experts to find . . . *massive* leaks . . . (a) around the metal frame of a door supporting a gasket (caused either by door buckle and weld separation and/or rust-induced deterioration), (b) around hinges and handles, (c) through an *unplugged hole* (former pipe hole?) and/or around an existing pipe penetration, (d) around adjacent spots behind a "dust deflector" on top of a wall-mounted cabinet, (e) through cracks in the wall plaster, (f) through an electrical outlet. . . .
>
> Major problems in the air movement in the four rooms examined . . . suggest that few rooms at [Plum Island] provide satisfactory containment.

Was this known to Dr. Callis, Plum Island's director? Given his earlier sense of care and dedication, had Dr. Callis lost his grip, his handle on the situation? It seemed illogical that the esteemed veterinarian would stand by and let his island crumble beneath him. He seemed preoccupied often with seminars and conferences around the world, neglecting the management of Plum Island. By then, "Jerry Callis had his own agenda," says a source. "As long as he could take off to Portugal, Thailand, Caracas, or somewhere deep in Africa or South America—these were some of his favorite places— he was happy." The safety panel found widespread belief among employees that management was lacking, communications had deteriorated, and there was a "dictatorial" atmosphere. Employees expressed fear of reprisal if things were brought to the attention of management. The findings of the safety review board were only the latest in the unbroken string of ill-fated events over recent years taking their toll on Plum Island.

AIDS AND AFRICAN SWINE FEVER

Before human immunodeficiency virus (HIV) was discovered in 1986, the cause for the emerging disease dubbed acquired immune deficiency syndrome, or AIDS, was unknown. Dr. Jane Teas, a pathobiologist at Harvard University's School of Public Health, compared the symptoms of AIDS with those of known animal diseases, and hit upon a striking similarity between AIDS and the African swine fever virus in 1983.

Amid the hype over Teas's proposed link between AIDS and African swine fever, forty-seven Plum Island employees were secretly tested for the presence of the African swine fever virus, normally not infectious in humans. Six tested positive. When asked about the findings, Dr. Callis said they were "nonspecific," or unreliable test results. Though they didn't contract the disease, six people had accidentally been exposed to the virus in Plum Island laboratories. They were virus carriers—capable of spreading it to other animals through tick and insect bites—and reservoirs of natural virus mutation, possibly into a strain that *could* affect people, like swine flu had decades ago. With no cure, the six Plum Island workers would have to live, symbiotically, with the virus. And without ever knowing they had it.

The subjects were tested without their knowledge or consent, since the samples were taken as part of routine physical examinations. Dr. Callis said he couldn't recall whether those that tested positive for virus were notified. To this day, no Plum Island worker interviewed remembers being told they were carriers of African swine fever virus. Management apparently deemed it unimportant.

Finally, after being stalled for months by the USDA to study the AIDS connection, Teas's colleague Dr. John Beldekas got a call from a secretary at Plum Island, telling him a shipment of African swine fever virus materials was being sent to him overnight via Federal Express. The next day, Beldekas waited in his laboratory all day for the package to arrive. It never came. Checking his answering machine at home in the late afternoon, there was a message from a neighbor saying a package had been left with her, and there was very little dry ice that accompanied the delivered parcel. It was from the USDA. He had no idea how they had his home address. Beldekas asked his neighbor to open the package and put the vials inside her kitchen freezer while he raced home. "So instead of delivering it to my laboratory where I had the proper freezer, it was left on my door-step. . . . Then, when I called Plum Island back to tell them I received it, and everything was fine, they were like 'Oh. Okay.' They did everything they were supposed to do, but they also did everything in their power to screw us up."

The joint discovery of Dr. Robert Gallo and French researchers of a new retrovirus called HIV settled the debate over AIDS and African swine fever. Today, HIV is widely regarded to have been an African monkey virus that jumped from chimpanzees to humans at some point between 1926 and 1946 (by virtue of humans eating raw, virus-infected chimp meat) and fully established itself in epidemic proportions by 1981.

Of the work, Dr. Teas later said, "It was viewed not as an interesting question but as a stupid thing I shouldn't be doing. . . . You would think the world could use a few ideas, especially with a problem like AIDS." The limited research aimed at finding a link between African swine fever and HIV ended.

CATS

Plum Island Lighthouse keepers always kept a few cats around for company, fenced inside the lighthouse reservation, and the cats greeted the USDA when they first arrived in 1952.[1] When the Coast Guard automated the light and vacated the lighthouse in 1978, the abandoned cats broke through the old fence and foraged the island for food. Slowly but surely, the wild cats bred and bred, until there were a "zillion cats," as one observer recalls. "Everywhere you looked, there were cats." For decades, the feline and human populations coexisted; most of the cats had become feral and kept the rodent population in check. Occasionally a cat would sneak into one of the outdoor cattle pens and eat from the feed trough. But when the island's cat lovers fed the felines, many became tame. A guard fed a gang of cats with food he brought over on the ferry periodically. "This group hung around the dock guardhouse," recalls another. "Consequently, there were dozens of cats hanging around where visitors entered Plum Island."

Management decided the island's large cat population created a negative image when outside scientists visited. Instead of discouraging feedings near the guardhouse, a memorandum came out in 1981 that said, according to one employee, "We aren't worried about the cats swimming to the mainland—we just can't allow them to be here and we're going to

---

[1]Soon scientists were experimenting on "imported" cats (domestics from southwestern Missouri) while researching a new cat disease called feline cytauxzoonosis, and its possible link to African East Coast cattle fever.

poison them." Scientific and support staff alike vehemently protested. *"You will not kill our cats! This is an agricultural facility, made up of veterinarians, for heaven's sake—we will do the next best thing. We'll neuter them!"*

Workers laid out scores of Have-A-Heart traps near the guardhouse, on paths and in the meadows. As the cats were collected, the vets took pause from their scientific research to neuter them, shaving their rears just forward of the legs to identify which ones were fixed. "For weeks, it was just the strangest thing, to see these cats darting around the island with nude bottoms," remembers the worker. But the neutering didn't work—many of the cats were too smart to be lured into the traps.

So they were poisoned instead. The "Animal Welfare Committee" ordered a technician to mix sodium fluoroacetate—one of the deadliest poisons known to animal or man—in rusty tin cans with hamburger meat and set them out for the cats. One of the cats, the harbor mascot Harry—named after dock guard Harry Sinuda—was spared execution. Just before Dr. Walker laid out the poisonous food traps, Harry the Cat was secreted inside the emergency power plant, where he escaped a gruesome fate. The others wouldn't be so lucky. "*Ahh*—the way they did it," says a former worker, slowly shaking his head. "Maybe they thought it would put them to sleep. Instead, the wicked dose of poison fed these cats—well they were walking around the island, choking and making terrible coughlike sounds." A day later, six cats were found dead. Before long, dozens of felines were being pitched into the incinerator.

Someone got mad enough to blow the whistle on this inhumane act. The USDA was snagged, it turned out, on a technicality: the use of sodium fluoroacetate was prohibited by both state and federal law, and Plum Island hadn't obtained the special permission required for its use. Asked about a "cat overkill" by a reporter, Dr. Callis said: "I'm embarrassed."

It wouldn't be long before he embarrassed himself again.

END OF AN ERA
What happens when he who makes the rules breaks the rules? Over one hundred microbiological safety rules were in place on Plum Island. At times they were painstakingly tedious. Cutting corners here or there may not seem so bad to the casual observer. But that one missed shower, an improperly installed air unit, or a forgotten air-lock door—each inched toward disaster. At the beginning, Dr. Callis wrote in the three-inch-thick safety manual distributed to all employees that the safety program "should be backed up with provisions for enforcement and penalties for willful violations of the regulations and instructions." And furthermore, "Personnel

shall report promptly to the director, through the safety officer, any safety violation that occurs or that is observed in the acts of others."

The federal law allowed foot-and-mouth disease virus to be *shipped* over the mainland to Plum Island, but not *kept* on the mainland. During a summer's day in 1982, Director Jerry Callis left a container of glass ampoules filled with a highly contagious Brazilian virus strain and related antiserum in the refrigerator at Orient Point for three days, before finally bringing it to Plum Island. An employee reported the infraction. Callis admitted guilt, and the Washington USDA office suspended him for two weeks without pay. "I unthinkingly left the container on the mainland," he later said. "I made a mistake. I'm not proud of it."

The director believed his suspension was far more severe than the violation warranted. After all, the viruses were wrapped in a self-destructing mechanism; that is, if the glass vials somehow broke, acid surrounding the container would rush in and annihilate the virus. At least that was the way it was designed to work. "There was some kind of fiasco," recalls Dr. Jim House, a recently retired Plum Island scientist. "And I don't think we'll ever know the details—it was unfortunate, really unfortunate."

A former veteran employee remembers Callis coming back from Brazil with more than cans full of viruses, and doing more than salting the germs away in the warehouse fridge. "The safety officer at the time was Jonathan Richmond. He was at Orient Point when Callis came back with the cans . . . [and Callis] opened them. Absolute safety violation. But before he opened them, Jonathan reminded him of the safety rules and regulations, and not to open them—that it must not be opened. Jonathan reported it to Washington to protect himself."

"I think [Callis] was getting a lot of flak at that time from the Washington area," says another former employee. "I recognized his voice when he called in [during the suspension], and he was calling from his home. It was very sad, because it hurt his reputation on the island and in the community. And guess where Jonathan Richmond is today? He's one of the heads of the CDC!"[2]

The following July, more virus safety rules were broken. Two Plum Island employees with a container of viruses scheduled for gamma irradiation, ostensibly en route to Radiation Technology Inc. in Rockaway, New Jersey, decided to make a two-day detour out of a three-hour drive, and stowed the biologicals in one of their garages at home overnight.

---

[2]One of the more prestigious Plum Island alums, Dr. Jonathan Richmond was until recently the director of health and safety for the Centers for Disease Control and the CDC's national "hot zone" expert, responsible for writing the rules for biosafety levels one through four.

In 1983, the National Academy of Sciences weighed in. Their two-hundred-page in-depth report repeated what lay people had been suspecting for years—Plum Island had run its course. "It is time to reconsider PIADC's [Plum Island Animal Disease Center's] future," it began. Though they commended Dr. Bachrach's pioneering work in genetic engineering, in the eyes of the nation's top scientists, Plum Island was through. "[I]ts isolation and high costs of operation, construction, and maintenance make it unsustainable in the long term. As soon as possible, USDA should proceed with construction of a new, highly secure, mainland laboratory to succeed PIADC as UDSA's principal center for research...."

While mainland research cost $175,000 on a per-scientist basis, comparable research on Plum Island cost $328,000. Of the island's total annual budget of $11.7 million, $7.6 million was invested in salary and personnel benefits, and only $16,000 was earmarked for new equipment—hardly enough for an electron microscope. In a dig at the locals, the academy found "the education level of technical personnel seems to be lower than what might be available near larger urban centers with university communities." Other than Dr. Bachrach's revolutionary work, Plum Island's virology program was "at least twenty years out of date." The experts also detected "a sense of isolation between different workers within the facility as well as between PIADC scientists and the larger scientific community.... [Plum Island] suffers from a paucity of mid-career, established investigators." Put another way, the place was languishing without direction by disinteresteds nearing retirement. And its ability to handle an exotic animal disease outbreak in the United States: "precarious at best," said the experts.

Jack R. Dahl, a hearty, mild-mannered North Dakota rancher and former president of the National Cattlemen's Association, helped write the report. "We identified a lot of problems that existed on Plum. We knew what the problems were—and we knew of the research work and the accomplishments. The problems outweighed the accomplishments."

The academy scoffed at the importance of its island location, in light of its populous surroundings and available modern technology. "Better biosafety containment is possible. An island location is not a guarantee of safety...." Twenty-five years ago, Labs 101 and 257 were state-of-the-art; now they were archaic. A full five years had passed since the virus outbreak, yet they wrote that "a potential for agent escape is still present." Walls weren't sealed, air-lock doors weren't properly gasketed, and exhaust air filters weren't installed. Personnel were allowed to move between support areas, laboratories, and animal rooms, despite the facility's own requirements. Though most labs prohibited eating, drinking, and smoking within containment, "surprisingly, the new renovation [at Plum Island] includes a lunch facility within the barrier zone."

A self-described "independent rancher" (the twenty-first-century euphemism for the near-extinct cowboy) who today runs a ten-thousand-acre ranch, Dahl recalls that closing Plum Island and moving operations onto the mainland was the "crux" of the study. The report urged the facility be torn down and closed.

All the negative press, the PR missteps, the biological bungling, and the outright mismanagement had impacted Dr. Callis's reign as Lord of the Manor. No longer was he the twenty-seven-year-old kid scientist who, at Doc Shahan's side, helped build the "World's Safest Lab."

One source described the situation on the island at the time as "stagnation." Another, "benign neglect." "Jerry had a helluva vision about Plum Island," says his successor, Dr. Roger G. Breeze. "But that vision sort of ran its arc. And I'm not trying to be critical about where that arc comes to rest." But rest it did.

A former official cites the director's penchant for travel as a contributing factor. "I'd say he was one for the ages, one of the few scientists who was a good leader and a good scientist. But by then his mission wasn't running Plum Island—it was running around the world, visiting other countries doing disease work, addressing meetings, and attending seminars—that was his agenda. And that's one of the reasons why Plum Island went to hell."

All the negative newspaper stories, egged on by Karl Grossman, *Newsday,* and others—much of it warranted, but not all of it—weathered the old veterinarian beyond repair. "There's one thing I think Jerry failed at," says Dr. Jim House, "and every director after him has also failed at—public relations. There was and is no program to explain the benefits of the place, the fact that it is a national treasure, that *it is needed.* Nobody does that—nobody goes out and gets the community involved and lets them understand it."

Much of that failure was an extreme desire for secrecy. "I think Jerry had this old-school cold war mentality," House contemplates. "You must remember, the Russians thought Plum Island was a really wild site out there. They knew it existed, and they played this game, they built it up in areas, saying it was a biological warfare site, or at least something of cold war value. And this just carried on." Dr. Carol House, also a retired Plum Island scientist (and Jim's wife), adds, "I think Jerry liked that." Though the outside advisory committee—now headed by Dr. Robert Shope—urged Plum Island to hold frequent press events, the hurtful, enduring sting caused by Grossman's 1971 AP germ warfare story caused Callis to lock up the gates tighter. Problems and difficulties on Plum Island were met not with candor and public recognition, but with suppression. Ironically, the lack of sunlight on Plum Island's activities over the years would lead the public—and rebuffed curious news reporters—to always assume the worst, fanning the flames of mystery, intrigue, and rampant speculation.

In one of his final communications, an uncharacteristically dry memo lacking the usual pleasantries, Dr. Callis told Plum Island employees that federal budget reduction legislation called for 5 percent cuts across the board, further depleting Plum Island's shoestring budget. To fight skyrocketing costs, he was considering a proposal to consolidate the two laboratory buildings into one, close the old Army-vintage administration buildings, and make Laboratory 101 a self-contained operation. This might—might— require cuts in staff, he said, but lest anyone worry, he would reduce positions through natural attrition, not through layoffs. By this time, Callis twice had stopped a move to privitize the federal workforce and reduce salaries, benefits, and the total number of staff positions. Callis may have faulted in many areas, but he remained doggedly loyal to his people, and to the promises he had made to them a quarter century ago: *"respect personal dignity . . . recognize work achievement . . . provide work security. . . . Believe in the Golden Rule and always practice it."*

Just days after his memo, the USDA hammer came down heavily. Regime change came quickly on Plum Island—Washington called up Jerry Jackson Callis, and told him his time was up.

"Let's just say retirement was not his own idea," says a source. "It happened abruptly—there were a lot of people and politics involved. He had really been there too long. We needed much better progression. We didn't need one person in one place—no matter how good he was." While Dr. Robert Shope maintains Callis's retirement was his own decision, other sources confirm he was "pushed out."

The USDA "elevated" him to the post of senior research adviser. This allowed him to maintain regular contact with his beloved island and offered another perk. Says one administrator familiar with the offer, "They said to him, 'We'll give you an office on the island, and carte blanche for several years on travel.' And Washington did that by dragging in money from other agencies and hidden places. They took real good care of him. It was a payoff, a golden parachute, if you will—they had to give it to him."

Dr. Roger Breeze, his successor, elaborated. "Jerry Callis is a very exceptional person. The United States never had a facility like Plum Island, until the debacle in Mexico hit between 1946 and 1952. People had to go out there, find it, and build it. Meanwhile, Jerry went over to Holland to learn about viruses from the Dutch. Now if you said to someone today, 'Hey this is the federal government and we want you to go to a foreign country to learn this whole new thing, and when you come back, we're going to build a laboratory for you,' people would laugh in your face. 'Yeah—sure—bullshit—can I get that in writing?' But Callis did it."

Lucrative offers from private laboratories had come his way for years, and he turned them all down. Plum Island was his life, and the island's life was in many ways his—it was his first job out of school, the only job he'd ever worked. Like a parent to his child, he could never see fit to abandon it. And there he remained for four decades.

Most of those interviewed for this book have the highest respect for Jerry Callis. "Jerry built Plum Island into a very prestigious place," says Carol House. "I think [his successors] had a hard time following in his shoes—anyone would, because of his memory and his depth of knowledge on everything that was Plum Island. A good—no, a *very good* manager." Dr. Shope acknowledges that "every leader has some disgruntled employees," but he found Dr. Callis nothing but "a real straight arrow." John Boyle, a former budget director at Plum Island, called him "a brilliant man, at one time a great scientist, and probably still that way today." Over time, a fondness and respect bonded the staff to their beloved director. "It wasn't always perfect," says building engineer Stanley Mickaliger. "There were discrepancies, and sure there were gripes—which were all normal. But we all understood something very important—Jerry Callis took care of us."

When asked about his retirement, Callis says simply and humbly, "I've enjoyed my career tremendously."

Plum Island librarian Frances Demorest—second in seniority to the now retiring director—saluted her director:

> *Dear Doctor Callis:*
> *Our thoughts return to 1953 and the local turmoil over the decision to locate the Laboratory at Plum Island off Orient Point. The representatives from USDA were not received warmly but, as time went on, the turmoil and anger subsided. It has been a long time and we have crossed a lot of water together!*
> *Through the efforts of two fine Directors, Dr. Shahan and yourself . . . the Center was recognized both nationally and internationally. Also, the Center made a tremendous economic impact on this end of Long Island.*
> *Harrison and I trust you will experience a busy, fruitful, productive retirement, and enjoy GOOD HEALTH and fond MEMORIES for a "JOB WELL DONE."*

Not everyone lamented his retirement. Some expressed glee over the change in the guard and looked forward to replacing the dusty old administration with fresh new leadership. "A lot of people thought of him as an emperor who treated Plum Island as his personal fiefdom," one former scientist says, "and to a degree, I believed he did. Whether your [research] program was funded depended on whether he liked you or not." But those who cheered Callis's departure didn't realize how good they had had it. "I tell you what," says one of them. "Six months later we were making novenas to Saint Anthony that he'd come back."

Because what came next would be far worse.

"You know, there's this old saying on Plum Island," says one worker. " 'When Dr. Callis went, the island went with him.' Once they moved him out, everything went downhill.

"In my humble opinion, this is when the whole island started to fall apart."

PART 3

THE DECLINE

10

The Kingdom and the Glory

> *Clarice:*
> *Best of all, though—one week a year you'd get to leave the*
> *hospital and go here . . . [shoving a map into the prison*
> *cell] . . . Plum Island. Every afternoon of that week you can*
> *walk on the beach or swim in the ocean for up to one hour.*
> *Under SWAT team surveillance, of course.*
>
> *Dr. Lecter:*
> *"Plum Island Animal Disease Research Center." Sounds*
> *charming.*
>
> *Clarice:*
> *That's just part of the Island. It has a very nice beach. Terns*
> *nest there.*
>
> *Dr. Lecter:*
> *[later, after Dr. Lecter discovers Clarice's offer was a hoax]*
> *No! It's your turn to tell me, Clarice. You don't have any*
> *more vacations to sell me on Anthrax Island . . . !*
> —THE SILENCE OF THE LAMBS (1991)

R oger Breeze wanted to be a vet because the local veterinarian was the most successful person who came by his family's sixty-acre dairy farm in the north of England. The Breezes milked cows and raised chickens in the 1950s. They delivered creamy milk each morning to their customers' doors, along with fresh eggs and chickens, eking out a living by profiting on both production and delivery.

Roger's idea was to follow in that country vet's footsteps. When he was seventeen, he attended vet school at the University of Glasgow in nearby Scotland, one of the oldest universities in the world. Though it wasn't his original plan, upon graduation he was given the coveted opportunity to teach at Glasgow. It was a prestigious appointment he couldn't turn down. To help make ends meet, he opened a local vet practice and worked nights and weekends. "I had some crazy nights," Breeze told *Outside* magazine:

> Once I had just finished pulling a newborn pig that was stuck in its mother's womb when I get another call about a sick dog. I go right over, knock on the door, and a bunch of Hell's Angels answer. They're all looking at me kind of funny, but I'm too worn out to care. I examine the dog and see right away that it's too far gone with distemper. So I take that dog out back and shoot it. The bikers pay me my fee, but they're staring at me wide-eyed, like I'm some kind of lunatic. It's not until I'm back in my car looking at myself in the rearview mirror that I see that my face and hair are all blotched and matted with pig placenta. I looked like the psycho vet from hell.

Two years into his professorship Breeze emigrated to America—a bold and unconventional move. He saw that young go-getters like himself, no matter how bright they were or how hard they worked, would be shunted into Glasgow's faculty caste system. There were eighteen veterinary pathology positions in all of Britain, and a slow thirty-two-lockstep ladder of advancement. Every professor parked on the same step was paid the same meager salary. "It didn't matter whether you taught Sanskrit or law," he remembers. " 'As long as there is still death, there's hope,' we used to say."

Roger Breeze disliked the stuffy peerage. *What does your father do? Who do you know? What high school did you attend?* No one seemed to care about what was really important—one's talent and ability to perform. "Can you imagine people in America asking, 'Where did he go to high school?' " Breeze asks. "How the hell would anybody know where to look, let alone care?" The religious discrimination also bothered him. Once a man "of great power" at Glasgow asked him to return to his alma mater to teach a discipline he didn't know. When Breeze explained his lack of knowledge in the discipline, the man said it didn't matter, saying, "There are too many English Catholics up here teaching and we need more Scottish Protestants." Ironically, Breeze was neither a Scot nor a Protestant—but with his University of Glasgow pedigree, they assumed he was.

All those stodgy trappings—and the royal mahogany paneling and leather furniture—never aroused him. No one in America cared what the laboratory lobby looked like and what the scientists' last names were, so long as they produced great science. In America, Breeze was more at home than he ever was in Britain. "Here it's 'On with the job!' That's what I like."

THE BREEZE REIGN

For all but a few of today's scientists, it's not about the money. It doesn't matter if you earn a million per year if five years down the road you are less employable than you were when you started. It's about science, but also, apparently, *the glory*. "They all want the glory," Breeze asserts. "That's what the whole thing is sold on today, and as director, your job is to provide that." To reach scientific brilliance, you need the proper tools, and that means expensive equipment. Most scientists do not question their ability; what they do question is whether a potential employer can offer the equipment and support they need.

Dr. Breeze's theory, honed in his first career post in the New World, went like this: institutions like Yale and Columbia, because of their reputations, have no problem recruiting postdoctorates for their three-year apprenticeships with lab chiefs. In this regard, America was little different from the Old World. Elsewhere down the ladder of prestige, that isn't quite the case; competition is fierce among the rest, and to prevail, a lab director must deliver the resources.

In 1984, after just a few years on United States soil, Dr. Breeze rose to chairman of the microbiology department at the Washington State University vet school, where he oversaw twenty-five faculty members. The school had a unique approach to its faculty recruitment that fit well with Breeze's management style. "It was a very entrepreneurial university," he recalls. "You could walk into the president's office and tell him you had a problem and you could cut a deal." He could reduce a salary or eliminate a position altogether and use those funds to buy an electron microscope for a star faculty member. Utilizing this flexibility, Dr. Breeze built the best department of its kind in the nation. In the mid-1980s, however, a new provost arrived. Out went the freewheeling philosophy. Now, over half of Breeze's powerhouse faculty was being courted by rival universities and bombarded with lucrative job offers.

On a cold January morning Breeze went to the provost's office to plead his case. "Listen," he said in his deep Scottish-sounding brogue. "You have to give us the flexibility we once had—please—or we're going to lose our people!" The provost leaned over his desk and lectured his spunky department chair. "Roger, you need to understand something. WSU is never going to be a great university. We're going to be a place where people pass through to achieve their best elsewhere. You've done a fantastic job developing these people—really you have—but you'll have to accept that." Breeze could hardly believe his ears. What the provost was really saying was that there's a cap on one's success, that one could go no further. And in America, no less!

"I told him, 'I have never heard so much bullshit in all my life.'" Breeze turned and walked out of the provost's office, feeling like he was back at Glasgow. He trudged back to his campus office, darting between waist-deep snowdrifts. He spied an orange slip of paper on the windshield of his car parked out in front of the quaint brick vet school building. "What the hell is this?" he yelled out. It was a $10 parking violation: Parking Without a Permit. "When it snowed," recalls Breeze, "they only brushed off the windshield on the driver's side looking for the permit." His parking permit had slid to the other side of the dashboard.

The ticket was enough to push Breeze over the edge. "I absolutely became unglued." Breeze stormed up to his office and resigned his chair. Previously, the USDA had asked him to run the Plum Island Animal Disease Center. He had responded by asking, "Are you crazy?" The gig came with a $20,000 pay cut, made worse by a cross-country move and a huge cost-of-living increase. Now *he* was the crazy one. Breeze renounced his British citizenship, pledged allegiance to the United States, and cast his lot with Plum Island, an island monarchy all his own. To Breeze, unless one's science was "the very best in the field," one's career was finished. He left WSU, taking with him the "entrepreneurial style" he perfected there.

"The whole management style I have is this—you have to go in, not just with money, but with resources." He could help the USDA deliver those resources at Plum Island. Or, as one familiar with Breeze puts it, "He could go up there and kick some ass."

When Roger Breeze arrived, he found Plum Island's laboratories "literally falling into the sea." Turning Plum Island into a gleaming new facility would take a tremendous financial effort, and money was unavailable. "Nobody said to me, 'Come to Plum Island and we'll pour money into it. . . . ' Washington thought the money they poured in gave them more of the same—the same science that wasn't good enough." The goal then would be to take Plum Island's small budget and stretch it as far as possible, to lure the best scientists with new lab equipment, and keep them happy once they arrived. "I told [USDA headquarters] we either have to do it this way or recruit only Czechs, because this place looks like Prague in 1956 and they'll feel right at home here." Meanwhile, the National Academy of Sciences, having excoriated Plum Island, said it couldn't be fixed and urged it be closed down. While Breeze spoke persuasively to the skeptics about how the island would turn around, privately he harbored doubts. He knew all eyes were on him, looking for his tenure to end in disaster. His first failure could very well be his last.

Taking over the Plum Island kingdom, Roger Breeze had put himself up against the proverbial wall. He had some snake oil to sell. The question was, could he sell it?

The first thing Breeze's "science first for glory" regime did was tear apart Plum Island's budget. Since the Army had left Plum Island in 1954, ongoing funding problems were par for the course. Dismayed over early cuts in his budget, Doc Shahan said, "The project may stand or fall, depending on adequate support in the beginning. . . . [USDA] cannot properly discharge its responsibilities without such a facility as Plum Island which is at present only an infant facing gargantuan tasks."

When Dr. Callis first took the reins in the 1960s, he too felt the money crunch: "For the last several years, the salaries of wage board employees have increased annually . . . [but] our budget has not been increased by an amount equivalent. . . ." And in the 1980s, Dr. Shope's advisory board proposed Plum Island prepare two budgets—one outlining the costs of a mainland laboratory, and another showing the additional costs of operating a facility on an island, in an effort to sway Congress into properly funding Plum Island. This way, Congress could understand why it had to fork over such large chunks of money to support functions as compared to research. Without the necessary funds, said the advisers, Plum's mission could be "jeopardized."

To raise Plum Island from its depths, Breeze first focused on where the money was going. For this, he needed a crack accountant, a brilliant numbers guy. Enter John Patrick Boyle, known as one of the best number crunchers in the business. An Irishman of medium build, with a frosty white beard, Boyle in his heavy Boston accent recalls Breeze's offer: " 'You'll have to build it up from nothing, John,' he said to me. 'It presents a real challenge.' " As they discussed the opportunity, something about the Scot (though he was really a north-Englishman, everyone thought him a Scot) captivated Boyle. He saw in Breeze a rarity in government service—a visionary with an ambitious plan, backed up with the enthusiasm and an uncanny relentless drive to carry it through. Boyle signed on.

Boyle's first visit to Plum Island was like something out of an episode of *The Twilight Zone*—to him, it was an eerie isle that time forgot. The once impeccably groomed and abundantly flowering "plantation" was now a wildly overgrown jungle. Paint had chipped off Lab 101's walls and huge chunks barely clinging on were flapping in the ocean winds. "Scientists were just sitting around doing nothing." Or at least Boyle thought they were scientists. "You couldn't tell the difference between a scientist, a tech, or the guy who swept the floor." The place had gone to pot—it was filthy and carried a rank, musty smell. The floors, caked with stains, hadn't been mopped in ages. On laboratory bench tops, which required an ultrasterile surface for virus research, Boyle spied dried soda pop, crumbs, and papers strewn about in piles. He took his index finger and swiped it along a pipe that ran along the wall, removing a quarter-inch layer of dust. "It was just ugly," says Boyle. "And everyone had the same attitude—'fuck it.'"

Plum Island was also a mess in other ways, especially when seen through the assiduously burning eyes of an accountant. "There was a lot of what we call indirect research cost [IRC]," remembers Boyle. "Scientific money that is eaten up by administration, support, building maintenance, grounds crew. And this side of Plum Island was essentially a Golden Cow." John Boyle explained how it worked during the reign of Dr. Callis: "Engineering would say, 'Well, we have this hurricane season, and we need [power] redundancy up here because we may lose a 50,000-kilovolt transformer that might go out on a Saturday night, when there's no one around.' They would use scare tactics like these, and add, 'Now if we lose that 50-KV transformer, then we lose power to Lab 101, we lose our negative airflow, and there's danger of what we have escaping out into the atmosphere.' And it would scare the bejesus out of management, so they'd say, 'Here, take this $75,000 and purchase a generator.' They dazzled them with footwork."

No longer would the tail wag the dog. To Breeze and his man Boyle, the primary need was not support, but rather Plum Island's pièce de résistance, its holy grail: the science. Without science, there could be no *glory*. "The IRC was outrageous—it was 78 percent [of Plum Island's yearly budget] when we got there." At a normal lab, Boyle says, it would be in the low twenties. To improve science and keep the island from being shut down? "Shave the IRC down to a bare minimum," says Boyle. No longer would engineering and plant management be the Golden Cow that got whatever it wanted. Boyle sharpened his pencil and got to number crunching, putting in late nights and seven-day workweeks, huddled over stacks of spreadsheets and financial data. With a stroke of his pencil, he slashed the animal supply contract and whittled down oil and new vehicle purchases. Next, he started counting up all the widgets—electric saws, toilet paper, circuit breakers, circuit boards, soap, hammers, nails, lightbulbs—and figured out where costs were coming from. "We bought thirty items, used twelve, and they reported six in inventory," he recalls. "People were carrying stuff off the island on Sundays, typical stuff people try to steal from the government—except here, they were doing it wholesale."[1] He installed a phone

---

[1] On one occasion—after seeing numerous instances of employees walking off the ferry on quiet Sunday afternoons with large boxes under their arms, dashing for their cars—Boyle spotted three men hauling a large machine off the ferry into a van. Rubbing his eyes in disbelief, he looked out the window again and swore; his hard work was again being undermined right before his eyes. Exclaiming, "That's it! I am not putting up with this anymore," he charged out of the warehouse toward the trio when one of the boat crew grabbed his arm and stopped him, saying, "Whoa, wait, John, you don't have to—they're just borrowing that. It'll be back in a few days—you'll see." The big machine was a floor-washer and the employees were taking it off to wash down the Greenport VFW hall, something they were doing every six months or so.

switchboard that provided reports of outgoing calls. "The day after I announced that one, there were cries I was violating people's privacy." But it worked—the monthly Plum Island phone bill plummeted in just one month from $7,000 to $3,000.

"John, I am hiring a doctor and I need an X-ray crystallography machine," Breeze told Boyle early on. "It costs half a million dollars—now go find me the money for it." Coming up with that kind of dough stumped even the uncanny number cruncher, who had just completed chiseling together a budget. He tore the whole thing apart again and found Breeze another $500,000, securing a renowned doctor's transfer to Plum Island. "He was very demanding to work for," Boyle says, smiling fondly, relishing working on the Breeze dream team. "I had to pull a string here and there and unravel it all and put it back together—and of course I found the money." But from where had it come?

"If you watch the pennies," Boyle's parents had taught him growing up in hardscrabble Dorchester, an Irish Catholic inner-city Boston neighborhood, "the dollars will take care of themselves." Now he was watching nickels and dimes, too, and to Breeze's delight. "We cut a substantial amount of money," says Boyle. "We cut well over a million dollars." He believes even more could have been slashed. After six years on the job, says Boyle, "I still didn't know where all the waste was. There was still some shit going on that I couldn't put my finger on—which tormented me to no end."

With the budget in Boyle's capable hands, Breeze next brightened up the physical appearance of the laboratories, taking special interest in the toilets. "Nobody had cleaned these rest rooms—there would be no toilet paper, and the sinks were filthy. I wasn't prepared to live with people not doing their job. Everything at Plum depends upon a handful of scientists who are pulling the train." At a minimum, those scientists had to have sparkling bathrooms, perhaps the most basic of all provisions. "I had to completely remove all of the furniture in the toilets because it was so bad—so dirty—it couldn't be cleaned." Breeze ordered brand-new porcelain, metalwork fixtures, shiny faucets, and gleaming tile, and refinished the rest rooms. Then he explicitly set forth the frequency and manner in which they were to be cleaned. He also conducted unannounced spot checks throughout the week to ensure full compliance.[2] "It was a personal tour de force," Breeze gloats.

But that was only child's play compared to what he did next. Within a

---

After that episode, employees knew better than to mess around with the new budget director. "Roger says tact and diplomacy are not my strong suits," says Boyle.

[2]Director Breeze also took a similar interest with the hallway floors. In a moment of candor, he said to me, "It's kind of silly when you hear a scientist talk about the hallways—is it waxed, is it shiny—why would you give a damn? But you are *not* going to create a top-class facility if you are only top-class about one thing."

year, the new administration had become a lean and mean machine, deaf to the rants and raves of the support staff and their supposed needs.[3] So when Walter Sinowski, the building foreman of Lab 257, told them in early 1991 that an emergency backup power cable needed repair, management had a different answer this time around: no.

Roger Breeze had become a slave to science.

Soon after taking over the reins, Breeze set out to acquaint himself with the island's three hundred daytime inhabitants. "When he first came in, we really thought he might be one of us," says one worker. Whereas Jerry Callis wore a suit and tie to work each morning, Dr. Breeze scarcely looked like a doctor, let alone the director. Sporting a plaid shirt and jeans and smiling broadly, Breeze was slapping workers on the back as if they were old friends. Veteran employee Martin Weinmiller recalls Breeze coming ashore on his first day. Heading home after the graveyard shift, Weinmiller watched the ferry bearing the new director tie up at the harbor dock. "He walks off the boat, and looks out at the island for a time. Then he says, 'Either I'm going to make this place or I'm going to break it.' "

Easing himself quite comfortably into the chair occupied by Dr. Callis for decades, Breeze quickly set the tone with employees with his newsletter, the *Plum Island Diary*:

> I am sure that the last many months have been very frustrating and disheartening. . . . This Center is not going to close, we are not moving to another location. . . . Together, we are going to plan and build the next proud 35 years of our history.
>
> The impression potential recruits gain of [Plum Island] must be diminished by the shabby state of most of our buildings and grounds resulting from years of neglect—and I know that our loyal E&PM [engineering and plant maintenance] staff who have tried so hard to do so much must also be discouraged.
>
> If we complain about federal salaries, crumbling buildings, and bureaucratic inertia, it should be no surprise that people lost interest

---

[3]All the deep cost-cutting apparently didn't extend to the creature comforts enjoyed by Dr. Breeze's cabal. Boyle first worked in the Orient Point office, but Breeze soon moved him onto Plum Island. "He put me in possibly the best quarters on Plum Island," remembers Boyle. "Talk about an Irish promotion!" From his desk in the old Army jail, he gazed through an all-glass solarium and viewed the parade ground and aqua-blue waters beyond. He complained to Breeze that the greenhouse effect made the office unbearably hot. "I told him it might be seen as elitist if I install an air conditioner, and he said, 'Yes, John, it will—but do it anyway.' So I put a big-ass 12,000-BTU unit in there—and E&PM [engineering and plant management] didn't like that at all."

in working here. Let's be proud of what we do and speak positively wherever we can.

Those with problems and suggested solutions can help me more than those with problems alone. . . . Thanks for your input in advance.

The whispers began: *Hey, look out for this guy*. When Jerry Callis looked out upon his island from his office perched over the Army parade ground, he saw an honorable band of three hundred loyal, dedicated employees. Breeze saw support staff and veteran scientists as a motley gang of serfs, a drain on science funds that did little more than punch the clock and collect an oversized paycheck.

In Breeze's first big move, he fired three scientists and forced a fourth into early retirement. He called the African swine fever virus team from the lab to his office and told them they had two weeks to pack up and leave. They were dumbfounded. "We had experiments with years of work gone into them," says Dr. Richard Endris. "With Roger, there was no kinder, gentler way," said another source. The team complained to the swine industry and the local congressman that taxpayers' work was "going down the tubes." That bought them a six-month reprieve. But once that time was up, Drs. Endris, Jerry Pan, Gertrude Schloer, and their leader, Bill Hess—a forty-year Plum Island veteran—were out on their rears. "He's seen fit to change directions," the seventy-two-year-old Dr. Hess drily told a reporter, "which is his privilege."

In Breeze's opinion, the four scientists were repeating the same science, or as one Breeze official said, "Reinventing the wheel, over and over again," testing viruses on one tick species after another. Said the official of the sacked four, "They were the worst of the worst—one of them hadn't published a research paper in nine years. Another one hadn't written one in five years. Sat there on their dead asses leaning on their elbows. They knew it was coming."

Contrary to what others said, Dr. Endris and Dr. Hess *had* published scientific papers in recent years. Breeze's reasoning that it was time for a genetically engineered vaccine for African swine fever made little sense to Endris, who says African swine fever antibody proteins (like Dr. Bachrach's VP3) just didn't have the prophylactic properties necessary. "[Breeze] did not understand the biology of it, didn't have the grasp of it—but he didn't let that get in the way of his politics." Ironically, Endris was one of those few on Plum Island who thought Dr. Callis's departure might be a good thing. A new director could rejuvenate the crumbling laboratory and reinvigorate its mission, he thought. Now he knew just how wrong he was. "Personally, after I saw what replaced [Callis]," he admits fifteen years later, "I wish he had stayed."

Breeze then confronted the remaining scientists. Sitting them all down, he berated them with facts and figures—the number of test animals purchased compared to the number of scientific papers produced. The number wasn't nearly high enough and the few papers published were terrible. "This was really sad," Dr. Jim House would growl years later. "The part that's bothersome is that he tried to play down the accomplishments of those who had been there before him—not one time, but numerous times. He tried to make their accomplishments look petty."

There was another reason the scientists were pushed out—to make room in the budget for a new ferry.

MELTING SNOWBALLS IN THE ANTARCTIC

Just when it seemed like things were clicking into place—the budget being tightened, unwanted scientists pushed overboard, and new equipment for new scientists under requisition—the roof caved in. Or rather, the dock caved in. The ubiquitous autumn storms that pounded Plum Island smashed the harbor dock to pieces. The $700,000 cost to replace the bulkhead blew a hole straight through Boyle's crafty budget and jeopardized monies earmarked for the new crystallographers, spectrographs, and electron microscopes. Washington told Breeze that Plum Island had to pay for the repair, putting him in the unenviable position of reneging on promises to his recruits. Says Breeze, "That would be like telling recruits at Columbia University, 'I can't buy your scientific equipment for you, because I have to go fix a piece of West 168th Street.' They would look at you like you were crazy." With the dock standing in the way of scientific glory, he had to do something.

In its long history, the secretary of agriculture's advisory committee, made up of representatives from livestock industries, had never met outside of Washington, D.C. Breeze proposed to Washington that the committee meet this year in Old Saybrook, Connecticut, to see Plum Island. As a friendly gesture to their new exotic disease lab director, the aggies agreed to organize it—but on one condition: he would not, under any circumstances, ask the committee for money. "The Department was going to put $1 million into the island and that was going to be it—and I was not to ask them [for any more]," says Breeze.

They all came out one morning—the National Cattlemen's Association, the Pork Producers' Council, and the dairy and poultry associations—all the groups that had a multibillion-dollar stake in what some might label "corporate welfare." Plum Island had always been painted as essential to the American people—wholesome, apple pie research to protect the food supply, defensive research that private industry just wouldn't do. But when the twenty dark suits crowded onto a boat bound for Plum Island with "Cap'n" Breeze, the scene looked less vital to the taxpaying public than to the billion-dollar agribusiness conglomerates the captain's passengers represented.[4]

In evangelical tones, Dr. Breeze walked and talked the group through the deteriorating laboratory, preaching how he planned to shore up its crumbling foundations, rid the place of the driftwood, clean the bathrooms, shine the floors, and bring science—glorious science. "He was very charming," says one observer. "So charming he could melt snowballs in the Antarctic." But meeting again the next day, the Pied Piper's smile had turned down at its corners. He talked hard and fast at them in his distinct accent.

"This place is never going to fly. You can't keep doing this business with the island falling down. We need $40 million here. We need it. I'm delivering the program, I'm hiring the people—but it won't work without $40 million." Breeze had previously been told by his engineers that it would cost about $25 million in total to repair the island into tip-top shape. He paused and looked at them, head cocked for effect.

"Now if you don't want to do it, that's fine—and we'll all walk away. No problem—I'll go find another job." But heads in front of him were nodding. *We're with you, Roger, we are—we are!*

They were. Thumbing his nose at his superiors, Breeze had secured, in a matter of days, a construction program that Plum Island had failed to get under way for decades. The livestock groups got behind it, and leaned heavily on reluctant USDA officials and Congress to give way. Though infuriated with their insubordinate new director, there was little Washington could do about it. How did his superiors respond? "Well," says Breeze, smiling, "everybody changes. If there was no money, it was all over, and they might as well have known that. You cannot make bricks without clay." Breeze soon had $22 million of clay in his hands ($3 million shy of his personal goal, but still enough) to build some 58,000 square feet of new space and renovate another 45,000 square feet. It was a truly remarkable feat.

Successful in dividing Washington officials from their industry advisory group, he trained his sights on the island, and on how to grab the animal kingdom for himself.

---

[4] A few years later, Tom Cook, president of the National Cattlemen's Association, would tell the *New York Times* that when it came to Plum Island, "Our bottom line . . . is we want to see the most dollars made available for research."

ong Island's North Fork was far too remote a place to attract good scientists, said Breeze. The cost of living there was far too high for government-salaried postdocs. Opportunities for spouses were also limited, except maybe to wait tables or work retail cash registers. As far as raising children, some scientists thought the school system fell short. "You'd have to go way up the Island to find any real sophistication," says one. And another, "The area is dead . . . it is not the type of place to attract young, upwardly mobile professionals. God knows how many miles you are from the nearest quality health care." Locals couldn't fill the scientific need, Breeze believed; they lacked the requisite education and experience. Recruiting would soundly fail. The laboratory would indeed fall into the sea, and with it would go Cap'n Breeze.

But there was a solution. By running a ferry across the Sound to Connecticut, Breeze could provide the scientists with a better place to live, one with a lower cost of living and better schools and real employment opportunities for spouses. At the same time, Plum Island would be connected with the Amtrak station there, allowing a link to universities like Yale and the University of Connecticut. "This ferry will turn Plum Island around," the new director boldly predicted. "It will help make us the number one research center in the world."

The Connecticut ferry's first customer was none other than Director Roger Breeze. He moved with his wife and children to a house in Cheshire. "I had to be on [the Connecticut boat]," he said. "It doesn't matter to me where I live. I just didn't think people would take it seriously without me being there."[5] The ferry's heaviest load was a mere seven or eight passengers. Former ferry engineer Ed Hollreiser says the boat often ferried a single passenger. Once, one of the new scientists Dr. Breeze recruited (all of whom "chose" Connecticut as their place of residence) realized he'd forgotten an important book while en route to the island. The ferry was sent all the way back to Connecticut, where a deckhand went and fished the book out of the scientist's trunk and brought it back—to the tune of $400 in marine fuel.

It was an open secret that new hires must hail from the Nutmeg State. "Anyone that wanted to work on Plum Island at that time—even on support staff—had to be from Connecticut," says a worker. Among the scientists and support staff interviewed who ventured opinions on the ferry, all of them suspected Breeze's wife was the motivation behind it. They say she

---

[5]But Dr. Breeze didn't even attempt to live on Long Island. The director lived alone on Plum Island in an old Army barracks until his family moved east from Washington State and joined him in Connecticut.

disliked the countrified North Fork, and the closest university teaching positions for her were miles away at Stony Brook. With its many colleges and universities, Connecticut offered far better opportunities for her and the children. While all of this may be true, the director's own motivations seemed far deeper than pleasing his wife.

As ridership on the Connecticut ferry increased, professional camaraderie at the laboratory began to decline. People were beginning to fraternize based upon which boat—and from which state—they hailed. The move "set up two classes on the island," said Ed Hollreiser, "[Breeze's] people from Connecticut and us peons from Long Island." "I don't have a personal problem with him," notes Dr. Jim House. "But he caused a divisiveness between the New York and Connecticut people. He created the Connecticut people because to him, no one smart would live on [the New York] side. But for forty years it had worked without it." Dr. Carol House agrees. "He would have conversations with people on the Connecticut boat that [the New York staff] wouldn't be privy to."

The ferry was a top-of-the-line, luxurious $1.2 million, 540-horsepower, 110-foot-long boat—but there were flaws from the start. Hollreiser said there were design problems with the engines, and the exhaust noise ran afoul of local town ordinances. "It was like they bought a Yugo," says a former worker. "The engines constantly blew up and it cost the government big dollars in repairs." When they finally got the leaky, noisy, shuddering craft running, it cost a hefty $100,000 a year to operate—a significant chunk of funds for a "cost cutting" regime to bear.

Dr. Breeze maintains that "[i]t had been difficult to fill jobs in the past," but nothing in the records indicate that local hiring over the previous four decades had been problematic at all. As veteran Fran Demorest wrote, "The professional staff moved to Long Island, bought or built homes, raised their families and used our school systems. These families joined in the many local community activities, services, churches, and other programs." Breeze's detractors, large and small, would always say the Connecticut ferry was a colossal waste, a sham, all the way down to the nifty uniforms he dressed the marine crews in. Even the local congressman decried the move. "It sounds like a total waste of money," snapped George Hochbrueckner, a Democrat representing New York's First Congressional District. "It sounds like a few people—including Mr. Breeze—decided they wanted to live in Connecticut. The taxpayers shouldn't be paying for this. . . . This does not make any sense to me." But the USDA had paid for it, and the congressman's objections were for naught.[6]

---

[6]Hochbrueckner would tell the *New York Times* a few years later regarding the Connecticut ferry, "If we realized the situation earlier, I would have made a point . . . but now the die is cast . . . we're too far downstream."

Even had he tried, Hochbrueckner could never have reversed the ferry. Because Dr. Breeze saw to it that undoing his Connecticut boat would be akin to unscrambling eggs. In a masterstroke, he did the one thing that would grant his new ferry perpetuity. He lionized the one man whose name was inextricably entwined with Plum Island, invoking in a simple step the potent feelings of a warm and distant past that instantly hushed would-be critics.

He named the boat after Jerry Callis.

The opinions of two of his allies are enlightening. "Yes, it was Roger's idea," says John Boyle, when asked who named the Connecticut ferry. "In my heart of hearts, I would never tell you what my honest opinion of that particular thing is. There are still a few things about him I can't quite figure out. He can be inscrutable—even to his closest friends."

The definitive word comes from Dr. Breeze's dear friend Dr. Robert Shope, with whom Breeze had lived after he legally separated from his first wife.[7] "He lived in Connecticut and wanted the ferry for himself. The excuse was that it would bring in a broader group of scientists—and it was a very controversial move." What did Shope think of the naming of the ferry? "It was very clever," Shope mused. "The *J. J. Callis*—he had it painted on the boat before anybody knew it—[after that] you couldn't take the boat away and nobody could counter the move." For no other reason than his own self-interest, Breeze traded on the good will his revered predecessor's good name, and in the process sliced the island's workforce into two parts.

DIVIDE AND CONQUER

By this time, in the words of one observer, "[Employees] were praying that Breeze would die and Callis would come back." No longer was Dr. Breeze the rustic faux Scot, turned backslapping American cowboy. He now charged around the island with a flashy imperiousness. Says a retired Plum Island engineer, "You know, a lot changed when Breeze came in. He seemed real worried about climbing the ladder. He could care less about us Americans—or American labor, for that matter. The island was all for his own benefit. I remember one time, he was coming over on the early morning ferry, and we'd been coming off from the night shift. He shook his finger at Walt [Sinowski, Lab 257's building foreman] and said, 'I'm gonna get your job. I'm gonna have your job one day. It's not long now. Not long.'

"Meanwhile, Walt had been retired already from another job, and

---

[7]"He and I have had a very good social relationship," says Dr. Shope, when asked about Roger Breeze. "I had separated from my wife and was living in Connecticut at the time and had a bachelor's house. He knocked on my door one night and said, 'Bob, do you have a spare room for me?' And I said, 'Yes, but I don't have a bed in it.' He said, 'Don't worry—I have a mattress in my car.' This was at a time when he was having marital problems, and he lived with me for about three months. So I got to know him well."

working at Plum was something he was good at and enjoyed doing. Walt said to him, 'You can take it right now, Breeze. Go right ahead.' We'd sit there and take that crap from him, day in and day out.

"And then one day—it happened. It really, truly happened."

The A-76 federal privatization program had visited Plum Island two times since its inception in 1980. The rules stated the government had to step aside when the private sector could perform the government's nonprofessional tasks, provided no "overriding factors" required the government to keep those functions federalized. On both previous occasions, Dr. Callis invoked the "overriding factors" exception and staved off privatization, arguing that Plum Island functions were far too sensitive to be contracted out. Placing two high-hazard biological containment laboratories in the hands of a private company would shift the emphasis from safety to profits. And it would kill employee morale. Why take the chance? Plum Island remained under federal control. Until Dr. Breeze arrived.

In 1988, a year after Breeze took control, a performance work statement (PWS) was prepared. This was essentially a chart of all the tasks that private contractors (and the government's own "in house" team) would bid on. From the start, there were signs that the PWS was prepared incorrectly. In an internal memorandum, the government review panel noted "many misleading statements concerning the real situation with underground storage tanks, existing fuel spills, the chemical management program . . . etc." The panel said the PWS had been "sugarcoated" and charged it did "not reflect the reality of the true pressing problems" at Plum Island. It seemed like the PWS was being rigged to lure an unsuspecting outside contractor into snatching the bid.

"It was a political decision to do it," Dr. Breeze says today of the Plum Island privatization. "It had nothing to do with me or anybody else in USDA—that's just the way it is." But at the time, responding to the review panel, Dr. Breeze exploded. "There are no misleading statements," he wrote. Instead, the panel had "misunderstood several issues. . . . There are no obvious deficiencies we know of. . . . If [the review panel] feels this PWS is 'sugarcoated,' I am very willing to make their specific concerns known." Definitive words from someone who supposedly had nothing to do with it.

On Thursday morning, February 21, 1991, four sealed envelopes were opened in Washington. Burns & Roe Services Corporation of New Jersey was awarded the five-year Plum Island contract, with the lowest and best bid: $16.3 million. The government's in-house team had bid $23.7 million on the same exact PWS. The private contractor had underbid the in-house bid by $7.4 million—more than 30 percent lower. The disparity meant one of two things. Either the private contractor grossly underestimated the costs of the Plum Island project, or the in-house government team engaged in the unfathomable: it tanked its own bid.

Breeze explains his point of view. "There is no secret to this. There's no way you can shuffle cards around, if you are giving people decent jobs with full medical benefits and all the other benefits that accrue with having a job with the federal government. You cannot compete with people that are being paid minimum wage, with minimal benefits—it's not possible. It's actually not something I agree with. I don't think it's the right thing to do by any means. But that's the process."

Breeze says the two bids during the Callis regime beat the private bids because they combined tasks. Those bids said: "Firefighters would act as janitors, and take away the garbage and clean the toilets and mop the rest rooms . . . the boat crew will maintain motor vehicles, and et cetera. And if that would have been clear up front, that firefighters weren't going to clean rest rooms, there would have been janitors and the government would have lost." John Boyle, who admits he was "heavily involved" in the privatization process, said, "You have to be really, really creative to have the government submit the winning bid. We put together a good bid—a competitive contract. To preserve it, [the union] had to give something up, and they didn't want to give up some jobs. I was quite frankly afraid we were going to win it. I didn't want to win it."

Most Americans instantly recall exactly where they were and what they were doing at the moment they heard President John F. Kennedy was assassinated, or that the space shuttle *Challenger* had exploded. For Plum Islanders, May 3, 1991, was one of those days.

John Boyle recalls the moment he heard Plum Island would be privatized for the first time in its history. "I was sitting in my office, staring out on the lawn, out on Gardiner's Bay, just daydreaming for a few moments, when Ernie [Escarcega, Breeze's facilities manager] calls me and says, 'A-76 is coming in!' And I thought, 'Well, thank God.' I personally considered it a blessing. . . ." Walking out of his office, Boyle saw men wringing their hands and women crying in the hallways. "Crestfallen is way too mild a word. For these people, it was like getting news your son was killed in a car crash. I am not overstating it, either." Dr. Breeze also remembers the scene. "Chaos—it was the first time any of the employees had an inkling they might really lose their jobs."

Seizing the moment, the very next morning Burns & Roe executives called an all-hands 8:00 a.m. meeting in the old Army chapel. Exhausted, Stanley Mickaliger had just finished an eight-hour graveyard shift in Lab 257 and trudged over to the chapel. "When I got to the meeting hall, they were already introducing each other, giving all these speeches—all these men with suntan lotion on all over, you know? It clearly wasn't the government anymore. They were these big contractor types." As the crowd slowly filed in, they were handed sealed envelopes holding the documents that determined the fate of their careers. Mickaliger walked over to an official-looking woman for guidance. She told him they were not taking questions, not now—"You have to catch the ten a.m. boat because we're not going to be paying overtime anymore. You can look at your letter on the boat."

Shunted onto the ferry, Mickaliger and the others tore open their letters. "It had all this information, on a bunch of papers, but I didn't understand a lot of it." One thing was clear, though—he'd been fired. And the severance package was nil. The fifteen-year-veteran's remaining choices were a bit limited. "I could have went to Calverton [National Cemetery, a nearby federal facility with openings], but that wasn't my cup of tea. See, I had been a master plumber for forty years. Now I'm going to dig graves and bury people? My wife and I lost our health insurance at age sixty—I had to pay 102 percent of the cost. The letter said if I retired, I got only 40 percent of the eligible Social Security benefit I paid into all those years. I remember complaining to them, saying, 'Hey, I know guys here have thirty years in. But I've put in a good fifteen years—do you think you can help me?' They gave me COBRA for six months."[8]

The days of *"respecting personal dignity . . . recognizing work achievement . . . providing work security"* were over. So was the old feeling that working on Plum Island was like being part of an extended family, that it was a *career*. For the roughly one hundred support workers left behind (which would decrease to seventy-five by 1995), employment at the Plum Island exotic germ laboratory would now simply be *a job*. And thereafter it would have all the dedication to mission, attention to detail, pride in workmanship, and camaraderie that accompanied *a job*.

A newspaper editorial published at the time captured the moment: "How would you feel if you worked for somebody for 15 years and a new boss came in and told you that you were losing your seniority and most of your benefits? What's more, you will be paid half of what you were paid to do the same job and your annual vacation days will be reduced from 25 to zero in the first year." And that was only if the contractor kept your position—"Remember," says one employee, "those that were allowed to stay were offered '*a*' job, not '*your*' job." And those were the lucky ones. A deeper sympathy was reserved for the seventy-five or so that were fired or pushed off the island into the bitter seas of early retirement. The negative

---

[8]When the contractor came in, a Burns & Roe corporate vice president had big news for his new charges. "All employees will receive, in addition to their base salaries, $2.07 per hour. And they can do whatever they want with it." Of course, nearly all were forced to apply the extra dollars to the health insurance premiums that tripled, just as their annual salaries plummeted.

effects on the community of the privatization layoffs at Plum Island, the east end's largest employer, were real.[9]

Even the USDA grudgingly admitted it had been a "challenging transition period." A former Plum Island official described the transition with a bit more flavor. He calls it "the biggest clusterfuck you ever saw in your life."

Like he did with the ferry debacle, Congressman Hochbrueckner complained again, saying privatization "was handled poorly. The workers were treated shabbily." For a politician who publicly bemoaned the woes of Plum Island numerous times, he accomplished surprisingly little. Says Dr. Breeze of Hochbrueckner's rants, "He said, 'I'm going to block it,' and that was just disingenuous at best. He knew quite well the way privatization worked. He knew exactly what was going to happen."

The congressman wasn't the only one feigning powerlessness. "I wanted the government to hold a series of meetings where you explained to the workers the process under way," explains Dr. Breeze. "Absolutely nobody would do that and it wasn't up to me to do it. I actually couldn't get it done." Those meetings never happened. Breeze says he had nothing to do with the privatization or the transition, though the record clearly indicates the opposite. He could have, like Dr. Callis before him, warded off the A-76 guillotine by invoking the "overriding factors" exception, based on the island's unique risks and unfathomable dangers among federal facilities. After all, Plum Island was no run-of-the-mill federal office building, where you'd apply for a passport or pick up a Social Security check. "When Dr. Callis was there, we had the rigid safety standards that almost made it impossible for the contractor to bid," says a longtime Plum Island scientist. "And they could have applied for an exemption this time around, but they didn't."

Retired Plum Island scientists Drs. Jim and Carol House believe without question that Breeze had the power to halt the privatization steamroller. "He could have stopped it, prevented it," says Jim. "Oh yeah, absolutely." Carol adds, "The next time [privatization] came up, [Breeze] threw in all of the inside support services, so it became a big enough 'plum' to bid on—he reengineered the [PWS] so that more was included. He personally did that." "We said from day zero that this is a place that should *not* be contracted out," continues Jim House. "It can't be—there are just too many concerns." Despite their beliefs now, Plum Island scientists remained

[9]In time, east end engineers and skilled tradesmen found diminished opportunities, serving thousands of wealthy summer colonists on the North Fork and in the Hamptons in the fine arts of pool cleaning, tennis court maintenance, landscaping, and golf course maintenance. A few dug graves at Calverton National Cemetery.

silent, instead of speaking out to fight the process. As a result, many of the support workers, says union leader Ed Hollreiser, felt betrayed.

As one employee said, "All it took was the swipe of a pen" to prevent Plum Island from being contracted out. But Dr. Breeze kept his pen in his pocket protector.

When asked why he wanted to see the government lose the bid, John Boyle replies, "Because I felt at that time—the neophyte that I was—that getting into the private sector would save money and that [the contractor] would get rid of the riffraff, and it would run more efficiently for less money." The USDA, indeed, trumpeted the Plum Island privatization. They said contracting out to Burns & Roe would save taxpayers $1 million a year, and that support costs would decline by $5 million, or 20 percent, over the length of the contract.

Not everyone thought it would be a financial windfall, however. One prescient senior employee told a newspaper, "There will be cost overruns, and eventually it will cost more than it did with the people they laid off." Later, it would be learned that the PWS grossly underestimated the required material and manpower to run Plum Island.[10]

After Burns & Roe came in, John Boyle was let go, the contractor having no need for his services. But eight months later, the contractor begged the financial wizard to come back because things were spiraling out of control. When he returned, Boyle didn't find a mess—he found nothing. "They had no accounting system to speak of. They had hired a bookkeeper who did not know what cost accounting was. I built a new accounting system there, from the ground up." And when he finished his task, he delivered some bad news. "They were over $1 million in the red. I knew about it two or three weeks before, when I started seeing the signs, and I was thinking, 'Jesus, this isn't going to turn around.' They wanted to shoot me."

In most organizations, taking out the trash, tending the grounds, mowing the lawns, and providing clerical support seem like responsibilities that should be handled by a contractor. But Plum Island is not your typical institution. Positions like firefighters, security guards, ferry operators, engineers, incinerator operators, ventilation system operators, electric operators, nitrogen freezer caretakers, power monitors, and sewer decontamination plant operators hold seriously heightened responsibilities. At the home of the most dangerous germs known to man, even the grounds crew and typists must be specially trained and abide by complicated guidelines. Ed Hollreiser comments, "My take is that management was so poor, con-

---

[10]For example, the PWS estimated 43 vehicles, 15 daily ferry trips, 50 biological air filters, and 75 meals. In actuality there were 51 vehicles to operate, 24 ferry trips to make, 98 air filters to regularly maintain and replace, and 135 meals to prepare.

tracting out would be an easier way to go. This way, they could yell at someone now besides themselves. They could hold [a contractor] accountable
and blame them for all the mishaps."

A USDA laboratory in Ames, Iowa, was also put out for privatization
at this time. Though the Ames lab houses far less dangerous germs, the staff
there, surprisingly, remained under federal government control. And it
remains private today. In August 2003, Iowa Senator Tom Harkin fought
off a renewed push to privatize Ames by introducing legislation barring
nongovernmental workers at the laboratory. He called Ames's research "a
vital function of the federal government, and it should remain the responsibility of federal employees." There seems to be no other reason for this
glaring inconsistency between the two laboratories than the personal
wishes of a director looking to slash nonresearch costs and yoke his workforce. Breeze muses, "It's sort of surprising that Ames hasn't gone out to a
contractor. I think it will eventually go to a contractor—it's just a push that
various administrations do to different degrees."

Privatization resulted in federally trained, highly skilled workers with
decades of experience being bartered away in exchange for a cost-cutting,
efficiency-driven private contractor. But on paper, it looked good, and the
credit redounded to Plum Island's director, Dr. Breeze. Most important for
Breeze, it freed up more funds for science, his primary goal. At least that's
how it was supposed to go.

Back in 1965, Jerry Callis described his island laboratory at an international conference by saying, "Safety is uppermost in our minds in
everything we do." If that meant less money for science, then so be it. Some
three decades later, costs, not safety, were uppermost in management's
minds. With the support staff slashed by 40 percent, and the Breeze team
cutting every support line item, some other details—like biological safety
and security—were compromised. For starters, the two-day Plum Island
orientation course, after which new employees were ordered to study a
three-inch-thick safety manual as if it were the Bible, was now boiled down
to a forty-minute VHS tape and a two-page flyer. The "Nothing Leaves"
policy was abandoned. Vehicles and items trafficked among the two laboratory buildings, Long Island, and Connecticut without being decontaminated at all. Contractors, who once had to be escorted ("The escort had to
take a shower with these men, take their cigarette breaks with them, even
go into the bathroom with them," remembers one), now roamed free.[11] The
color-coded badge system was discontinued, and the identification num-

---

[11]For the record, I was escorted during the last of my six visits to Plum Island.

bers disappeared. Against the advice of an outside consulting firm, the five-man full-time professional fire department was converted to one firefighter and a bucket brigade of volunteers.[12]

For years, security on intruders was "like white on rice," as one veteran employee describes it. The once thirty-four-man-strong armed guard platoon—checking off ID numbers at the ferry dock, manning the lab compound gatehouses—was dispensed with. One "safety technician" now stood in place of all the guards. He wore an empty gun holster to scare away intruders. One of the fired security guards, Phillip Zerillo, told a reporter there was "a total disregard for security. . . . The place is just going crazy. It's running by luck now." Today it's no different. "You could walk onto that island right now," says an employee. "Two Eskimos in a kayak could invade and take Plum Island."[13] There are two private security firm guards, one at Orient Point and one protecting all of Plum Island. The U.S. Coast Guard patrolled the island's surrounding waters until 1977, when it decommissioned the Plum Island Lighthouse. After a brief Plum Island detail following September 11, 2001, Coast Guard cutters again retreated, leaving Plum Island without any marine patrol.

An internal memorandum dispensing with and erasing away decades of safety closed on an bizarre note: "We welcome any and all suggestions, recommendations, criticisms, and attaboys . . . as well as fishing tips, a good joke, and restaurant recommendations."

In perhaps the most egregious of the safety lapses, the sentinel animals, Plum Island's "canaries in the mine," were eliminated. These test animals were kept outdoors and tested periodically to ensure no germs had escaped the lab. Said a USDA safety director of the move, "From a biological safety perspective, the best thing that ever happened to Plum was the discontinuation of Animal Supply," because he believed it eliminated the threat of disease transmission. But it also eliminated the island's last line of defense. After all, they had successfully alerted the Plum Island scientists to the

---

[12]An outside consulting firm hired to evaluate the Plum Island Fire Department recommended three options to Dr. Breeze: buy new equipment and retain the full-time professional department; retain the outmoded equipment and the department; or buy new equipment and institute a volunteer brigade of workers performing other functions. The firm stated that the volunteer brigade, while the cheapest, was "the worst of the three options." Dr. Breeze chose that option. After one fire alarm went off under his reorganized fire department, Dr. Breeze was thoroughly disgusted—with his own choice—as reflected in an internal memo blasting a supervisor: "An ill-assorted group of people drift up wearing various items of clothing, but usually not the firefighter turnout gear we went [through the trouble] to provide, and mill about without apparent discipline or leadership."

[13]Indeed, on August 1, 1997, vandals from Connecticut came over to Plum Island by boat and vandalized the Plum Island Lighthouse. By chance, the Plum Island marine crew at Orient Point noticed them and shooed them away. On several occasions I was offered the opportunity to be smuggled onto Plum Island: "We could get on my boat, put you in a lab coat, put you ashore, and you can walk right into the laboratory. You can go in and walk out with whatever you want." Though the offers seemed both convincing and tantalizing, I declined.

virus outbreak. Having the control animals outside "keeps you honest," says Plum Island scientist Dr. Doug Gregg. It was akin to a mother removing the smoke detector from her baby's bedroom. The laboratories, now more than ever, became ticking time bombs and the public, unknowing sitting ducks.

But one thing remained clear: the new way was certainly cheaper. And the cost-conscious attack on safety soon became literal—the Plum Island biological safety office itself suffered. First it was divided into contractor safety and government safety. Then contractor safety was eliminated, and the government safety department was slashed down to three staffers. The money saved went toward recruits' salaries and their new scientific equipment, and amenities like touring bicycles and a high-tech exercise gym.

Veteran scientists became concerned over the hacking away of four decades of carefully thought-out biological safety procedures. Former Plum Islander Dr. Ronald Yedoutschnig told a reporter, "When I was there, every person on the island was the same. But today, there are fewer permanent employees. I would be more safety conscious because the [new] people are less safety conscious. The agents we are working with are highly infectious." Says Dr. Richard Endris: "When I was there, the safety was good. Now when it went from a system that is based on employee loyalty and integrity to one of the lowest bidder, I was very concerned that safety would be compromised. Little things, like the backup power generator going out during a storm. The redundant systems are the absolute key. You have to maintain air pressure, the airflow, the freezers. . . ."

Dr. Gregg ponders what privatization has wrought upon the island's morale. "Morale is not as good," he says. High turnover in the workforce contributes further to the problem. "The contracting out was a major blow to the unity of the island," says Dr. Jim House, "and it's still a problem today. A lot of the esprit de corps with the people is gone—they're not there for more than a year and there's a turnover. The turnover after two, three years had 90 percent of the people gone—people that had been there for years." Unfortunately, low morale often translates into poor performance on the job.

The USDA should best be able to assess Plum Island's safety. Out of a possible 100 points under rating system, the USDA scored Burns & Roe, the contractor they chose to run Plum Island, 54.3, 43.9, and 60. The score of 60 was the minimum acceptable performance number.

Sometimes poetry captures the human condition better than prose. Plum Island Lighthouse keeper Captain William Wetmore penned "Plum Island" in the mid-1800s:

There is a rock-bound Island off Long Island's shore;
Where you hear the music of the Ocean's roar.
There you see a light-house on a rocky bluff—
Tides there a roaring, surging; waters sometimes rough.

One hundred and fifty years later, this line and verse was pulled from a file in Lab 101:

What in the world's going on over there;
We heard that there is a "new Breeze in the air."
Who denounced the achievements of thirty-some years;
And announced that a time of success would appear.
Just what has he done in three years on the job;
See those new scientists that fancy boat bobs.
Back and forth, forth and back to Connecticut and
Of course the new carpet has served to expand
USDA's credit both here and abroad;
That use of our taxes we can all applaud.
It seems that there's only one thing to say:
That new Breeze just may blow Plum Island away.

One employee summed it up more succinctly: "Breeze?
"Breeze was a fucking doomsday machine."

Boomerang

> *You have to understand. If somebody took all this away from you . . . you wouldn't take a little revenge?*
> —PLUM ISLAND EMPLOYEE

The sprightly librarian Frances Demorest was now the longest-serving employee on Plum Island, and her seniority carried a lot of weight among the Plum Island family. When the sky started falling, many of them ran to her, shared their problems, even cried away their woes on her shoulder.

Well spoken and outspoken, Fran resolved to do something, and in 1990 she wrote Congressman Hochbrueckner a six-page single-spaced letter carefully detailing everything she heard. "If I should be harassed and then forced to take my retirement," Fran wrote, "I shall do so. But I feel that someone has to attempt to stop what is being done to and at Plum Island. I started employment on March 15th, 1954, and am still very active, in good health, maintain an impeccable leave record, and desire to continue working at PIADC." She proceeded to describe the firing of the four scientists, the ferryboat escapade, the inequitable perks afforded to new scientists, and now the privatization debacle. "This is a lengthy letter," she closed, "and I hope it has provided fuel, facts, and additional information . . . looking forward to you taking the 'bull by the horns.'" Following Fran Demorest from that point, one had to wonder if Dr. Roger Breeze had somehow come upon a copy of her courageous letter.

Breeze was enamored with *The Silence of the Lambs*, the thriller that mentions Plum Island as a summer getaway for the evil Dr. Hannibal Lecter. He actually decorated his office with a life-sized cutout of the cannibalistic Dr. Lecter. Breeze's next move seemed like a page right out of Thomas Harris's novel. Soon after she wrote the letter, he ordered Fran to take a new assignment: the sixty-seven-year-old grandmother was relegated to the murky basement of Building 14 for a special project in Central Files—the obliteration of Central Files. In a cruel twist, the longtime assistant librarian was forced to discard Plum Island's entire library of records. "I was ordered to destroy everything—*everything*! I was to clean out all of it, because he said there was no more need for it." Exiled to the musty catacombs of the hundred-year-old Army hospital, Fran destroyed forty years of historical records—they were shredded, bagged, and lined up in the dank hallway awaiting the incinerator. What about the preservation of Plum Island's past? "Breeze didn't give a damn about the island's history," says Dr. Gregg. "He wanted everything thrown out."

"When he took me off my job and put me down in that basement," says Fran, "it just felt humiliating." Barely finished with bagging the shredded records, she was transferred again, to the telephone switchboard in the guardhouse. While trying her best to handle her new responsibility over the next two months, Fran became gradually lethargic and had to be hospitalized. Her doctor diagnosed her with bilateral severe maxillary sinusitis.

She was unsure of the cause of that condition—the otherwise healthy Fran had never before felt tightness in her chest and sinus pain—until a scientist checking in with her after her hospital stay put his finger on it. Unknowingly, Fran had been inhaling insidious formaldehyde vapors seeping into the switchboard room. When notified of this, Plum Island's brass had the solution. She was transferred to the rusted, corrugated steel warehouse at the harbor, to chug around on a grimy forklift and record inventory. Physically weakened and mentally drained from the abuse, Fran Demorest quit her job ("I took my retirement," she prefers to say with pride). She wasn't surprised by the antics—she had predicted them when she wrote her letter.

"Fran needed to be fired because you couldn't change her," remembers John Boyle. "Fired or transferred into some other job where she wasn't involved in everybody's business—I think we eliminated her job. I'm not certain."

And, thought Breeze and his minions, that was one less voice they had to contend with from the Plum Island peanut gallery.

THE BROKEN PACT

For decades, little was known about Plum Island, owing in large part to the blind loyalty of those who worked there. The initial tumult over the USDA's "choice" of Plum Island had long ago faded from memory. The three hundred strong, locally hired workforce put bread on many a table on the bucolic North Fork, increasingly hampered by an evaporating rural economy in need of jobs with good pay, security, and retirement benefits. A close-knit Plum Island family grew, strengthened by the ferry ride each morning, like children riding to school together on a big yellow bus. Employees' children—children of scientists, secretaries, and steamfitters—went to east end schools together and played sports together. Workers—of both white- and blue-collar extraction—went fishing together, and barbequed together on the weekends. If you didn't work on Plum Island, chances were two or three of your friends or neighbors did. An unspoken pact evolved, an implicit bond between the Plum Island workforce and the brass: attacking Plum Island wouldn't be attacking your place of work—it would bring shame upon your *family*. Better to be tight-lipped about our Plum Island, went the thinking, because the press and general public doesn't understand us and always discredits us, and makes us look bad— and that's not good for business. That's not good for *our family*. "It was a great-paying job with great benefits," said one worker, "when others around here were digging clams and planting potatoes—so we kept it quiet. And management basically said to us: 'We'll keep this place safe, you keep your mouth shut, and we'll take care of you.' "

But many now felt that Dr. Roger Breeze had shredded, like with Central Files, that pact of loyalty, stuffed it in a big black bag, and sent it to the incinerator. And unlike a typical corporate downsizing, here the employees knew the downsizing and vile treatment were preventable, had the director really cared about them or the nature of the work being performed.

Dr. Breeze would pay the price for his disloyalty. The floodgates opened and revelations poured forth. Federal agencies started receiving an unprecedented number of anonymous phone calls about the Plum Island Animal Disease Center. So did the major newspapers, and the television stations. The Federal Office of Special Counsel, which runs the toll-free whistleblower program, received more anonymous tips about Plum Island than about any other federal facility in the nation.

Workers called the U.S. Environmental Protection Agency (EPA) in Washington, D.C., and told them about large sewage spills and massive landfills and Army bunkers filled with drums of hazardous chemicals. They reported leaking underground oil tanks, a sewage plant that didn't work properly, and a faulty wastewater treatment plant that dumped untested sewage into Long Island Sound. They told them fellow workers were ordered to toss unmarked plastic bottles and Clorox bottles filled with unknown liquid chemicals into the incinerators. Workers were told to run the incinerators only when the winds blew due east, out toward the Atlantic Ocean, to avoid mainland detection of noxious emission clouds.

An incredulous EPA came out to Plum Island to investigate the claims. They couldn't believe their eyes. They came upon dump sites and bunkers stuffed with scores of rusting (and ruptured) metal drums containing solvents, oils, ethylene oxide, creosote, hydrochloric acid, and paraformaldehyde, and still more drums that contained unknowns. They toured the wastewater treatment plant and found it out of service, along with effluent that wasn't chlorinated to kill germs and a storm pipe that ran straight into the sound without going through treatment. The EPA then witnessed thick black smoke billowing out of the Lab 101 incinerator, and asked for samples of the ash to test for toxic content.

The EPA cited Plum Island for multiple violations of the Clean Water Act and environmental laws prohibiting the storage of hazardous materials. Threatened with legal action, the USDA agreed to stop polluting, and to spend $150,000 on remediation efforts.

The USDA knew for years about these environmental conditions and that they might come to light one day. A letter by the assistant secretary of agriculture, dated years before the EPA investigation, noted Plum Island's "improper handling of asbestos-containing material [forty thousand pounds worth], sanitary landfills in violation of state standards, and improper management of hazardous wastes." But they never told the EPA. Dissatisfied with Plum Island's snail-paced remediation progress, the EPA brought legal action against the USDA, seeking $111,000 in civil penalties for the environmental and hazardous waste law violations. Years after the action was brought, much of the waste still remained. Again, the EPA charged the USDA with violating federal environmental laws. In 1999, the USDA agreed to pay $32,500 in fines (a fraction of the earlier fine) and promised, once again, not to violate laws protecting the environment.

Whistleblowers also called the U.S. Occupational Safety and Health Administration (OSHA), telling them about employees becoming strangely ill from exposure to unknown animal viruses and bacteria. OSHA learned about the lack of a biological workplace safety program, and of inconsistent vaccine programs. The agency also learned about radioactive isotopes being handled without monitors and protective equipment, locked emergency exit doors, and about the professional fire department being replaced with an unqualified volunteer bucket brigade.

OSHA was far less skeptical about problems at Plum Island than the EPA. Plum Island had made OSHA's "High Hazard List" of unsafe workplaces back in 1988, when workers' compensation claims on the island were found to be triple the national average. Investigating the island, OSHA inspectors quickly discovered exposed electrical cables, shoddy electric outlets, open incinerator pits, untested fire alarms, expired fire extinguishers, locked emergency exit doors, unmarked hot water pipes, and workers bitten and trampled by animals. OSHA cited the USDA for no less than 139 workplace safety violations. When asked by a local newspaper about the violations, Dr. Breeze explained that $50,000 would be redirected from important road-sealing and building painting projects to correct the OSHA safety violations. The money would not be coming out of scientific funds, he said, including those earmarked for new equipment, the fleet of bikes, or the scientists' exercise gym.

On national average, OSHA on-site inspections occur about once every eighty years. But OSHA returned to Plum Island five years later. This time around, OSHA not only found many of the previous violations uncorrected, they also encountered numerous new violations: no emergency procedures or safety training for handling ethanol, formaldehyde, or radioactive compounds like Cobalt-60; disposal of used virus syringes in penetrable containers; no vaccinations for employees; improper storage of compressed gas cylinders; unlabeled and mislabeled hazardous chemical containers; no protection against carcinogen exposure; and no safety training for handling blood-borne virus and bacterial pathogens. OSHA apportioned the blame for the 124 *new* violations between the USDA and private contractor Burns & Roe. Of those, Plum Island was cited with 67 "Serious" violations, those which can lead directly to death or serious injury. Burns & Roe was ordered to pay $54,375 in punitive fines; those costs were ultimately borne by their client, the U.S. government. As a repeat offender, the USDA should have warranted—at the very least—a heavy fine. Yet because it was a sister government agency, OSHA could not legally fine or shut down Plum Island, like it would any private business. Instead, the USDA's entire penalty for 67 potentially fatal safety violations—an immense 263 violations overall—was a letter listing the violations sent to the secretary of agriculture's office.

When asked about the OSHA violations, Dr. Breeze scoffs. "The OSHA thing is a rat to me, I think. What does it really mean? The exit sign needs to be above the door, there isn't a sign warning this or that. When people think of OSHA, they think of small children next to rotating circular saw blades or furnaces, but it's usually not like that. It's like a vehicle inspection—people find bits and pieces and they need to get things changed, but they're not structurally very important."

Though the USDA's fellow government agencies investigated Plum Island and meted out punishments, nearly all of it was deflected. After all, Plum Island was not a private company that could be closed—neither the

EPA nor OSHA had the authority to shut down the USDA. So a harsh reality ruled on Plum Island. The federal government was brazenly ignoring its own laws.

If the first round of whistleblowing was limited in its effectiveness, the second round held great potential. Current and former employees called every media outlet they could think of. They told reporters about herds of deer roaming the island, and a doe and her two fawns being gunned down by soldiers in Army Blackhawk helicopters, and about an employee who caught Nairobi sheep disease.

A media siege erupted. CNN and ABC, *Newsday* and the *New York Times*, national and local reporters all bombarded Plum Island for information on the allegations. Their faces blurred and voices distorted, workers sat before television cameras and told what they knew.

CNN ran a report with footage of Pete Swenson, a former Plum Island safety technician who left the area in disgust and moved to Virginia. Seated next to his wife at his kitchen table, the stocky, middle-aged Swenson told a harrowing tale. Outdoors he had been bitten on the leg by a horsefly and scratched the bite, thinking nothing of it. He entered Lab 257 to decontaminate a room that held animals infected with Nairobi sheep disease. Days later, his leg swelled and ulcerated, and he developed severe flulike symptoms. His illness got so bad he checked into the Long Island Veterans' Hospital. Swenson told the attending doctor where he worked and that he suspected he had contracted Nairobi sheep disease, at which point a foreign disease specialist was brought in. According to Swenson, the specialist told him, "That disease is not here in the U.S.—I'm not sure how you could have it." Swenson recovered, but not without a memento for his troubles: CNN cameras focused on an oval-shaped scar about two inches wide, burnished deep into in the middle of his left calf. When reporters asked Dr. Jerry Crawford (a new Plum Island official handling the media) if Swenson had been infected, he replied, "Probably, no." Possibly, they asked? "Possibly—yes. Some of the diseases—humans can catch a form of it, and get flulike symptoms."

Two stories involved more disturbing occurrences. In the winter of 1992, three cardboard boxes filled with biological samples blew off the back of the ferry into heavy seas. An emergency call went over the radio that morning, and a special boat crew was called to duty. A captain and two able-bodied seamen plowed back and forth through Plum Gut, and then around Plum Island's perimeter, looking for the missing cargo. An island official called off the search after only a few runs. The three boxes were never recovered. Three employees confirm the accident. "What if it didn't sink?" says one of them, shaking his head. "What if some kid hooked it on his fishing pole? What if it floated to a beach in Southampton, or Sag Harbor, or Montauk?" The germs were taken away by the currents, bobbing up and down inside a Styrofoam cushion inside a cardboard box. But management didn't seem too concerned. " 'Nah, don't worry about it,' they said to us—'Don't worry about it.' "

The second most damning incident had scores of dead birds lying outside Laboratory 101. This allegation had a document written by Dr. Breeze that went along with it:

> Concerns have been raised about dead seagulls in the [Lab] 101 light court. The most recently dead bird was frozen by [Burns & Roe] and has been passed to [Lab 257] for pathological examination. . . . I would ask that [Lab 257] confirm that infectious agents used in [Labs] 101 and 257 are not avian [bird] pathogens . . . [and] report on postmortem findings in this bird.

"Seagulls used to land on the 101 roof all the time, with no problem," says one employee. "We have contractors building all over the place, including on top of the roof. Then, all of a sudden, these birds are dying, and they didn't attribute it to anything." Fearing a problem, a contractor presented one of the birds to Dr. Breeze with a memorandum for the record. "It's the fox watching the chicken coop," the employee says of Plum Island's self-analysis of a biological containment leak. "Take the seagulls and bring them to someone who doesn't have an agenda." Asked about the event, a USDA official said, "The veterinary pathologists on Plum Island are about the best around. They'll find out what killed the birds." Under the Freedom of Information Act, a request was made for the results of the "postmortem findings" on the nine birds reputedly autopsied (sources indicate there were many more dead birds). As of this writing, the USDA has refused to respond.

When the gaggle of news reporters first learned of Plum Island's many woes from the disgruntled mob, they turned to the island's director for answers. Dr. Roger Breeze told the *New York Times* they had the wrong guy. "I'm not responsible for the operation of the support services and facilities at Plum Island," he told them. As one knowledgeable source says, "Breeze had thought, 'Whoa, wait a second here—I did all this work for USDA, got this lab built in four years when they couldn't over the last twenty-five, and now *I'm* taking the heat?' And that is when they brought in Crawford. He was the fall guy to take the flak."

Enter Dr. Claude G. "Jerry" Crawford. Breeze's moves to divide and conquer Plum Island had met with plenty of controversy and criticism; he had somehow earned a reprieve from this next fiasco, the Plum Island

Workers' Rebellion. Washington granted him cover in the form of the forty-nine-year-old Jerry Crawford, a USDA career man previously working in Peoria, Illinois, whose specialty was crop seeds. Many suspected an ulterior motive for giving Crawford the newly created position of deputy area director, or "DAD," as he was known in correspondence (and referred to derisively by the workforce). The plan was to have Breeze do research full time, while letting DAD "run the buildings"—another way of saying "take the flak."

Ed Hollreiser remembers Crawford's first day on Plum Island. "When he came in, he gave this big speech to all of us, where he explained his whole background, and how it was on plant research, and how it had nothing to do with animal research. Then he discussed the vacations he took to Europe with his wife. Meanwhile, half of us are about to lose our jobs." Crawford was a master of malapropism. When asked about cost overruns, he said, "There are no golden toilet seats on Plum Island." Asked about all the safety problems under Dr. Breeze, he said, "We don't search handbags or strip search. I believe we can show [security] is warranted for the type of research we do here. We're not making nuclear bombs." And when asked about the OSHA violations, he noted it took six months for the safety agency to issue the violations and said, "If these violations had been life-threatening or extremely hazardous to employee safety, one would think OSHA would have done something immediately." Then realizing what he had said, Crawford added, "Uh, I don't mean to say these violations are not important. . . ." Finally, when asked about the consequences of an employee's refusing experimental, non-FDA–approved vaccines, "We will try to find another position. If we don't have one, they will be terminated. Simple as that."

When it came to public relations—ostensibly the reason he was on Plum Island—a running joke was that Crawford himself had been infected by the virus that caused foot-*in*-mouth disease. One top Breeze administration official said, "We all independently formed our own opinion of Crawford's capabilities in a relatively short period of time, and realized the man was a complete fool."

Dr. Breeze, spared this particular media feeding frenzy, says the media issues were handled terribly. "It was very reactive." Another adjective he uses is "comical." In fact, John Boyle recalls Breeze "sitting in the bleachers, laughing his ass off on that one."[1] Within the next year, the USDA transferred Dr. Crawford to a position near Washington. On his way out,

---

[1] Realizing the dearth of Crawford's PR savvy, USDA headquarters begged Breeze to speak to a *New York Times* reporter to cool things down, and the two hit it off. That reporter ultimately became Dr. Breeze's second wife.

he told a local reporter, "My goal, by opening the island up to the press, was to show the people that we were good neighbors."

RAW SEWAGE, RECYCLED

"If you flush it or drink it, it comes to me."

"Joe Zeliff" worked in Building 102, the wastewater decontamination chamber for Lab 101, known to those on the inside as "The Big Shithouse" (Lab 257's chamber was the "Little" one). One of Joe's responsibilities included pulling liquid samples from the sewage tanks after they went through the decon process—that is, after the goo had snaked through a series of tubes that heated it up to 214 degrees Fahrenheit for one hour. The samples were handed over to the safety officers, who would inject them into test animals to ensure that systems in Building 102 were operating correctly. It was crude safety, but it worked.

For years, Joe would prepare a safety sample each Thursday morning. He doesn't pull samples now. "We haven't done that, well, since Breeze came in," Joe says. Now, Joe has no idea whether the liquid waste coming out of Building 102 is alive or dead.

"They've recently added new gauges, but I don't think even those gauges read right." Sometimes the low temperature alarm rings at 203 degrees Fahrenheit on the new system; other times it goes off at 198 degrees, and still other times it buzzes at 206 degrees, the temperature which is supposed to set it off. Sometimes he doesn't get a low temperature alarm, like the one time he realized too late that the heat exchangers had failed and the sewage went through cold. The old system had ten tiny heat probes inside the booster, heat exchange, and retention tubes that relayed the temperatures within up to the control panel. The new system only has two temperature readings: going in and coming out.

When Joe and other operators in Building 102 hear the low temperature alarm, they are first supposed to shut the system down. Then they gradually bring it back up to 214 degrees, and only then do they restart the pumps. But it doesn't work that way. Why not? "Because frankly, it's a pain in the ass—you'd be shutting the entire system down every single day, every fifteen minutes. It's not that the operators don't want to do it. It's just that the gauges malfunction constantly." The tanks, including the 27,000-gallon main tank ("You could honestly get lost inside it," says Joe. "It's the size of my house") have large cold spots, like when one swims in a warm lake and hits a chilly pocket of water. In these cold slugs, billions of viruses and bacteria stay alive through the contaminated sewage treatment.

A February 1993 incident form noted a "critical alarm sewage spill" in Building 102. Joe remembers, "I was directly involved with this." Since the building opened in 1956, the plant had always run twenty-four hours a day, seven days a week. But the private contractor ran it only ten hours a day. "The collection process will be unattended, with 100 percent reliance upon the automatic and emergency collection equipments," stated an internal USDA letter on this decision. "They did it to save money," says Joe, shaking his head in disgust. "What else?"

One winter night, an operator left his shift after recording the 27,000-gallon No. 2 tank as only 7,000 gallons full. High winds and rough sea conditions prevented the next shift from getting to the island the following day, so it remained unattended. When the weather finally calmed, the shift operator opened the door to Building 102 and came upon a putrid smell—something had gone terribly wrong in the tank area. A single 1,000-pound cow excretes from 75 to 115 pounds of manure a day, and a single 200-pound pig, 35 to 65 pounds; it isn't hard to see how quickly raw, infected sewage accumulates on Plum Island each day. Second only to the incinerator charging room, Building 102 is the biologically "hottest" area, and is therefore kept painstakingly clean. On a normal day, "that building is as clean as a whistle," says Joe. "It is cleaner than my house—you could eat off the floor in there." But that morning, the operator and his assistant, who came upon the large pool of gushing brown mess, were gagging within seconds for clean air. "I saw nothing but shit all over the place," says the operator. "Raw, no decontamination, viruses and who knows what—everywhere." The men witnessed a macabre sight: a continuous flow of contaminated sewage, vomiting out of the overflow valve, after being pumped from the No. 2 tank, through a daisy chain of pipes. Underneath the pool of sewage, fluid was emptying into the floor drains, where sump pumps sent it back into the No. 2 tank; from there, it cycled through again. "We were essentially recycling raw sewage," says Joe. First, the men processed some contaminated sewage out to make room in the tanks. Then, donning Tyvek protective jumpsuits, hip boots, and face respirators, they drained and mopped up the mess, flushing the room down with Roccal solution. Research in Lab 101 seized up for two days. "They couldn't so much as flush a toilet in 101, because we couldn't handle it in 102."

The spill form cautions its reader "NOTE: This bldg. was UNMANNED for 32 hrs." Under the old way of doing things, it would have been manned—the overnight shift would have worked overtime on the island through the next day and ridden out the storm. Of course, it would have cost more money than running Building 102 at one 10-hour shift per day. Joe observes, "See, the contractor said, 'We need more money to pay for more men.' The government said, 'We're not paying you any more money,' and the contractor said, 'If you're not paying us, we're not putting operators in there twenty-four hours and you're going to have another spill.' " At this time, the contractor's vice president of operations told a reporter that safety on Plum Island was "really governed by the client," and his client—the USDA—was under "very serious budget constraints." Another executive privately admitted exactly where his company stood, saying, "We can't staff the facility to the level we'd like to."

This round-robin routine is a prime example of why certain government functions at Plum Island should be kept in federal hands, and ought not be privatized.

TURTLES AND GEESE

After exiting Building 102, effluent travels to the nearby wastewater treatment plant, which uses a process called activated sludge. It enters a pond at about 150 degrees (assuming Building 102 is functioning properly), and aerators froth the fluid up like a bubbly hot tub. The hard stuff drops to the bottom and a set of six-foot-long weirs skim the "cream" off the top and push it into a clarifier, where more settling takes place. The bacteria are routed back into the aeration tank, and the remaining cream goes under a set of ultraviolet photochemical radiation lamps before being pumped into Plum Island Harbor, which empties into Plum Gut, whose waters disperse north into the Long Island Sound, and south to the Hamptons. The cream used to be chlorinated, but was replaced with UV lights to cut costs.[2]

The most troubling part of this whole process is found at the end of the pipe. Despite the particularly hazardous content of the Plum Island wastewater treatment plant's outfall, it is tested monthly, no different than any other municipal sewage plant. Following New York State and EPA laws, Plum Island tests for suspended solids and coliforms. Coliforms are harmless bacteria, not normally in water, but found in warm-blooded mammals. The presence of fecal coliform bacteria in the water is a bellwether for other microbes; that is, it indicates that the water contains many other bacteria and viruses not being tested, including harmful disease-causing germs. When traced back to a municipal sewage plant, this means the water is transmitting typhoid fever, diphtheria, hepatitis, dysentery, gastroenteritis, or other domestic virus and bacteria infections. High levels of fecal coliform out of the exotic animal germ labs on Plum Island present a unique and highly dangerous situation.

For water to be acceptable for swimming, it should have fewer than 200 colonies of coliforms per 100 milliliters of water sampled; fewer than 1,000

---

[2]The jury is still out on which method of final treatment, chlorine or UV radiation, is preferable. While chlorine is considered highly toxic to marine life, UV radiation may not be inactivating all of the billions of viruses and bacteria contained in sewage. UV is far cheaper than using chlorine, which must be replenished regularly.

colonies are acceptable for fishing and boating.[3] Plum Island's legally permissible limit is 700 colonies per day. Required by law to be tested once a month, Plum Island's coliform levels were above that limit for seven months during a twelve-month span between 1996 and 1997. In October 1996 and March 1997, the levels tested at 900 colonies, well over acceptable standards for swimming in the waters of the Hamptons, America's elite summer playground. But that was the best of it. Other monthly counts ranged from 1,000 to 6,000. And during one fateful month, February 1997, Plum Island spewed out an enormous *23,000 colonies per 100 milliliters sampled*—115 times higher than the acceptable limit for swimming, 33 times higher than the island's legally permissible limit, and 23 times higher than the fishing and boating limit. Additional documents show this was not a fluke. Back in 1990 and 1991, Plum Island's fecal coliform counts held steady in the 2,000 range—three times the legal limit—and peaked to a shocking 20,000 during *four* separate monthly testing periods. Plum Island has been—and still is—flushing deadly exotic animal viruses along with the standard typhoid and diphtheria germs into coastal waters enjoyed by millions of people.

In 1999, hundreds of thousands of lobsters in Long Island Sound mysteriously died, causing the New York–Connecticut lobster harvest to fall off virtually 100 percent that year. The die-off, which ended New York's reign as the nation's second largest lobster-producing state (behind only Maine), prompted the federal government to declare the event a "marine resource disaster." Researchers pinpointed a parasitic paramoeba that kills off nerve tissue and a bacterial infection that causes "shell disease" as culprits, as well as large amounts of malathion from the 1999 West Nile virus pesticide sprayings seeping into the Long Island Sound. Surprisingly, no scientific efforts have been made to study the link between the two microbes and the fouled sewage flowing out of Plum Island.

Dr. Floyd Horn, the official in charge of all USDA research, said in 1998, "We're not polluting the sound in any way. We heat the discharge to sterilize it and it goes into a treatment plant before it goes into the sound." But one employee extensively involved with the process has a different response. "That stuff wasn't killed off in the old system, and I'm not sure it's killed off in the new system." When the *New York Observer* questioned Dr. Jerry Crawford about the high levels of fecal coliform, he answered, "The location where the samples are taken has a tremendous number of turtles and geese. . . . They shit in the aeration pond . . . the shit has *E. coli* in it. Normal sewage treatment plants don't have animals."

---

[3]These are generous thresholds. Vermont, for example, sets 77 colonies of fecal coliform as that state's acceptable maximum level for swimming.

Building 102 and the Plum Island wastewater treatment plant are not performing their critical functions. Instead, they pump thousands of gallons of virus- and bacteria-contaminated waste into the water. Consider this: a single bacterium that replicates itself in twenty minutes will be, within twelve hours, the parent of 16,777,216 new offspring. And that's just a single germ out of billions and billions that make their way out of Plum Island each day.

PATH OF LEAST RESISTANCE

After the EPA came down on Plum Island, an environmental company came over and drilled a number of monitoring wells twenty feet down in which it installed a pipe with a cap. Every month, "Steve Bosley" totes his collecting buckets and portable pump from well to well, unscrewing each cap and carefully snaking clear tubing about sixteen feet down each pipe. He then switches on the pump—"*p-p-p-p-p-p*"—which brings up what's been accumulating way below the island's surface. "I'm still pulling oil out of the wells," Steve tells me. "I'm supposed to pull the sample until I run clear water. I've been running about four and a half gallons of No. 2 and No. 6 oil before I run clear water." The No. 2 oil oozing out of Plum Island's sandy underground is essentially diesel engine fuel; the No. 6 oil has the viscosity of tar. "Yes, I would say it runs pure oil for four gallons, then it turns creamy, then to white, then to swamp gas, then almost clear."

Steve surmises, "We've had a bunch of busted pipes and leaking tanks. The oil must have permeated the whole ground." He isn't exaggerating. A 1992 internal USDA memorandum from the assistant secretary of agriculture notes "Over 40 underground storage tanks. . . . Most of these tanks are probably at least 30 years old. While tests of the drinking water have not indicated any problems to date, this could change overnight." The memo urged action "as soon as possible," but seven years later, nothing's changed. The oil pockets below Plum Island coexist with pockets of clay and sand, as well as aquifers, or vast pockets of fresh water from which people obtain their drinking water through fourteen shallow wells. The aquifers, and other liquids like oil, do not remain fixed in one place. They move and often merge, according to geology experts. One drop of oil, dabbed on a blade of grass in the center of Long Island, will travel south and reach the Atlantic Ocean within twenty-five years. The pockets of oil Steve pumps from a well that's sampling only a tiny portion will do the same thing. "That oil, in my opinion, is going to take the path of least resistance, go through the sand to the potable water," says Steve.

THE GRAVY TRAIN

One of the ways Dr. Breeze kept his prized scientific recruits happy was through special nonsalary compensation. For example, during overnights and weekends, a scientist would be stationed on the island as a duty officer to respond to animal or human emergencies. During the Callis era, duty officers were paid $60 for a weeknight and $160 for the weekend. "Basically, if there was a problem in the lab, you had to go in there, turn off a centrifuge, move stuff from a freezer or drop feed down the chute to the critters [test animals]," explains Dr. Carol House. But under Breeze, new scientists were assigned duty officer shifts and paid "some $600 to $700 a weekend to get in on the gravy train," says one scientist. The role went from being the most dreaded to the most coveted.

"Danny Clewis" remembers a problem that occurred in Lab 257. Around 2:00 a.m. on Sunday morning, an alarm rang in the control room: the Box 6 freezer had gone down. As one of the overnight building operators, Danny went into the containment corridor to check on Box 6, and found the compressor shot and the freezer all but cooked. Normally stored at minus 158 degrees Fahrenheit, biologicals have to be shifted from the broken icebox into other freezers quickly or the germs will start to get pesky. Danny phoned the duty officer and told him Box 6 was out. "Well, he got very ornery with me," says Danny, describing the reaction on the other end of the line. "He didn't want to come down to assess the situation. He yelled at me that I woke him up and told me to do it myself." Tact isn't one of Danny Clewis's strong points, because he shot back, "Listen, buddy, this is your responsibility, not ours—and you're messing with the wrong guy." He was right. Operating the building's negative airflow systems and steam plant was Danny's domain, but when it came to viruses and germs, they all looked the same to him.

The duty officer then asked, "Well, what do we usually do in this situation?"

" 'What do *we* do?' I asked him back." "What *I* do is usually call *you*— the duty officer, for Christ's sake!"

The duty officer slogged down to Lab 257, and the two of them began unloading the contents of Box 6 to transfer them into other freezers. But it wasn't all that easy. "We had biologicals in that freezer that were frozen and dated in the 1950s," Danny recalls. "We opened the door and about half of them fell onto the floor." Luckily, only a few tubes and flasks broke on the tile floor and spilled their contents; a Roccal disinfectant solution was brought in for the kill. The move then faced another obstacle—the other freezers were already overcrowded. One freezer was literally bursting at the seams with germs, so Danny wedged a two- by four-inch beam of plywood between it and a wall to keep its door shut. Each freezer has an attached log that records the exact contents of the materials kept inside. But Danny noticed the duty officer, all hot and bothered from being woken in the first place, haphazardly separating the carefully grouped test tubes and stuffing them into whatever freezer open space he could find. Without logging any of them. "He's the boss," Danny said to himself and shrugged.

Three weeks later, Danny got a phone call. "This one scientist wants to know where everything is, particularly all his samples. How the hell am I supposed to know? I told him to call up the duty officer from that night."

The stories that circulated turned out to be more than grumbles of disgruntled workers. The facts proved that the USDA was grossly negligent, the workforce had looked the other way for years, and the public had been in the dark—in short, it proved that Plum Island was in disgraceful shape. Though some said the disclosures to the EPA, OSHA, and the news media were unwarranted, other workers and scientists believed the door needed to be opened wide. One employee laid his thinking bare when he explained, "You have to understand. If somebody took all this away from you, what would you do? You would be really pissed off and annoyed— you wouldn't take a little revenge? I covered up for these people for years, and it would have cost them millions then. So I—and all the others—blew the whistle." Says the former union leader Ed Hollreiser, "I never had a vendetta to close Plum Island down, but I wanted it run right. The contractor took shortcuts to save money, made more profits, and got lean, but then safety was compromised. Safety doesn't need to be lean—it needs to be fat."

For all the whistleblowers' efforts in bringing these dreadful conditions to light, Plum Island remained—and remains—a real biological and environmental hazard.

12

Meltdown

> *Look, I'm not a scientist—but to me, it was a biological meltdown.*
> —PHILLIP PIEGARI, FORMER EMPLOYEE

Friday, August 16, 1991, 12:00 p.m.— A tropical depression formed from a cluster of tightly packed thunderstorms just east of the Bahamas, and it gradually intensified throughout the day and overnight. By late Saturday afternoon, it became a "tropical cyclone with winds that exceed seventy-four miles per hour and circulate counterclockwise about its center," otherwise known as a hurricane. When a storm officially becomes a hurricane, the National Hurricane Center gives it a name. The names are handed out in alphabetical order to delineate how many have come before it that hurricane season; this second hurricane of the season was simply named "Bob." Like many hurricanes originating in this region, Bob plowed north by northeast, following a path parallel to the East Coast of the United States. With each passing hour, the hurricane's intensity accelerated. As Saturday slipped into Sunday, Bob came within 30 miles of Cape Hatteras, packing winds over 115 miles per hour. Minimum barometric pressure was 957 mb. Out of five possible levels, Bob was classified as a Category-3 (CAT-3) hurricane, a storm capable of extensive damage. The hurricane spiraled on, gaining strength as it went.

Saturday, August 17, 6:00 p.m.— Phillip Piegari sat down for dinner with his family in his modest Jamesport, Long Island, home. Jamesport is one of ten tiny hamlets along Route 25, a countrified area dotted with truck farms, vineyards, colonial bed-and-breakfasts, and ubiquitous yard sale signs.

Phillip's wife, Zyta, born and raised in Ecuador, set out a delicious dinner for Phillip and their two young boys, Peter and Matthew, who immediately dug in. Phillip halted their progress by loudly clearing his throat. He then recited a solemn blessing, like he did before every meal.

On this night, however, talk at the dinner table was not the usual chatter about school, weekend fishing plans on their modest boat, or the upcoming family vacation. All talk, and quiet thoughts, centered on the hurricane. The television in the family room was on, blathering faint sounds about the violent storm spiraling up the East Coast, headed straight for Long Island. Because the 118-mile long land mass extends out at a right angle from the coastline, it has forever borne the brunt force of tropical storms that come north and hug the coast. During one recent December nor'easter, for example, the ocean flooded over the Hamptons barrier beach island and met the bay. When it retreated, it gouged two new inlets and over forty million-dollar beachfront homes were destroyed; the road now comes to a screeching halt at a rushing ocean inlet, and then continues on the other side. The Piegaris listened from the kitchen as the nightly news anchors instructed people to board up windows, take fragile items off of shelves, and stand in the frames of doorways when the storm hit. Heeding the warnings, earlier in the day, Phillip took the boat out of the water and picked up large bottles of water. He packed the freezer with extra ice. Zyta pulled her fragile crockery off the shelves.

The Piegari family was particularly concerned because the hurricane was headed directly for the Plum Island Animal Disease Center. Phillip was due to report there in just a few hours for the graveyard shift. For twelve years now, he had been a ferryboat seaman and building maintenance engineer on Plum Island. Phillip had experienced rough weather conditions on the island before; he felt confident he and the men on the late-night shift would see this storm safely through, just like all the others. He assured Zyta and the boys he would be fine and admonished them to stay indoors during the storm.

At forty-four years old, Phillip cut a slight frame at five feet, six inches tall. But his small size and reddish-brown mop-top belied the man on the inside. From 1967 to 1973, Phillip sailed around the globe six times as an able-bodied seaman with the U.S. Merchant Marine, and saw heavy combat action in the Vietnam War. After he returned home, he enlisted in the marine division of the Army Corps of Engineers. Others might shrink from the risk of that night's assignment, but Phillip was hardly afraid. He had a job to do, an important responsibility on Plum Island, and a sacred obligation to provide for his family. To a combat veteran, it was just another day on the job.

Sunday, August 19, 12:00 a.m.— Piegari and the men of B Crew arrived on the island. They were shuttled over on the *Plum Isle*, one of three government-operated ferries used to transport scientists and staff to and from the facility. Unknown to the passengers, but likely known to the USDA, the old *Plum Isle* had recently failed a Coast Guard inspection and was not certified to carry passengers. The other two ferries were tied up in safe anchorage at the New London, Connecticut, naval submarine base. Lack of certification notwithstanding, the *Plum Isle* was the only vessel available to transport workers during the storm. It made the trip over rough waters without incident.

B Crew, three men and a foreman, normally pulled an eight-hour graveyard shift. But unlike previous routine shifts, that night they had to support and maintain Laboratory 257 while a hurricane passed overhead. Piegari recalls the crew's mission early that Sunday: "We were operating the building as usual. We had certain responsibilities. For example, we had to maintain steam in the building. We had a constant flow of water that had to be treated as sewage. We had to regulate and monitor the air pressure in the building, to see that we hold negative pressure biocontainment. We had to monitor the freezers to see they were the proper temperature. We constantly patrolled the building to make sure we weren't losing anything, in the way of our systems."

Upon their arrival, the crew drove along narrow winding dirt trails to Lab 257. Lab 257 was a long, rectangular building with three-foot-thick reinforced concrete walls and a flat, black tar roof. Tiny porthole windows cut into the whitewashed walls and circled near the top of the building every twenty feet. The windows were recessed deep within the building's walls. Little light shined through, for they were caked with the grime and coarse sea salt that had accumulated over eighty years' time. The building was surrounded by a four-foot high cement wall, and a rusted barbed wire fence encircled the compound. Entering the building through the air lock, B Crew was greeted by the shift's first problem—the hot water coil broke. The coil runs on electric power and generates steam for the building. Second in importance only to electricity, steam is an essential resource for the lab's smooth operation; it is needed to heat the building, treat contaminated animal sewage, and produce hot water for decontamination showers. It must constantly be kept on-line. Within a short time, the crew repaired the coil, and steam pressure was restored. The next few hours were relatively quiet.

Sunday, August 18, 7:00 a.m.— Hurricane Bob's tentacles stretched from Maryland's eastern shore of the Chesapeake all the way up to Cape Cod. Storm surges of more than eight feet rocked boats clear off their moorings, knocked down homes along the shoreline, and shaved off fifty feet of coastline. Seas were far beyond navigable. They were awash in choppy white-caps, tall wave crests, and deep troughs.

The storm swirled around the eye of the hurricane, a core about twenty miles wide. Within it, conditions were relatively calm, with a lull in wind velocity. But this brief pause lasted barely long enough for victims to brace themselves for the calamity on the other side of the eye.

Around 9:30 a.m. on Sunday, Bob's eye unpredictably twisted further east, heading straight for Block Island, about twenty-five miles due east of Plum Island.

At this time—with the worst of the storm still yet to come—Lab 257 went dark.

The laboratory's basement held the boilers, sewage treatment plant, and main mechanical room containing air-handling and monitoring equipment to prevent viruses from escaping. The sewage plant "cooked" raw, contaminated wastewater to kill off any live biological agents. On Lab 257's first floor were the laundry and glassware storage rooms, glassware sterilization unit, incinerator, decontamination rooms, and air locks, through which everyone passed in order to enter and exit the lab. It also contained the 120-Area. The 120-Area was the "hottest" part of the building, the site of the most dangerous, infectious virus research. This area was equipped with a series of freezers filled with liquid nitrogen, pumped in from an outdoor tank that kept temperatures at a frosty minus 158 degrees Fahrenheit. The freezers housed biological materials used in the lab—tissue samples riddled with viruses, bacteria, experimental cultures, and vaccines.

Research laboratories and the animal holding rooms were on Lab 257's second floor. When the Army refitted this former mine storage building, it slapped a sheet metal livestock chute onto the side of Lab 257 to herd animals into the second-floor animal rooms. By the hundreds, cattle, pigs, sheep, horses, goats, llamas, and other quadrupeds were herded up the ramp, through the doors, and into the second-floor holding pens. This floor housed the autopsy room, where animals were dissected and experiments performed on carcasses. When the dissection was through, lab workers shoved the diseased animal carcasses into a disposal chute, which dumped them into the incinerator one floor below, where they were cremated at temperatures exceeding 1,800 degrees Fahrenheit. The hot cremation exhaust ran through a filtered chimney on the roof and then poured out of the building in thick black tufts.

Lab 257 was a self-contained operation that functioned separately from the rest of Plum Island. For the animals, it was a one-way operation—animals that entered the lab never left alive. Engineered to prevent harmful viruses and other biological toxins from escaping, the lab's sustaining lifeblood was electric power. Electricity was needed to run the sewage decontamination plant, to create steam in decontamination rooms, to power the freezers, and—most important—to run the building's negative air pressure system, the cornerstone of Lab 257's biological containment system.

Simply put, electric power was the building's most precious resource. Without it, Lab 257 morphed from a safe research facility into an extremely dangerous biological threat.

As the hurricane swirled closer, strong winds blew down power lines on Long Island. Normally, Plum Island's power was supplied by the Long Island Lighting Company, via an undersea cable on the ocean floor. But the LILCO power grid shorted out and mainland power to the island laboratory failed. Fortunately, there was a backup plan. Oil-fired power generators kicked in at Building 103, the Plum Island emergency power plant, and supplied the island with electricity. The huge generators in Building 103 were old, but they were well maintained and effective.

Building 103 supplied Lab 257 with power through overhead power lines and through underground cables that provided "redundancy." Typically, if one power conduit failed, the other was switched on as a backup so power could continue uninterrupted. This Sunday, however, redundancy wasn't given a chance.

Hurricane winds, gusting over one hundred miles per hour, toppled the island's overhead electric poles. "They fell over like they were as light as toothpicks," recalls Dave Stakey, in a scratchy smoker's voice. A genial, slightly disheveled man, he sports a long ponytail and diamond studs in each ear. As one of the island's longtime electrical engineers, Stakey knew the island like the back of his hand.

With the wires down, workers in Building 103 threw the wide jumper switches that engaged the emergency generators. After a series of flickering lights, all of the island's buildings were back on-line except one—Lab 257. Earlier that morning in 257, the power went down a few times. Explains Phillip, "It was intermittent and then it came back steady. But then the lights started to slowly dim." At this point, B Crew member Stanley "Shine" Mickaliger, a twenty-year veteran of Plum Island, was not sure if power was coming back. But he was hopeful. A sixty-one-year-old master plumber, Mickaliger joined Plum Island after years in the trade. Of Polish extraction, Shine—called "Sunshine" for his ever-grinning as a boy—grew up in Riverhead, the youngest of nine. "For this short period of time," says Shine, "we were thinking the [emergency] generators would come on-line.

'Why are they gone?' we all wondered. In the past, we had power failures where we'd been out for an hour or two at most."

But this time, things were different. After two hours without power, their worst fears came true. They radioed the power plant and learned that emergency power was being pumped into all buildings on the island, yet the lights were still out in Lab 257. B Crew gathered together in the boiler room in quiet disbelief. At around 8:00 a.m., Shine broke the silence and delivered the grim news. "Guys, the overhead is gone—that's it, guys, we're out." The members of the team shifted uncomfortably and exchanged troubled glances. They knew they were on their own.

Just a few feet beyond the eerie yellow glow of their flashlights lay the end of the universe. Everything past it—all of Lab 257—was pitch-black. Ordinarily, working overnights in the lab was akin to "being in a tin can," says Shine. "After everyone went home, it was just the few of us in that big building, all alone." The building usually had a low, steady hum to it, sounds of boilers heating, steam pumping, fans turning, and light fixtures buzzing. That was the sound accompanying the men on graveyard shift every night since Lab 257 opened for business way back in 1954. Now there was a different sound. The distinctive "sound" of utter silence in the dark. And man's worst fears were all around.

Three months prior to Hurricane Bob, in a flurry of sparks and a wisp of gray smoke, one of the underground conductors shorted out; with it went the underground cable as a source of electricity. Dave Stakey, who was working on the island that night, restored power. "I was on duty the night the cable went. It was an 'uh-oh,' and then we put [Lab 257] on the overheads." Standard operating procedure called for using the underground cable as 257's primary power source, while the overhead lines were used as an emergency standby. So for the past three months, Lab 257 had been running on its emergency feed, the overhead transmission lines. "We no longer had an alternative means of supplying the building," says Stakey. The emergency backup had become the only option.

In full-time use, the overhead power lines went down about once a month. "We had this continuous problem on the overheads, because you had birds flying into them," Stakey says. "We would get Canadian geese—and occasionally an osprey—flying into the lines and shorting them out, either from phase-to-phase or phase-to-ground." In the orange-purple tint of dusk, the great fish hawks swooped down, grasping sharp sticks to build nests on the cross-arms of the poles, and sometimes the sticks sliced into the wires, causing short circuits. To rectify the problem, New York Telephone erected nesting poles back in the 1950s to lure the birds away from telephone and power lines, but it never worked; it actually increased the number of nesting places and expanded the flocks. An engineer from the early days said that during 1954, the year Lab 257 opened for business, adult ospreys had flown into power lines and momentarily knocked out the power to the laboratory no less than three times.

The island's top brass knew about the damaged underground cable because B Crew's foreman, Walt Sinowski, Stakey, and the other "electrics" told them. According to Shine, every time Sinowski began the shift, he penciled into the Lab 257 logbook, "UNDERGROUND STILL OUT—NOT REPAIRED." Underground power and telephone cables had blown out before and were promptly repaired. But this time, management flatly refused to repair Lab 257's electrical umbilical cord.

Though the 1991 budget was invested in items like a new ferry service to Connecticut and a new gym, Dr. Breeze and his facility manager, Ernest Escorsica, thought replacing the cable was too expensive. The cost: $70,000. It would have to wait for next year's budget. An appeal to Washington for emergency funds could have been sought, but no such appeal was made. In a letter after the hurricane to Congressman George Hochbrueckner, Dr. Breeze wrote matter-of-factly, "We made the decision not to replace the underground cable which provides power to Lab 257 immediately after it shorted out . . . because overhead power lines to the building were available and all [fiscal year 1991] funds had been committed." Apparently, Dr. Breeze was unable to appreciate the damage hurricane winds could have on overhead power lines, or the need for backup electric power to ensure biological safety.

Plum Island deputy Jerry Crawford conceded the government's misjudgment when he later blundered to reporters that "in an operation of this size, something can fall through the cracks." But, Dr. Crawford added, he did not feel that the lack of backup power was critical to the maintenance of Lab 257. To electrician Dave Stakey, however, the power going down was beyond inexcusable. "We had our generators at the power plant," he laments. "We certainly had enough fuel. We could've given them power if we had a way to get it to them."

So without a working underground cable or on-site power generator, Lab 257 remained at the mercy of flimsy overhead lines and their questionable reliability.

For this gross negligence, a terrible price would be paid.

Sunday, August 18, 8:00 a.m.— Bob's winds howled louder with each passing hour. Rain poured down like a shower of falling nails. The ground shook with each thunderbolt. Waves that normally rolled gently onto the island's beaches now crashed high upon the shore, spraying water thirty-five feet into the air and flooding low-lying areas. Just three hundred feet from the coast, B Crew huddled in total darkness inside Lab 257. They heard low, rumbling sounds and felt the muted vibrations of the hurricane, protected by the lab's three-foot-thick masonry walls. But inside the building, a deadlier storm was brewing.

After the power went out, each crew member grabbed two heavy-duty flashlights, stuffed one in each pocket, and the men split up and patrolled the building. To maintain biological containment in 257, B Crew needed to preserve sewage treatment, storage freezers, steam, and negative air pressure. Plum Island biological safety regulations call for a much larger crew during a storm of Hurricane Bob's magnitude, yet there were no reinforcements. Even with electric power, maintaining the building during the storm wouldn't be easy. Without power, it was futile. But there was no time for the men to dwell on that. They rose to the occasion, strong, handy men, resolved to handle the crisis as quickly and efficiently as possible. Lightning bolts flashed through the third-floor windows like strobes, briefly illuminating the hallways and lab rooms and casting ghostly shadows just before darkness set in again. Muffled through thick animal room doors, grunts and groans became noticeable to the ears of B Crew. The test animals, already moaning from the painful symptoms of deliberate viral infections, wailed loudly in fear, sensing the turmoil within the lab and the wretched storm outside. With the chillers down and no working ventilation fans, temperatures climbed over 100 degrees.

Sunday, August 18, 9:30 a.m.— By now, B Crew should have been relieved by the next shift, but travel between Plum Island and the mainland was completely cut off. Trapped by the raging storm, the crew continued to work feverishly inside Lab 257. While patrolling the ground floor of the building, Phillip's and Shine's shoes suctioned to the floor and made squishing noises with each step as they progressed down the hallway. Though he'd rather not have known what lay beneath, Shine instinctively moved his flashlight to the ground to investigate. He saw a thick layer of sludge coating the floor. It was rising rapidly. "Leaking wastewater," Shine figured. "There was overflowing sewage on the floor," Phillip remembers. "It was awful—because there was no place for it to go."

The sewage treatment in Lab 257 is based on a simple principle: heat kills. When cooked to a temperature that no living organism can withstand, bacteria and viruses are killed or permanently deactivated. Contaminated animal waste—blood, pus, sputum, saliva, vomit, feces, urine—from the second-floor lab rooms is drained through the beveled floor and piped downstairs to a thousand-gallon sewage holding tank. While that's a lot of sewage, the tank's capacity is still finite. "When you lose your power, your tank still fills up," notes Shine. "The animals in the labs are continuously throwing off water and waste." A single cow excretes over a hundred pounds of waste in a single day, and there were droves of them in Lab 257, all ailing with viruses. If the waste fills this primary tank, additional fluid is routed automatically to a smaller secondary overflow tank. It is transferred by means of an industrial pump that is fueled by electric power. "Regardless of power—this emergency tank is critical," Shine says.

As Hurricane Bob raged outside, Lab 257's sewage tank filled rapidly. With no emergency backup power, the overflow waste could not be redirected to the second tank. Shine read the gauges on the tanks, which remained operational despite the loss of electric power. "We're monitoring the gauges—they're still working—and it's getting up there. Way up there," Shine remembers. "We realized we needed to make provisions. We had to do something about this tank." If the crew didn't act fast, contaminated sewage would seep everywhere. For the workers trapped inside the lab, there would be no escaping the consequences.

But it was too late.

Instead of bursting, the primary tank released contaminated waste through an emergency drainpipe. The waste spilled over, onto the floor. Shine and Phillip, wearing only gloves and rubber boots, mopped up the raw, infectious muck and worked frantically, trying to keep it from accumulating. If they paused even for a moment, they knew they would end up knee-deep in . . . God only knew what. They had to get the overflow problem under control or the basement of Lab 257 would soon be flooded in a sea of dangerous viral and bacterial biological matter. Contamination would escape the building, release into the outside air, and seep into the ground.

Searching in the dark through an equipment room upstairs, Phillip found an old, small gas pump and rushed it down to the basement. Phillip and Shine attempted to start it, but each time they pulled the cord, the pump sputtered, letting out a "*p-p-p-p-p*" and stopped. Taking turns, one held the flashlight while the other tinkered with the pump. Sludge continued to flow out of the drainpipe and onto the floor. The pungent odor emanating from the floor began to overwhelm the men. They stepped back for a break, gagging, and cupped their hands over their noses and mouths. Government safety requirements mandated that face respirators be worn for this type of work, and respirator units were available upstairs in the utility closet. But B Crew and the other workers had never been trained by the safety office to use them.

Standing in a torrent of rising sewage, they willed themselves to withstand the stench and returned to the pump. Resourcefully, they ran a hose from the primary to the secondary tank, using the makeshift pump to bypass the normal electrical system. They fired up the pump, and it coughed and splattered liquid in their faces. Then it slowly kicked into gear and began to suck the sewage into the overflow tank. Phillip expressed relief with a thumbs-up, and Shine gave him a high-five. As Phillip and Shine admired their rapid patch job, two of the other crewmen entered the room. *It's beautiful! It's running beautiful,* thought Shine, viewing their successful handiwork; while wiping beads of contamination off his face, he looked behind him and saw his co-workers. "Get the hell out of here!" he yelled at them. He and Phillip had already been in the room too long as it was. There was no reason why anyone else should be exposed to the puddles of sludge.

Sunday, August 18, 10:30 a.m.— The eye of Hurricane Bob passed directly over Block Island, twenty-five miles east of Plum Island. Inside the eye, winds were relatively calm, but the storm reached its maximum intensity just past the edge of the eye, in an area known as the "eye wall." On Block Island, wind speed at the eye wall was recorded at 105 mph, but that was inaccurate. The equipment couldn't record any higher. As the outside of the eye wall battered Plum Island, wind speed at the hurricane's most powerful point could only be imagined.

A long, deafening alarm screamed out in three bursts, paused for a long second, then screamed out again. It echoed through the dark hallways of Lab 257: "*AAANNN ... AAANNN ... AAANNN ...*" It was the freezer alarm. "*AAANNN ... AAANNN ... AAANNN.*"

The freezers were melting.

With the power out, temperatures in the electric freezers rose from their subzero levels. Inside the freezers, virus samples and experiments are kept in vials and Petri dishes at minus 158 degrees Fahrenheit. If the temperature rises high enough, biological samples stir in their dishes, activate, and multiply into lethal predators. Highly infectious viruses attach themselves to air particles and travel freely through the air before attaching to a host animal. Inside these freezers, on this cold, dark night, were cultures containing the germs that cause African swine fever, foot-and-mouth disease, Nairobi sheep disease and Rift Valley fever, among scores of other plagues—many infectious and communicable to humans. Some of the cultures were five decades old. All were living pathogens. For many of the germs, all it took was one short breath to become infected.

"We had freezers thawing out all over the place. Puddles from freezers were going down floor drains," remembers Phillip. During a previous era, this accident would have been easily rectified. "We used to have this canister," says Shine, "a special single canister, and go to the outside tank and fill it up with liquid nitrogen. We would hook that canister up to the freezers and get the temperature down if we needed to." But now, no such luck. The portable canister was missing—it was removed from Lab 257 some time ago and never returned. Like emergency power and the safety respirators, the island was loaded with resources. And the workers had no way to use them.

All the men could do to counter the oozing biological freezers was mop up the puddles of liquid thaw into wastewater drains, keep the freezer doors shut, and pray for the storm to subside.

Sunday, August 18, 12:30 p.m.— A pungent odor began to permeate Lab 257. The doors leading into the "hot" areas—the 120-Area, the other lab rooms, and the animal rooms—are surrounded by rubber gaskets. The gaskets form a hermetic seal between contaminated air inside the hot rooms and clean air circulating throughout the rest of the building. An electric air compressor powers the gaskets; but since the power outage, the door gaskets were slowly deflating as air seeped out of them. While the freezer alarms continued screaming loudly through Lab 257, new buzzer alarms signaled a breach in the air-lock seals. Polluted air from within the lab and animal rooms began to mingle with clean air circulating through the rest of the building. "Walt Sinowski was after management for years and years," Phillip remembers, "to install an outside air compressor or an outside tank, in the event this might happen. You know, to prevent this type of thing."

With the door gaskets deflated, the rank odor of infected air wafted through the hallways. It flowed under the nostrils of healthy animals and the men of B Crew. The stench was even more potent near the animal rooms. Normally, the crew checks on the animals periodically by entering the animal wing and peering through the doors' thick glass windows, protected by the inflated gaskets. Entering the animal wing of Lab 257, the men were overcome by the smell of decaying, diseased animals. "We had this odor all over us," Phillip says. "It was just awful—it can't be described." The animals let out guttural wails and cries for help when they sensed the workers trying to check up on them. To top it off, a failed ventilation system contributed to the air-lock breach. "There was no venting over these animals, and the door gaskets were gone," recalls Phillip, who pauses and blankly stares into space with a look of disbelief. "On the floor there were animal droppings. There was urine. Everything was there to see, right there on the floor. We were breathing in all of this stuff." Shine concurs: "It was coming through the doors and ductwork and there was nothing to prevent it, nothing." The men were forced to turn back.

The men of B Crew were defenseless—with no means of protecting themselves, except to stay clear of the animal wing. Outside the building was no better, with a CAT-3 hurricane pummeling the island. "What could we possibly do?" Phillip asks, defeatedly. "We did everything we could to save this building." In the midst of the escalating calamity, crew members summoned the strength to remain levelheaded, at least on the outside.

A 1956 USDA film trumpeting Plum Island called the then-revolutionary negative air control system "an ingenious device" and a "wonder in itself." Because of this, "viruses stand no chance of getting out." Negative air can be best described as a series of supply and exhaust fans. One huge supply fan draws air from the outside, then circulates it through the building with a series of vents. At the other end of the building, a large exhaust fan expels the air through an enormous battery of air filters. A series of air dampers regulates the flow of air in and out of each contaminated room. Slightly adjusting a room's air dampers can throw the entire building out of balance; miscalibrate it and you can actually draw water out of a toilet. Lab 257's "hottest" room, the 120-Area, has the strongest negative air pressure, and the animal and other lab rooms share similar high-negative pressure. For this elaborate system to function properly, it requires two things: outside air vents to allow air to flow freely in and out of the building, and an electric current to turn the fan motors.

Sunday, August 18, 3:00 p.m.— Since their arrival at midnight, the crew had taken gauge readings every hour. The gauges monitor Lab 257's negative air pressure system, the foundation of the building's biological containment system. The dials are supposed to display the amount of negative air pressure in each lab room, like a speedometer measures speed. But the gauges inside Lab 257 were hit or miss. "Two-fifty-seven was a mess—it was always breaking down," says an employee familiar with the system. "An older-type gauge and a newer-type gauge measuring the same room would give two different numbers." Tapping the glass only made the gauge needles sway from side to side, before resting at a randomly inaccurate number.

When the power went down, the normal hum of the fans declined in pitch to a chopping noise that faded with each revolution until the fans ground to a halt. The needles on the monitoring dials began to fall. The twenty-four-hour-a-day, 365-day-a-year containment system failed. With the negative pressure system off and the door gaskets blown, air particles escaped and traveled the path of least resistance, from contaminated areas to the outside.

The Army engineers who built Lab 257 came up with a fail-safe mechanism for such an emergency. The air vents that allow air in and out of the building were fitted with outside "air dampers." If the system failed, these air dampers were designed to close, locking "bad" air inside the lab. But like so many other controls on the night of Hurricane Bob, the outside dampers faltered. "They were frozen in an open position—and we couldn't move them," says Shine. Given the USDA's abysmal record on Plum Island safety, they weren't checked periodically or adequately maintained. The strong hurricane winds might have broken them off or simply prevented them from closing. Biological containment was completely jeopardized.

The men, realizing with horror that containment was entirely lost, rubbed their eyes in disbelief. "There were insects in the building—we saw them," Phillip recalls with a pained expression. With the system down, airborne insects—mosquitoes, flies, moths, all in abundance on the feral island—burrowed through the air vents into the laboratory building to escape the storm. Inside, the insects mingled with disease animals, spoiled air, and contaminated raw sewage. They flew freely in and out of the building during and after the hurricane.

"We had no power—there's no on-site generator at 257. They had everything we needed, everything in 103 [the emergency power plant]. We had our incinerator, our own boilers—and"—Shine's voice is trembling, and the husky engineer is choked up probing the memories—"give us something, an old pump outside, even a lawn mower engine. Something— something for us to help—ya know?"

Phillip interrupts to cool him down. "Look, I'm not a scientist, but . . ." he confesses, head in his hands. Advanced degree or not, Phillip can relate what he saw, what he heard and smelled, how he felt. And he knows in his heart and in his head it wasn't good.

". . . to me, it was a biological meltdown."

Sunday, August 18, early evening— The eye of the hurricane slammed the coast of Rhode Island. On Plum Island, B Crew was physically exhausted and mentally defeated after doing everything humanly possible to maintain Lab 257 for eighteen consecutive hours, a period that seemed an eternity. Ravenous with hunger, after witnessing sewage spills, melting freezers, and the failure of the biocontainment system, no one was foolish enough to pick at the food in the lunchroom refrigerator, fearing they might be ingesting contamination. Then their radios chirped. It was a worker from the power plant. The sump pump that normally flushes out leaks was overwhelmed by gushing floodwaters. There was heavy flooding, and they needed help fast. Phillip agreed to brave the storm and lend a hand. He figured he could also pick up food for the crew, since the power plant's provisions were presumably safe. Preparing for the trek, he religiously followed the normal laboratory decontamination procedures. At this point it was a laughably futile exercise.

Upon entering a change room with his flashlight, Phillip stripped naked in the dark. He removed his wedding band and his crucifix. He entered the shower room, felt around for the knobs and the soap, and showered for about three minutes. "I took the standard shower," Phillip says, "but the water wasn't nearly up to temperature—it was barely luke-warm. Remember, we had no power and no steam." He scrubbed his skin vigorously, spat, and squeezed his nose and blew air through his ears. He cleansed his fingernails with a special nail file, then donned "clean" clothes in the change room on the other side.

"I was allowed to leave by the duty officer, who gave me permission," says Phillip. "He was in charge of the whole operation—he said go and eat and try to bring 'clean' food back to the men in 257. By that time, we had been in there for eighteen hours." That duty officer was Dr. William Lagreid. Only recently hired to join the island's scientific research staff, Dr. Lagreid was stationed in the duty officer's quarters, on the east side of the island, far from Lab 257. His new position put him first in command of the island with full responsibility over any laboratory accidents or mishaps. During Hurricane Bob, Dr. Lagreid was the only scientist on Plum Island. It is certainly puzzling that director Dr. Breeze appointed this rookie solely in charge after receiving a hurricane warning. A more experienced duty officer might have evacuated Lab 257 prior to the hurricane.

Outside 257, Phillip was thrust into the tail of the hurricane. Somehow, he clawed his way along the road through the island's forest to the power plant. The journey, which usually takes a few minutes, now required considerable time and effort. He threw his full body weight against the 100 mph winds and shielded himself from driving rains, which pelted his face. He hurled himself over large trees and debris strewn across the dirt path; the trail looked like a battle zone, a winding maze he navigated in the rainy dark. Phillip drew a deep sigh of relief upon reaching the power plant. Though he was soaked to the bone, he shook himself dry and got to work. Phillip and the others barricaded the east door to impede the water rushing into the electrical gear room. They bailed out the water, using small buck-ets, which prevented the generator from shorting out and avoided a power outage to Lab 101 and the rest of the island. Mission accomplished, Phillip moved on to his second task. In the power plant's well-lit kitchen, he pre-pared an uninfected feast of sandwiches, brewed a pot of coffee, and wrapped up the rations securely. Without wasting a moment, he made his way back to deliver the provisions to the B Crew. As he reentered the Lab 257 air-lock door, his co-workers cheered Phillip's triumphant return.

Monday, August 19, 12:00 a.m.— Hurricane Bob lost much of its strength after it left Plum Island and made landfall on Rhode Island. It continued on a northeastern path, past Boston, and up the coast of Maine. Eventually it spun across the Atlantic Ocean and broke apart off the coast of England.

Back on the island, the alarms died out. The frazzled men of B Crew heroically weathered twenty-four hours working inside Lab 257; it was now sixteen hours after their normal shift ended. Amazingly, they were not relieved. It would be another eight hours before they were. Ferry transportation was still impossible, as the *Plum Isle* could not brave the violent waters without risk of capsizing. High crosswinds prevented helicopters from flying safely. A tired and dazed B Crew spent the next eight hours repairing leaking steam pipes. "We're going around, making sure doors were closed, looking for water from contracting pipes after we lost our steam. Water is all over the place, we're tightening every pump packing and every flange." The utterly exhausted men, with weary eyes slogged about the cavernous hallways, armed with their now-dimmed flashlights and thick adjustable pipe wrenches. The sounds of the storm gradually dwindled. Except for a sporadic groan, the animals quieted down, and the building systems remained down. "We were in complete silence," remembers Phillip. "When the building runs normally, you never seem to notice the sounds of leaks—and then, with everything quiet, you can hear water dripping—*drip, drop, drip . . . drip. . . .*"

Monday, August 19, 8:00 a.m.— B Crew was finally relieved and the next crew, A Crew, took over. Shine remembers the scene. "When we left, there was no need for decon [decontamination]. The water in Lab 257 was ice cold. What would be the point? We took our outside clothes, put them on, went down to the guardhouse, took a cold shower there to be safe, and went home."

"We brought contaminated clothes off the island," says Phillip. It hardly mattered after the collapse of Lab 257's biological containment systems. After this harrowing work shift on Plum Island, the safety regulations were utterly pointless to the men of B Crew.

Tuesday, August 20, 12:00 a.m.— Management refused to offer the crew, held virtual prisoners for thirty-two hours in the chaos of Lab 257, any respite. The fatigued men reported to work for another graveyard shift. "Management didn't exactly say, 'Guys, you really broke your hump in there—great job,'" says Phillip. "They didn't offer to send us to a doctor to be checked."

B Crew's valiant efforts were belatedly praised in a letter dated September 17, 1991, by R. D. Plowman, head of Agricultural Research Service, the arm of the USDA that ran Plum Island. It commended the B Crew for its "quick and decisive action taken ... to prevent damage—possibly even a blackout—at PIADC [Plum Island Animal Disease Center] during Hurricane Bob." Possibly even a blackout? They had been using flashlights in the dark for thirty-two hours. Dr. Plowman then concluded, "I want to personally thank you for your hard work and courage in the face of such a dangerous situation."

That same day, the men received another letter.

"As a result of the A-76 [federal government privatization] process, it has become necessary to conduct a Reduction In Force. Your position has been specifically identified for abolishment, and you [are] to be released. . . ."

As part of the USDA's efforts to cut costs, Plum Island laid off the men of B Crew.

Great job, men.
You're fired.

13

The Aftermath

They won't need a boat to remember me by.
—PLUM ISLAND DIRECTOR ROGER G. BREEZE

Hurricane Bob's devastating effects ranged far beyond Plum Island. During its run up the East Coast, over seven inches of rain fell for twelve hours straight. A total of eighteen people, from South Carolina to Maine, perished in the storm, which left almost $2 billion in damage in its wake. Like a sports team retires player numbers, the National Hurricane Center retires hurricane names; when they cause enough death and destruction their names are never repeated. The name "Bob" was retired, and it went into the annals of history as the eighth costliest Atlantic storm in United States history.

Established when the facility began operations a half-century ago, the comprehensive Plum Island biological safety manual sets forth explicit "Emergency Hurricane Procedures." Lying in perhaps the most vulnerable spot in the northeast hurricane corridor, Plum Island played host to violent fall weather as soon as the USDA arrived—and long before. Storms in the 1700s and 1800s wrecked numerous schooners on its rocks, whose victims' bodies (often anonymous) were buried on the island; during the 1950s, said an old employee, every day after a rough storm hit, another person resigned from staff. A gale swept through the east end in November of 1953 as finishing touches were being put on Lab 257. Much like Bob, this storm struck Plum Island dead on with devastating force. Tidal waves rushed in from Gardiner's Bay, wrecking one of the Army's boats and flooding three feet deep against Lab 257's four-foot concrete barrier, recently installed and dried. The T-boat was found the next day, dragged across the beach and broken to pieces. Channels connecting the marshy ponds with the bay were cut wide open from waves. Seawater flooded in, creating a brackish environment that not only threatened the freshwater wells, but could shear off twenty acres at Pine Point and erase the land buffer between Lab 257 and the ocean. Awaiting the inevitable destruction of the next nor'easter, the Army Corps of Engineers sandbagged the channels shut and filled the area with tons of jetty rocks. Though Lab 257 was saved, the close call should have called into question—before its doors opened—the decision to locate a germ lab on the island's southwest shore. A Plum Island hurricane inundation table shows Lab 257 completely inundated with water during a CAT-3 hurricane's twelve-foot surge elevations.

Knowing firsthand the potential for disaster, the scientists who founded the animal disease laboratory on Plum Island drew up the hurricane emergency plan:

> The aftermath conditions of a severe storm or other natural caused disaster could severely limit or prevent the emergency operations of facilities. . . . Potential breaching of the agent contaminant aspects of [Plum Island] facilities and escape of disease agents could also occur in this type of emergency condition.

Upon issuance of a twenty-four-hour hurricane warning by the National Weather Service, procedures dictate securing laboratory buildings to protect against damage. This includes covering all windows with one-quarter inch plywood and sandbagging buildings to minimize water damage in low-lying areas. The emergency plan specifically mentions sandbagging Lab 257, but when Hurricane Bob was on its way, no one sandbagged 257 or the power plant. Because of this oversight, the power plant flooded. It was only through the ingenuity and hard work of men like Phillip Piegari that the generators were saved, averting the loss of power to the entire island and a far greater catastrophe.

In addition to securing the buildings, the emergency procedures mandated additional safety measures for approaching hurricanes:

a. Water tower must be filled to capacity;
b. All underground electrical feeders shall be utilized;
c. Stand-by generators must be operational and be attended by competent operators;
d. All sewage in Buildings #102 and #257 must be processed and tanks emptied; and

e. Employees must be advised to have food, water, prescription medicines, etc., within their respective work stations.

Management failed to follow these procedures after receiving the hurricane warning for Hurricane Bob. In fact, they didn't follow a single one. Item a. simply did not occur. Items b. and c. were impossible, thanks to management's disregard of safety when it failed to repair Lab 257's underground power cable. If procedure d. had been followed, sewage would not have spilled onto the floor and contaminated the building and the men of B Crew. Finally, the lack of proper provisions mandated by item e. forced Phillip Piegari to leave containment to obtain provisions for the crew. "I don't think they expected the hurricane to be of that magnitude, that it could do such damage. But they knew it was coming and didn't prepare for it," one crew member later reasoned.

The government broke Dr. Jerry Callis's cardinal rule. "Each person," Dr. Callis wrote in the introduction to his three-inch-thick island safety manual circulated to all new employees, "has a moral and legal responsibility for assuring that maximum biological safety precautions will be taken in all operations." A reckless disregard of Callis's edict and the standard emergency safety procedures caused Lab 257 to come apart at the seams during Hurricane Bob. Those responsible for Plum Island safety, notably island Director Dr. Roger Breeze, compromised the safety of both the island's employees and the public at large. It is only by a stroke of good fortune that contamination didn't noticeably spread to Long Island, Connecticut, and beyond.

The government refused to admit anything went wrong in Lab 257. Dr. Plowman's letter of commendation didn't acknowledge that a power outage actually occurred. Management treated B Crew's thirty-two-hour dance with a hurricane like a typical day on Plum Island. Over time, the men found other employment or retired. A few continued to work on Plum Island for the private contractor, at a fraction of their previous wages, without any meaningful retirement benefits.

Soon after the hurricane, Phillip Piegari developed flulike symptoms—constant nausea, severe headaches, and hot-and-cold flashes. His family physician requested the blood sample that Plum Island officials took when he began employment. The government refused to release it. After a *Newsday* article uncovered the government's stonewalling, officials gave in and released a portion of the blood sample. After batteries of tests, neither his doctor nor doctors from the State University Medical Center at Stony Brook were able to diagnose Phillip's illness. Like all standard medical centers, the facility was not equipped to check his blood against exotic "animal viruses," many of which can infect humans. One location, however, did have the ability to test for them: the Plum Island Animal Disease Center. They refused to help. Instead, the scientists insisted that Phillip was a disgruntled laid-off worker suffering from a typical Lyme disease infection. But his medical doctors flatly refused to diagnose Phillip's condition as Lyme disease. The symptoms went undiagnosed and unabated for six years before they gradually subsided, though he continues to suffer occasional mysterious relapses where he contracts viral-like symptoms. Over a decade later, Phillip Piegari tries to lead a generally healthy, normal life. Nonetheless, he is certain that he was infected with an unknown virus from his contact with contaminated sewage and poisoned air in Lab 257 on that fateful night. And Plum Island's determination to prevent an accurate diagnosis only furthers that likelihood.

There is reason to believe Phillip wasn't the only one who contracted something that night.

Two years after he helped rescue Lab 257 from certain meltdown, Stanley "Shine" Mickaliger came down with relentless arthritic conditions. First he had a difficult time bending his elbow. Then his legs hurt him when he walked. And then he couldn't walk at all. "For eighteen months, I was deadly sick—my wife would have to fit me into the car to go see the local doctor." Shine's country physician put him on a heavy regimen of cortisone steroid shots to bring down inflammation, which eased some of the pain, but replaced it with a dogged malaise and awful bruises from bumping into things around the house. Plum Island viewed Shine's illness the same way they did Phillip's and those of others—with denial and with contempt. Unlike Phillip, Shine couldn't offer his doctors a baseline blood sample, since Plum Island never took one from him during his twenty-year career. When he asked a Plum Island safety officer for help in finding out the cause of his condition, the official told him there was no money in the budget to do it, and that they didn't have the dime for it. "It's hard to prove," says Shine, "and I wasn't bled by Plum Island, so who knows?" He couldn't point to the Lab 257 meltdown as the definitive cause, however, because that fateful weekend wasn't the first time he was exposed to contagion.

On Wednesday morning, March 2, 1983, a sewer line leaking from Animal Room 264, which had cattle infected with the *Isfahan* strain of vesicular stomatitis, spilled into the equipment room. Sludge splashed on Shine and two other building engineers taking their coffee break. Shine and another worker immediately placed plastic tape over the gushing pipe and flooded the floor area with hydrochloric acid.

Ten minutes after the crew's panicked phone call, safety officers Drs. Walker and Richmond barged into the equipment room and ordered the area locked down and decontaminated. The crew locked the corridor door and sealed it with duct tape on both sides. Food was incinerated, and clothes were stripped and stuffed into the steam autoclave. The workers mopped the corridor floor with Vanodine disinfectant solution and then poured a full-strength gallon of One-Stroke into the drain in Room 264. An hour later, a safety tech, wearing heavy rubber gloves and a full-face respirator, removed the duct tape and drained the trap into a bucket. Engineers located a small hole in the pipe, and Shine patched it up with silicone. The room was deconned a third time and finally declared clean late in the afternoon. After ninety-six hours of close monitoring, miraculously neither Shine nor the ten others exposed to the contaminated waste came down with any disease. Following biosafety rules to the T, it required seven and a half hours for the engineering and safety department to control a tiny pinhole in a pipe during Jerry Callis's administration, with far more efficiency and concern than the next regime would devote to a full biological meltdown.

After the meltdown, with no help from Plum Island and no diagnosis from his doctor, Shine turned to the one person who could help him recover—his wife, Fran. "My best doctor was my wife. She took all these books out of the local library, and threw away my meds." Fran put him on a strict regimen of exercise, good food, and positive thinking, and he slowly regained himself. To this very day, he has no idea what his illness was. A possibility is one of one of the feared "slow" viruses, so named not for the tempo of virus growth, but for the protracted time of the disease's course, which can be months or years.[1]

The laymen tried to figure out what they had contracted on Plum Island, and pleaded for answers from their North Fork doctors, family general practitioners better suited to bandaging knee scrapes and prescribing antibiotics for ear infections. "You are exposed to so many viruses over there," says Shine from experience. "They say it won't bother you, that the germs aren't zoonotic, that they won't transfer to you. Then you become ill, you tell your doctors you are ill, and that you work on Plum Island. And look here—they have no tests for you. Hepatitis they have a test for— but for Rift Valley fever? And USDA, their response to us always was, 'Prove it to us. Show us what you caught here and how you caught it.' Now how the hell I am supposed to do that?"

In a newspaper article that appeared the day after B Crew was com-

---

[1] AIDS and Creutzfeldt-Jakob disease (CJD, the human variant of mad cow disease) are two examples of slow viruses.

mended for its "quick and decisive action" and then summarily fired, Plum Island officials stated there was never any danger to laboratory staff or to the public during the hurricane. Manuel Barbeito, an island safety officer, told *Newsday,* "There is no potential problem here . . . this is a safe facility," and stated that the laboratory air filters operated during the hurricane without power and prevented diseases from escaping. When Phillip, per his physician's instructions, asked for a list of what he was exposed to during the storm, an official handed him a letter. "It said the only things we were exposed to were paint, paint thinner, and oil—that's all." Apprised of the hurricane incident by Plum Island employees and worried local residents, Congressman George Hochbrueckner wrote a letter to the Department of Agriculture, demanding information on the storm's effects on the island's laboratories.

Though the government told the public nothing had happened, steps were being taken on Plum Island that reflected a different belief. A few days after the *Newsday* story, a portable generator the size of a tractor-trailer appeared alongside Lab 257. With the underground cable still shorted out, the portable generator provided the emergency power the lab had lacked for months, and covered the momentary power breaks when both the overheads and underground cable were restored. Though management claimed the biological containment system had worked properly through the hurricane, technicians replaced all of Lab 257's outside air dampers with new units. New procedures were adopted to regularly inspect the air dampers— which, according to the government, also worked properly when Hurricane Bob hit. Henceforth, after even a minor power interruption, employees said safety officers climbed atop Lab 257 to personally inspect the roof and ascertain that outside air dampers were closing properly. And just days before they were canned, B Crew was finally trained on how to use the face respirators. "They were afraid of lawsuits," guesses Shine, "so they did this to have it on record that we were all trained."

Today Shine and Fran spend their days working around the house, hustling firewood, and taking long walks along the beach, pointing out sea turtles and searching for washed-up fishing lures. Hurricane Bob, searing pain, and the recurring nightmares are now in the past. In his work cabin, out back behind his modest home, Shine keeps perhaps the world's largest collection of jigs and lures. Thousands of multicolored and feathered wooden, shiny metal, and plastic lures adorn the walls and ceilings of the dark brown cabin, warmed up in the winter by a small space heater. When he's not fishing with his two older brothers, Charlie and Eddie, aged eighty-six and eighty-one, he's hunched over the workbench fashioning the lures from broken ones found strewn on the beach. Shine's at his happiest hammering, sawing, gluing, and picking away at the rigs that will trick next season's blues and stripers into thinking they have fixed their mouths onto some-thing delicious to eat—only to realize it'll be Shine, not the poor fish, doing all the eating.

Phillip still lives out on the east end, working for the county now, spending his free time on his beloved boat with his black Labrador, Jezebel. A few years ago, Zyta contracted breast cancer, underwent extensive chemotherapy, and thankfully pulled through. The cancer survivor now works as a translator and was recently honored at a police department ban-quet for lending her bilingual skills to help solve a murder case that had gone unsolved for years.

THE BREEZE SUBSIDES

Dr. Roger Breeze left Plum Island in 1995 for a better career opportunity. His two predecessors had been honored by the ferryboats *M. S. Shahan* and *J. J. Callis*. "They won't need a boat to remember me by," Breeze told a *Newsday* reporter who asked him about his legacy. He was even more direct with me when we spoke. "My memorial has to do with the people I got there. I'm not interested in any damn boats and buildings. Facilities and boats don't do the research. People do. You can set out a stack of my scien-tific papers, and I'll be judged by those. I come back to this—it's the glory at all levels, and not in a negative kind of way."

Roger Breeze drew responses out of people, whether they were his superiors, his scientists, or his support workers. A head Washington-based USDA official said, "Some people just think he's the best—and some don't." From that distant official's vantage point, Dr. Breeze was "innova-tive and dynamic." He had reworked Plum Island's entire scientific pro-gram, rebuilt the facility, and saved Plum Island from imminent demise.

Ed Hollreiser sees Breeze as a "strange guy, very cunning—he'd call me in for little chats and tell me things that he said he didn't wanted repeated, but he really wanted me to spread the word." Plum Island safety officer Tom Sawicki says, "Breeze was here for a reason, he did what he had to do, and a lot of people didn't like it." Fran Demorest says, "It was his stepping-stone. And he made enemies there."

Dr. Robert Shope, who had lived with Breeze for a time in Connecti-cut, takes a middle view. "He did some things that weren't too popular with his superiors at USDA. And he may have gotten rid of some people at Plum who were deadwood, and in that sense, he wasn't very popular. But I think he was part of the driving force of the island." When asked to com-pare him to his predecessor, Shope thought of Breeze as "a totally different type of person—and still is. He's a wheeler-dealer type, and Callis was very conservative and played by the book and that sort of thing. Just two differ-ent people."

"My problem with Roger," says Dr. Carol House, "is that he still has an influence. He still shows up at town meetings and stands in the back, hovering. He still has a large influence over there, and he has pulled potential [Plum Island director] candidates."

"Roger's very hard to talk about," Dr. Jim House says, slowly, measuring his words. "Not one of the brighter moments in Plum Island history. Roger had a way of manipulating situations so he was always politically on top. No matter what he did, he would come out smelling like a rose. He was very, very clever.

"He did make strides, enhancing the amount of molecular virology done, but he even took that to an extreme. He was somewhat of a visionary, and he's into biological warfare, his new thing now. But he didn't have the vision or skills as a manager to run Plum in a smooth and productive manner. I didn't have a lot of respect for him scientifically. He was selling genetically resistant animals, and we didn't do genetics of animals. We had this genetically resistant cow, and transgenic pigs. Of course that never went anywhere.

"A lack of perspective—of all the things you'd say about him, that would be the one—a lack of perspective."

The two research groups at Plum Island are without question the best of their kind in the world," boasts Dr. Breeze, "and that wasn't true when I went there. If there's one thing I do know about very, very well, it's how to motivate scientists to go beyond what they think they can do—that's what I do best." But in some cases, the science on Plum Island may actually have been set back. Proof of that, says Dr. Richard Endris, is that some three years after the four scientists had been dismissed, one of Dr. Breeze's new recruits reestablished and set up—from scratch and at great cost—the same African swine fever tick colony research that Breeze disbanded upon his arrival. As for the new laboratory facility, it included a new animal isolation wing, and a fancy sandstone two-story brick office complex slapped across the front of Lab 101. The brown brick and shiny glass façade conceals the deteriorating 1956 laboratory facility behind it.

Retired from government service, John Boyle still follows the career path of his old boss. "You saw what happened after Roger left Plum Island—he became an associate area director, he then became an area director. Now he is a big-time guy in Washington." Dr. Breeze is the associate administrator for "special interagency programs" for Agricultural Research Service. With his boss, Floyd Horn, at the Department of Homeland Security, Breeze oversees a good part of the USDA's scientific research.

Undoubtedly, the steep trajectory of his career path in America trumps the sluggish thirty-two-step ladder he left behind at Glasgow University.

Is it possible that Dr. Breeze was blinded by his own ambition? "He is a very talented guy, and cares very deeply," says Boyle. "I think he cares so much that maybe it even overrides his talent, because he is so tenacious, once he sets out to do something, it *will* get done. But he really cares about science—good science. It's why he took a liking to me, because I worked so hard to get him the ferryboat." Blinded by the glory of science, or blinded by unadulterated ambition—or perhaps both—Dr. Breeze's curriculum vitae soared while the people of Plum Island tumbled and the island itself crumbled.

In the wake of that unbridled progress, people's lives changed, and not necessarily for the better.

"If this place wasn't going to be different," Breeze told a reporter, reflecting on his tenure, "it was going to be gone."

For certain, Roger Breeze had made Plum Island different.

"Living with success," he says, "is harder than living with failure."

PART 4

THE FUTURE

14

The Homeland

> *The further you get into this, the more mind-boggling it will become.*
> —PLUM ISLAND EMPLOYEE (1997)

Unfortunately, this story doesn't have a happy ending where the troubles work themselves out into tidy solutions. In fact, there is no ending. The island workforce walked out and went on strike in August 2002. The following June, President George W. Bush moved the laboratory from the USDA to the new Department of Homeland Security. The Plum Island saga gets more intriguing with each passing day.

Dr. Breeze physically departed Plum Island in 1995, but he continued calling the shots from his new office in Washington as a procession of faceless directors came and went. Breeze finally got his man in sixty-five-year-old Dr. David L. Huxsoll, whom he appointed Plum Island's director in June 2000. "Roger handpicked him," says one scientist familiar with the decision. "He has that biological warfare background that Roger likes. Breeze has always been into germ warfare. He loves the mystery, and the intrigue—he's really into it." Dr. Huxsoll grew up on a farm in the rural town of Aurora, Indiana, where he recalls being so attached as a child to his family's livestock, he cried for days when the fattened baby calves he had named and petted were sold at market. Like Dr. Callis, Huxsoll attended Purdue University, but the comparisons end there. After a brief vet practice stint in northern Illinois, Huxsoll was drafted into military service and embarked on a three-decade military career chasing diseases around the globe. Colonel Dr. Huxsoll was named commander of the Fort Detrick biological warfare laboratories in 1983, the first veterinarian to hold that command, bringing full circle the veterinarian connection to biological warfare that Plum Island founding father Dr. Hagan began in 1941. "The most valuable thing out there," says Dr. Huxsoll, "it's not the gun, it's not the tank, it's not the jet fighter—it's man. So we do whatever can be done to prevent illness, and should illness occur, restore that person to operational status."

As Fort Detrick's[1] commander, Colonel Huxsoll saw heavy action during the now-infamous Ebola virus outbreak in Reston, Virginia, featured in Richard Preston's *The Hot Zone*. He made the controversial decision to send the Army into a domestic matter that the Centers for Disease Control (CDC)—lacking any hands-on expertise—was having a difficult time managing. "At that time, I considered everything, the potential hazards, and the safety issues. . . . There comes a point in time when you see—and you know—the only logical response is to make that uncomfortable decision with the best interests of lots and lots of people in mind." The decision was the right one. Huxsoll's well-trained medical soldiers—led by virus hunter C. J. Peters, who stalked Rift Valley fever through Egypt a decade before—successfully beat back the Ebola virus.

After the 1991 Persian Gulf War, Huxsoll served on three U.N. biological warfare inspection teams in Iraq, leading two of them. In Iraq he interrogated directors and middle managers of medical and academic facilities like the University of Baghdad and the College of Agriculture. As with the United States, following the veterinarians revealed clues about germ warfare. Says Huxsoll, "Probably the greatest capability in addressing the biological villains, and those having a true understanding of it, was at the veterinary vaccine places." Special military satellite image maps enhanced by computer line drawings afforded the inspection team an uncanny recreation—down to exact shapes and sizes—of each suspected germ weapon manufacturing plant. Among the long sheet maps Huxsoll unfurled in front of his impressed Iraqi hosts ("I think it got to the point they probably thought we had more capability than we really did") was a single-cell protein facility at Al-Hakam. Ostensibly, that operation grew colonies of bacteria in big fermenter vats to be dried and milled into high-protein animal feed. But the team discovered a fair amount of evidence that it was used for biological agents, and ultimately the Iraqis admitted it. It wasn't easy to determine, however. "The plant that would produce biological agents for

---

[1] By this point, the Fort Detrick biological warfare laboratories had assumed a friendlier-sounding name: U.S. Army Medical Research Institute of Infectious Diseases, or USAMRIID.

weapons purposes," Huxsoll notes, "may not look too much different from the plant that produces biological agents for vaccine purposes or for making beer." Chemical agents are another story. Chemical plants have what is called a large "signature," while biological facilities have a far smaller footprint. "If you're going to dump chemical agents on a significant portion of Long Island," Dr. Huxsoll postulates, "you have to have tons of the stuff. Now, in the case of biological [agents], you can measure what you need for the same area on Long Island in kilograms, not tons. That's because you can disseminate them in an aerosol, and if you are good at this, you can spread it over a huge distance." Following those tiny footprints all throughout Iraq, Huxsoll's teams uncovered volumes of anthrax, botulinum toxin, ricin, gas gangrene, and other anticrop and antilivestock germs being prepared and weaponized on the tips of bombs.

Before the first Gulf War, says Huxsoll, little effort was put into biological warfare intelligence. But after he helped uncover the Iraqi program, America turned its eyes to the rapidly disappearing Soviet Union, and Huxsoll inspected weapons plants there, too. "Their programs were far beyond the imaginations of anybody," he says. "We're talking about big facilities. Huge. Huge beyond the imagination. Incomprehensible. Very plain, very stark. Desolate and unattractive. Nothing decorative about it." In the vanquished Iraq, Huxsoll's horde of UNSCOM (United Nations Special Commission on Iraq) scientists, interpreters, communications officers, and photographers could go anywhere and be as intrusive as it pleased. When it came to Russia, however, for every suspected facility the American inspectors asked to see, the Russians could see an American one in return. "And that really begins to create some problems. Not that you are trying to hide anything, but to have them go into Eli Lilly and Merck and take samples?"

One of the doors the Russians would pry open was the door to Plum Island.

The first thing Dr. Huxsoll did after accepting the Plum Island directorship (he refers to it as just "one more interesting type of experience") was to call a truce with the community. He raised the white flag by proposing to open the island to public tours, perhaps once a week, so the public could learn more about the mysterious animal virus kingdom. "There's nothing to hide," says Huxsoll. "No need to hide anything at all. I can be extremely proud of what we do here, and extremely proud of why we're doing it." But the half-hearted olive branch never materialized, and the September 11, 2001, attacks foreclosed the idea of an open-door policy for the foreseeable future.

With Huxsoll in control, Plum Island finally appeared to be in capable hands, under the leadership of this tested, experienced ex-military commander. That is, until one hears the island's previous directors talk about him. "Dave Huxsoll is a really good guy," says Dr. Breeze. "But he doesn't have the power that I did. The bean counters are running that in a way that actually isn't the right way to do it. . . ." And Dr. Callis, in a rare statement about Plum Island leadership, admitted, "Huxsoll is there just because the USDA thinks it can upgrade [to biosafety level four]."

What about Plum Island itself and anthrax? The USDA made it a priority after the anthrax attacks in the fall of 2001 to state repeatedly that it did not have and never had anthrax on Plum Island. If that really is true, it is quite curious that the FBI's lie detector tests for scientists suspected of the anthrax attacks included these three questions:

- Have you ever been to Plum Island?
- Do you know anybody who works at Plum Island?
- What do they do there?

As noted earlier, the Army's housewarming gift to the USDA upon bequeathing Plum Island in 1954 was 131 strains of germs, including 12 vials of "N," the now-declassified code name for the original weapons-grade anthrax designed by Plum Island founding father Dr. Hagan. If Plum Island never worked on anthrax or kept it in its freezers, then the USDA must explain when the code name "N" changed from anthrax to another germ. And if there is no anthrax on Plum Island, the USDA has both the FBI and its own founding fathers fooled.

The *New York Times* revealed in late 1999 that the USDA was quietly upgrading Plum Island from a biosafety level three to a biosafety level four laboratory. The story provoked a howl of protest from Long Island's east end communities. The structural difference between BSL-3 and -4 is minimal—face respirators are worn while working with "hot" agents in a level three facility, while level four requires full-body spacesuits. Here's the important difference: the level four germs. Once upgraded, Plum Island joins charter members CDC and Fort Detrick in an exclusive club whose membership benefits include working with agents, lethal ones that have no vaccine and no cure. These include mad cow disease, the Nipah pig virus (that killed a million pigs and 265 people in Malaysia in 1999), and the Ebola, Marburg, and Machupo hemorrhagic fever viruses (all related to Rift Valley fever virus). Biosafety level distinctions, however, never deterred Plum Island research in the past. For example, Rift Valley fever virus and glanders bacteria are classified borderline germs, lying between biosafety levels three and four. Both are biological warfare agents and have been studied intensively at Plum Island. Some literature classifies glanders, the bacteria used by Germany in World War I, as level three, yet a federal government document relating to 1989 Plum Island research is titled "Biosafety for Animal Rooms with Glanders—Biosafety Level 4."

Despite the USDA's secret efforts, Congressman Mike Forbes, Hochbrueckner's successor, who had eyed Plum Island skeptically since his surprise visit with Karl Grossman, intervened and killed the BSL upgrade line item in the 2001 federal budget. For the first time, the local community triumphed over the USDA's designs for Plum Island. The USDA again pushed to upgrade Plum Island after September 11, 2001, but it has been fought off thus far by Forbes's two successors.

When news of the BSL-4 upgrade came out on the heels of the 1999 West Nile virus outbreak, former employees and local environmental groups bombarded local officials—including New York State Assemblywoman Patricia Acampora—with phone calls. Acampora grew up on Long Island knowing little about Plum Island. "When you take the ferry over [to mainland Connecticut], you see this big island," she says. "All the other islands out there are small and don't have much to them. Then you wonder, 'Wait, what are those buildings over there?'"

In her five years on the job, Acampora had never concerned herself with Plum Island. Now, resolving to take the bull by the horns, she formed a task force to address the upgrade and all the strange things she was hearing. "Even though it's federal property, if something awful happens there, it affects the people of my district, no doubt about it. I had to speak out." Rather than a soapbox for disgruntled workers or a USDA admiration society, this would be, for the first time, a serious and inclusive effort to get real answers. It took political courage to do it, because the issue cut both ways. Many voters called to berate and accuse her of trying to shut the island down. "'Look, I couldn't shut the place down if I tried,'" she remembers telling them. "'I only want to start communicating. That's what we're trying to do here.'"

Acampora drafted political leaders, former scientists, engineers, USDA officials, New York State environmental officials, and county health officials. Seeking to legitimize her efforts with Plum Island officials, Acampora included experts like former Chief Engineer Merlon Wiggin and the respected former Plum Island scientist Dr. Carol House. "When you are dealing with the scientific community, right away they always get defensive," Acampora says. "They feel that no one but them can understand their work. Now I may not be a scientist, but I'm no dummy either."

The plan, she says, was to "make them talk to us about what was going on there." The group had five marathon-length meetings in which the

USDA was "pummeled" with questions, with the assemblywoman leading the charge. "I asked them point-blank, 'Can you tell me that you have a fail-safe facility?' 'Can you tell me that nothing can go wrong, and that nothing has ever gone wrong?' And they didn't want to answer these questions." Plum Island's leadership still seemed unwilling to come clean. Dr. Huxsoll sent the task force a letter stating that unlike his predecessors, he would be open to communication. "And that's the last we heard from him," says Acampora. Huxsoll also failed to appear at a March 2001 community meeting, prompting one politician on the committee to exclaim, "I am stunned. . . . It's these types of actions that have created this atmosphere of mistrust and suspicion. . . ."

Dr. Breeze hovered over the proceedings, and attended meetings here and there. "I spoke to him, too," says Acampora, referring to Breeze. "They're all the same." Except for another former director. Coming out of retirement to lend a hand was Dr. Jerry Callis, who was "very helpful" to Acampora and helped explain the science to the task force. At one open meeting, a boisterous crowd of detractors and supporters quieted as the white-haired Callis slowly rose from the back of the audience to speak. "I plead for your understanding," he began in low, measured tones. "These discussions remind me of 1953, the year I moved here. I wasn't welcome back then. I had my career here. . . . It was a very *fine* career." Then he spoke to the issues at hand. "We all make mistakes . . . we make poor decisions. All of us are becoming more aware about what we need to do to protect the environment. We can do dangerous work and protect the public. I can't address the need to improve the facility [to BSL-4], but they [the USDA] will listen to your concerns."

One of the few task force members supporting the upgrade, Greenport mayor David Kapell seemed "practically ecstatic about the prospects for local development," in the words of the *New York Times*. "There's always a concern," he told a reporter, "but it's got to be considered in light of the 50 years of history of a safe operation. . . . I have faith in the government." The *Times* noted that as a real estate broker, the mayor had already sold at least half a dozen homes to Plum Island scientists over the years and stood to sell many more after an upgrade.

Plum Island, in a rare twist, then produced some good news. In April 2001, Plum Island's Dr. Fred Brown developed a "rapid test" that shrank the testing time for diagnosing foot-and-mouth disease virus from two days to ninety minutes. "The rapid test we developed is the most important discovery in fifty years," adds Dr. Breeze. The test is linked to an Internet-based epidemic management system in real time. "I'm working on this with the military—all kinds of people immediately start to work on the problem, and that completely changes what's possible." Said USDA spokeswoman

Sandy Miller Hays, "This is not a magic bullet." But it's a commendable stride in the right direction, the kind of positive output the public would expect from Plum Island's animal disease research.

Hamptons-area environmentalist Robert DeLuca wasn't impressed. "Short of being on a geological fault," he said of the Plum Island upgrade, "I don't think you could find a worse place" for the laboratory. Thor Hansen, a World War II naval officer, stood up and said he'd been through three typhoons in the South Pacific and he knew rough weather better than anyone. In his opinion, the east end of Long Island had plenty of it and he was deeply troubled by the laboratory's location in the middle of a hurricane path.

"What disturbs me," wrote a state senator on the task force, "is the consistent flow of misinformation. . . . I feel that some of the misinformation borders on a cover-up. [I]t shakes the foundation of our very form of government." Committee member Dr. Carol House witnessed an all-too-familiar routine. "I think [the task force] went a long way. But Plum Island didn't help themselves on that committee. They were caught up constantly. The problem was that there was no continuity in leadership, so no one in management—those answering the questions—knew the answers. . . ."

One comment in particular stoked the community's ire, according to reporter Karl Grossman. USDA official Wilda Martinez told the task force, "We'll listen to you and the community. But in the end, the federal government will decide what to do upon its own and move forward." They did just that. Despite the community's uproar, Dr. Huxsoll went ahead with a Request for Proposals to construct a $125 million laboratory renovation. The USDA, of course, said the largest renovation in Plum Island history had nothing to do with a BSL-4 upgrade.

The aggressive task force gained some ground in other ways. They successfully lobbied the USDA to remove sensitive material available on its Internet site, like maps and charts of the laboratories. Yet other dangerous items still remained on-line, waiting to be preyed upon.[2]

The task force uncovered another disturbing finding. The USDA, after all these years—and after all the EPA environmental law violations—*still* had stockpiled hazardous waste and *still* flushed contaminated sewage into area waters. Unable to order in the EPA as a state official, Acampora turned

---

[2]Conducting Internet research for this book, I uncovered many schematic and other diagrams of Plum Island available on-line, including a cache of sketches and documentation from the U.S. Army Corps of Engineers from a study they conducted of Plum Island in the early 1990s. Ancillary to a free and democratic society, this and much other material is available on the Internet, including a virtual how-to guide published in 2000 by virus hunter Dr. C. J. Peters, titled "Are Hemorrhagic Fever Viruses Practical Agents for Biological Terrorism?" It reports the kill ratios of various quantities of Rift Valley fever virus, Marburg virus, yellow fever virus, and anthrax spores.

instead to the New York State Department of Environmental Conservation (DEC) for help. A July 2000 DEC inspection report uncovered what it called "very troubling" environmental pollution on Plum Island, including solid waste, incinerator, hazardous waste, and sewage discharge violations—the same multiple infractions for which the EPA had cited Plum Island a decade ago. Not surprisingly, the USDA hadn't bothered to ameliorate the mess. New York State Attorney General Dennis Vacco filed a lawsuit against Plum Island, and Vacco's spokesman scolded Plum Island: "If the federal government doesn't follow environmental rules, who does?" In a June 2001 court-approved consent order, the USDA admitted violating three sewage effluent limitations. It exceeded permitted discharge by *over 39,000 gallons daily* and flushed out high fecal coliform levels (the dangerous barometer of other toxic animal viruses and bacteria) to the tune of *60,000 gallons a day*. Once again, Plum Island agreed to stop fouling the water. Six months later, not much had changed. In December 2002, the National Resources Defense Council listed Plum Island second on its "Dirty Dozen" roll of the twelve biggest polluters in New Jersey and New York.

How does Plum Island get away with it time and again, and for so long? For one, as mentioned earlier, until 1991 Plum Island was an extremely close-knit family of workers, hailing from a band of sleepy colonial-era villages that preferred tranquillity to making trouble. "Loose lips sink ships," went the philosophy. Another reason lies with its harmless-sounding, and patently misleading, name: "The Plum Island *Animal Disease* Center." The 1999 West Nile virus outbreak and the 2001 anthrax attacks showed the public that humans caught animal diseases, too—and died from them. The USDA presented a formidable barrier of secrecy that newspaper and television reporters—all with ephemeral attention spans—could never fully cross. Whenever its culpability was called into question—which was more often than not—the USDA could always hide behind its shield of government secrecy and national security, no matter how insecure its facilities were. Finally, and perhaps most important, as its own island it enjoys total freedom, and no public accountability. A arm of the federal government, Plum Island has no constituents, or media, or shareholders to report to. And short of a congressional law or presidential executive order, no one can stop it.

The USDA fancies calling Plum Island "The Alcatraz of Animal Disease." This is terribly misleading. Unlike the prisoners of Alcatraz, live germs on Plum Island have made it out alive on two proven occasions, from each of the island's two laboratory buildings. And those are the known out-breaks; many more no doubt escaped detection. Employees have been infected with animal germs, and some have become violently and perma-nently ill. The USDA prefers to shrug its shoulders and demand impossible proof. Plum Island has proven to be an environmental catastrophe and a workplace safety nightmare, chronicled on multiple occasions by its own sister government agencies, the EPA and OSHA. Given the direct facts and circumstantial evidence, the insidious connections between Plum Island and the initial outbreaks of three infectious diseases—West Nile fever in 1999, Lyme disease in 1975, and Dutch duck plague in 1967—are far too coincidental to be dismissed. Liaisons between Plum Island and a top Nazi germ warfare scientist smuggled into the United States under a top-secret military program add a new slate of questions to the already murky formative years of joint USDA-Army biological warfare activities there. The most burning revelations lie in the consistently ineffective biological security measures, which let unknown viruses and bacteria find their way through supposed decontamination filters and systems and into the air and local waters to this day.

All of this should be terrifying science fiction, but it's worse. Because it's all real.

Clearly, Plum Island is no CDC or Fort Detrick. Except in one impor-tant way. Plum Island—funded and run more like a small town high school biology lab than a high-containment virus laboratory—works on germs as deadly as the other two: microbes that cause AIDS, Rift Valley fever, polio, West Nile fever, Japanese encephalitis, swine flu, mad cow disease, to name a few, and countless others in Plum Island freezers (not to mention anthrax).

Looking at the 1956 vintage Lab 101 in full operation, and viewing Plum Island's past to see its future, one thing is clear. Alcatraz is indeed a misleading comparison.

Where and why did it go wrong?

Money. Power. Duplicity. Politics. Glory. All had a hand in shap-ing this twisted animal kingdom. Political support for Plum Island ran from nonexistent to hotly antagonistic to almost laughable. From U.S. Senators Ives and Lehman's 1952 desire for public hearings on the lab's location, to a local congressman's plea to choose an island elsewhere, to Congressman Forbes's 1995 successful fight to halt the controversial BSL-4 upgrade—federal officials have viewed Plum Island as a pariah. Other than nuclear waste depots, Plum Island may be the only federal facility in America where locally elected federal officials flatly refuse to lend any support. Other facilities, like the nearby Groton nuclear submarine base, just across the Long Island Sound in Connecticut, can count on support from their congressmen and senators because of their boost to the local economies and the steady jobs they provide. Not so on Plum Island, where officials vacillate between investigating it (or pretending to) and ignoring it. The community never wanted the laboratory there to begin with, and though many locals—whose sons and daughters and brothers and sisters worked there— buried the proverbial hatchet, the politicians kept theirs in an easy-to-reach place. Had the USDA and Army established their labs in locales where the community and its political leadership actually supported it, like Maine, or Montana, or Washington State, or on St. Thomas in the Virgin Islands, it would have been different. But they didn't. They plunked a biological warfare lab down in a populated region that begged them, *"Please. Don't put it here. Not in my backyard."* Right on the edge of the largest population center in the United States.

When the Army left in 1954, it took its deep pockets and congressional war-hawk caucus with it. Left with no funding from the military complex and zero political support, Plum Island found itself, literally and metaphorically, out at sea, forever swimming in the red with never enough money to meet its skeletal budget. It became a forgotten orphan in an obscure research division of a federal agency that had its priorities—and political supporters— based far away from New York, nestled deep within the farmbelt. Contrast this with the Ames federal laboratory (the USDA's domestic disease version of Plum Island), always championed by Iowa Senator Tom Harkin, who also happens to be the ranking Democrat on the Senate's Agriculture Committee. During its half-century of existence, Plum Island has never had a Tom Harkin. But it sorely needed one.

Plum Island endures a vicious cycle of bottom-of-the-barrel annual appropriations that leads to decisions that cut corners and sacrifice safety and security. In the 1970s and 1980s, Plum Island's engineering and plant management department cited "insufficient personnel" as the prime reason for its lack of preventative maintenance, proper training, and multiple containment leaks riddling Lab 101's roof. In the 1990s, the private contractor Burns & Roe cited financial woes when it refused to staff the sewage decontamination plant twenty-four hours a day, even after a massive toxic spill occurred. Funding issues at Plum Island have not only impaired its performance, but have adversely affected the island's dated, crumbling infrastructure as well.

These types of priority decisions—untenable choices between science and biological safety—made each year on Plum Island are never forced onto Fort Detrick or CDC. The CDC received approximately $2 billion from Congress in 1998. In that year, the Fort Detrick laboratories received $25 million and millions more in in-kind support from the Army. Plum

Island skimped by on about $12 million. It appears that the 2004 Plum Island budget, now under the auspices of the Department of Homeland Security (DHS), will be approximately $16 million. Do animal and zoonotic science and health just not rate? Is it poorly understood or not explained well enough, compared to purely human health priorities? Is there no recognition of the susceptibility of humans to animal diseases accidentally or intentionally released on the population?

When residents protest Plum Island's geographic location as unfit, the USDA time and again points to the CDC. USDA officials say the CDC has been operating safely for years in downtown Atlanta—"CDC is immediately adjacent to a child care center in a very highly populated city," in the words of one official. By inference then, Plum Island, off the mainland in a less populated, immediate geographical area, should be acceptable. However, the USDA overlooks three important realities. For starters, Plum Island conducts its business entirely by itself on a private island off-limits to the public, where there is no opportunity for the slightest oversight of its activities, inside or outside the laboratory. The CDC's very location in downtown Atlanta under the public's watchful gaze keeps it honest, while Plum Island—out of sight, out of mind—is afforded ample opportunity for liberties when it comes to safety. Countless safety and security breaches go unnoticed every day because no one is watching. Second, what is now known about the USDA's privately contracted activities on Plum Island— poor biological safety and island security, a reprehensible attitude toward workplace safety, and a lengthy track record of polluting the environment—speaks for itself. Past events prove the USDA to be an unfit steward of the island's laboratories, bequeathed to it by the Army half a century ago. Finally, Plum Island is a Third World republic compared to the CDC and Fort Detrick in funding, in scientific talent, and in biological safety expertise. Neither of the other institutions has had a germ outbreak; Plum Island has had two verified outbreaks. Neither of them has been assailed for violations of environmental and workplace safety laws; Plum Island has repeatedly run afoul of such important laws.

The Plum Island task force surveyed the island's security, and found two unarmed guards. "A bunch of teenagers were over there—just before September 11th—and [Plum Island officials] didn't even know it," remembers Acampora. "The kids came back and told us they went onto the island." Acampora, a Republican, teamed up with Democrat U.S. Senator Chuck Schumer to issue a bipartisan call for the federal government to refederalize Plum Island. "The federal government is the only agency [sic] that has the expertise and manpower to ensure that [Plum Island] will be monitored by electric sensors and patrolled by boats and helicopters," said Acampora. The time for talk was over. "The health, safety, and welfare of our community and the nation are at stake, and we will prevail." But their plaintive calls fell upon deaf ears. As of this writing, the only change the USDA has made is arming the two private security guards, one at Orient Point and the other on the island, who secure the entire facility. DHS claims to have added some security cameras. As for the other critical island functions? "I think federal oversight has to be there," Acampora says. "At least federal background checks to ensure we have the right people there, like at the airports."

The task force had tried in vain to bring accountability to Plum Island, and disbanded after the BSL-4 upgrade was stopped. While the task force remained active, management placed people at the Orient Point ferry dock to check identification badges, just like they had when the island opened for business in the 1950s. But Acampora found the ID check gone by January 2002. It had all been for show. "There we go again," she says. "They need to have someone with their thumb on them. I don't really think it's my job, but I helped get the ball rolling." If it's not the local assemblywoman's job, then somebody else needs to mind the store where Plum Island is concerned—which is the strongest argument yet for moving the USDA's facility from the creaky old virus laboratory onto a bustling, conspicuous downtown street on the mainland.

THE USDA'S MAGIC TOUCH

You may live three miles, three hundred miles, or even three thousand miles from Plum Island. In every case, the same central question may be asked: would *you* want Plum Island in *your* backyard? Remember—if it isn't in your vicinity, what happens there still has the potential to reverberate throughout the nation: the 1979 meltdown at the Three Mile Island nuclear power plant in Pennsylvania is proof of that.

Plum Island's future has never appeared more nebulous. As part of the emergency appropriation doled out in the wake of September 11, 2001, the USDA received $23 million for Plum Island, which is less than half of what the Ames, Iowa, domestic-germ laboratory received. Despite the smaller appropriation, the USDA maintains the funds are solely for implementation of security improvements—good news, because that kind of money can buy a lot of new air filters, armed security guards, and marine patrols.

In 2002, the Bush administration began the largest reorganization of the federal government since President Truman formed the National Military Establishment in 1947. The new federal reshuffling moved two federal laboratories into a new Department of Homeland Security: the small Environmental Measurements Laboratory in lower Manhattan, which protects against potential radiological and nuclear events, and Plum Island.

There remains speculation that the homeland security funds and the reorganization were a mere cover to upgrade the lab to BSL-4. New York's junior U.S. senator, Hillary Clinton, raised concerns that the upgrade will allow Plum Island to research deadlier germs and endanger millions of New Yorkers. Yet Plum Island *already* works on deadly germs that pose genuine dangers. Of course, there are a few supporters of an upgrade. "Plum ought to be the leading site in the country for animal disease defense against these foreign threats—accidental or deliberate—it should be the leading lab," huffs Dr. Roger Breeze. "If the county doesn't want us to have [a BSL-4 laboratory], we won't have the defense, that's fine by me—I just need to get on with my business." Breeze envisions the new improved Plum Island boasting "ten times as many scientists than are there right now."

Plum Island officially moved into the new Department of Homeland Security on June 1, 2003. At the time of this writing, Dr. Huxsoll packed up and left, and there is no director in place on Plum Island. All media or public inquiries about the island are referred to the DHS public relations office in Washington. The homeland security legislation included a provision that the president must notify Congress 180 days prior to any change in the island's biosafety level. The clause is eerily reminiscent of the 1952 Plum Island public hearings provision inserted by Senators Lehman and Ives— and we know how that one turned out. It's all too clear from the law that once 180 days expire, Plum Island may be upgraded to biosafety level four at the whim of the federal government, no matter how strongly the people and their elected representatives protest.

In March 2003, Plum Island Assistant Director Carlos Santoyo told the Associated Press that the transfer of Plum Island from USDA to Homeland Security would be little more than a "paper transfer." Nonetheless, moving Plum Island out of the hands of the USDA and into the DHS might raise the island's profile and could get it the funding it so desperately needs. Throwing money at the problem won't solve it, but in this case, spending large sums reasonably and responsibly may be the only way to resuscitate the moribund island. Knocking down Lab 101 and building a modern containment facility from scratch would seem to be a good start, but a wrecking ball may not be possible. Lab 101 has become so virus-ridden over the years it may have to be fumigated, locked up, and mothballed, like its older brother, the ghostly Lab 257, and its first cousin at Fort Detrick, the boarded-up 1950s-era "Anthrax Hotel."

There is one reason above all why transferring Plum Island from the USDA to the DHS might be the best thing that ever happened to Plum Island and its threatened environs: the USDA itself.

Plum Island has taught us that veterinarians aren't all of the Dr. Dolittle variety, and that there's plenty more to the United States Department of Agriculture than USDA Grade-A eggs and the food pyramid (or, if you're a traditionalist, the four food groups). Most people know that agriculture relates to a most basic human need: food. But beyond that, what exactly is agriculture? Traditionally defined as the science of crop and livestock production, it represents a sea change in humanity—from that of primitive hunter-gatherer to organized planter and rancher. This science is relatively brand new; a study in 1968 noted that agriculture has been practiced for less than 1 percent of human history. Over the last half-century, the USDA has emerged as its greatest power and tireless promoter, taking modern science methodologies and applying them to foods grown and raised, processed, and shipped into the American marketplace. In a postwar fervor that can best be seen as a scientific "Manifest Destiny," the aggies aided and abetted the demise of the family farm in favor of agribusiness, trading localized production for the largest units of mass food production obtainable. Like a "boiler room" stockbroker, American agriculture has become enslaved to the incessant march of increasing numbers. In 1976, the average U.S. farmworker fed fifty-two mouths, while the Russian worker fed only seven; livestock production per animal was up 130 percent from the previous fifty years. That was a quarter-century ago—since then, production has grown exponentially.

The USDA has accomplished this growth through proselytizing the new religion of science with an uncanny blind faith, ignoring any fallout over the last half-century. It claimed with certainty that its chemical DDT would not contaminate the wildlife and marine ecosystem. Thanks in large part to Rachel Carson, the USDA now admits it does.

The USDA championed chemical pesticides—scientifically engineered to quell insects that threatened exponential agricultural growth—and maintained they were food-friendly. But now, it grudgingly admits the chemicals are infused into the food products people eat, and wreak havoc upon wildlife.

The USDA said that fertilizer nitrates used on large farms would increase crop yields, and they would never reach groundwater aquifers. Today, the USDA admits that nitrates have seeped into and blighted potable water supplies in dangerous concentrations.

The USDA promoted dieldrin—a compound twenty times more toxic than DDT—to eliminate the Argentinean fire ant in a $156 million federal crop spraying program. Over time, the fire ant grew more resistant to pesticides and tripled its original geographic range. Says Smithsonian historian Pete Daniel, "The lost war against the fire ants should have been a cautionary, a warning against hubris and unintended consequences. Instead, it typified a mind-set that substituted faith in science and bureaucratic expertise for common sense."

The USDA has always denied it had anything to do with biological warfare on Plum Island, but such assertions ring hollow. There is its early germ warfare work for the Army in the 1950s, and the joint USDA-Army work on the Rift Valley fever virus in the 1970s and 1980s. A July 13, 1992, Plum Island visitors' manifest lists fourteen visiting Joint Chiefs of Staff, Army, and Pentagon officials. According to the document, the purpose of the visit was "to meet with [Plum Island] staff regarding biological warfare." The visitors were part of the Arms Control and Disarmament Agency reviewing the dual-use capabilities of the facility.

The Soviet Union didn't buy the USDA line. Former Soviet biowarrior Dr. Ken Alibek wrote in his 1999 book *Biohazard* that Plum Island "had figured in our intelligence reports for years. [It] was used during the war to test biological agents." I ask Alibek to explain those reports about Plum Island. "We had these 'Special Information Reports' that came from KGB and GRU [Soviet military intelligence] that described Plum Island, Fort Detrick, and Dugway Proving Ground as biological warfare sites.

"There's no question that Plum Island was a threat—you know, at that time—I cannot even translate for you in Russian how 'Plum Island' would sound. Today I know what a 'plum' is and what 'island' is, but then," he says, laughing, thinking of the fear the strange-sounding name instilled. "Someone had told me they worked in exotic livestock diseases, foot-and-mouth virus and such." As part of a 1994 arms agreement brokered between Russian President Boris Yeltsin and the United States, each side was permitted to choose three suspected germ warfare sites for full inspection. When the Russian team arrived in February 1994, they visited their first two choices while aboard *Air Force Two*: Vigo and a Pfizer pharmaceutical plant in Groton, Connecticut.[3] Then the team announced their third and final snap inspection: Plum Island. Dr. Breeze ventures a guess on why they chose Plum Island out of all places in the United States. "It used to be a military base with all these old underground bunkers, right across [the Long Island Sound] from a nuclear submarine base. We were doing a lot of reconstruction, and I think they looked at their satellite photographs and saw all this construction going on."

Upon arriving, the Russians first demanded a tour through the Army bunkers, searching in vain for a trapdoor leading to an underground complex. Then they walked around Lab 101, blueprints in hand, pointing and insisting to be shown the basement, except that the portion of 101 they

[3]Vigo was a long-abandoned World War II–era Army germ weapons plant in Terre Haute, Indiana, that had since been sold to Pfizer.

were interested in didn't have one. The plans showed the area as an "unex-cavated" crawl space, yet the Russians refused to believe it. It was plainly apparent to the Plum Island staff that not all the Russians sniffing around were scientists. Though Dr. Alibek defected two years earlier, he confirms that inspection team members Grigoriy Shcherbakov and Aleksey Stepanov were two of his subordinates working on weaponizing genetically engineered anthrax. Dr. Carol House recalls, "A couple of them were green-beret types, who would take anybody out without thinking twice, and there were some intelligence agents, too. It wasn't well hidden."

Then there was a question-and-answer session on the viruses being studied there. Team leader Oleg Ignatiev, a short man with a low, booming voice, told *Newsday* reporter John McDonald afterward, "We find this is a very large center. It does very important work—but we have to work on it and think it over before reaching a conclusion." Days after the Russians returned home, the newspaper *Izvestiya* reported that the United States was in violation of the Biological Weapons Convention.

According to Alibek, the violation was mutual. The Soviet Union had a biological warfare program against livestock, code-named "Ecology," that boasted thousands of scientists in multiple facilities. One Ecology-designed weapon he heard being discussed was called R-40. "Whether it was some new fever or new foot-and-mouth disease virus, I don't know. What was interesting is that this weapon agent was cultivated in cattle; then, when the virus concentration was high, they bled them and made weapons from cattle blood." Primitive science, but deadly.

The Soviet threat remains strong. Many of the program's scientists, faced with starvation or working for rogue nations, chose the latter. And to this day, Russia still has three "former" biological warfare facilities that continue to be off-limits to the West. "Closed," says Dr. Alibek in his gruff Russian accent. "Top secret—and nobody knows what they do behind those closed doors." The former biowarrior now works for a Virginia biotech company with Pentagon contracts, developing defenses against the recombinant DNA germs he helped create, such as an antibiotic-resistant strain of anthrax. He's even visited Dr. Huxsoll on Plum Island to interest the USDA in some business. Dr. Alibek says of Plum Island, "It's a struggling facility trying to find adequate funding. This is what I know."

The catastrophes at Plum Island did not occur in a vacuum. The single-minded zeal of the USDA's science mantra ends in wreckage far and wide. What happens when the blind pursuit of science is taken too far? When scientific glory is pursued for glory's sake? In Plum Island's case, the not-so-invisible hand of government prioritizes an abundance of food (really the livestock from which food is derived) ahead of its human work-force and the public health. Plum Island runs at the sufferance of deep-pocketed agribusiness. One political scientist put the USDA's thinking in no uncertain terms. "Critics inside the establishment are not appreciated, and outside criticism that cannot be dismissed as malicious, romantic, or uninformed is viewed as trivial in the context of agriculture's record of increased food production." There are no signs that DHS will change this modus operandi.

Then there is the question of the animals themselves. Plum Island has sacrificed hundreds of thousands of them, in all shapes and sizes, in an ambitious effort to tame foreign animal germs. Yet fifty years later, there is still no cure or silver-bullet vaccine for foot-and-mouth disease virus—Plum Island's cause célèbre—or any of the other germs studied there. Is all this animal carnage necessary? If the modest success of the past is a barom-eter of future successes, productivity at Plum Island needs to be addressed.

The best solution may be not to rebuild on Plum Island, but to pack up instead and leave the mess behind. In this case, the USDA and DHS should consult those communities in Maine, Montana, Washington, and the Virgin Islands that *actually wanted* the lab during the first go-around in the 1950s. Generations from now, after the germs have long subsided, Suffolk County can establish that magnificent park preserve it wanted to build on Plum Island before the Army canceled its sale of the island.

In its final battle, Pat Acampora's task force faced down mad cow disease. At an early morning meeting in Acampora's office, with Dr. Breeze present, USDA officials were asked if Plum Island had ever studied bovine spongiform encephalopathy, better known as mad cow disease. This animal infection's human variant, Creutzfeldt-Jakob disease, gnaws away at brain tissue and has a 100 percent mortality rate.[4] The Plum Island officials told the group they had not worked on mad cow and weren't equipped to study it. But a few months later, when 255 sheep in Vermont were suspected of contracting mad cow, the USDA made plans to ship the carcasses to Plum Island for necropsy, tissue and blood samples, and incineration. Tipped off of the plan, Acampora went livid. When she confronted Plum Island about it, the USDA told her, *"Well, the situation has changed."* "They don't even have a state permit for their incinerator!" she roared, and threatened a

---

[4]Mad cow disease and Creutzfeldt-Jakob disease are caused not by live germs, but inanimate "prions," free-floating snippets of infectious DNA material that lack the structure that characterize viruses and bacteria. More resilient than any known germ, they resist the decontaminating processes of steam autoclaving and even incineration. Little is known about prions, but they are known to be extremely dangerous to animal and human life.

showdown: if they carted the sheep carcasses through Manhattan or Connecticut onto Long Island soil, she would get the police in Nassau and Suffolk counties to barricade the roads.

Fearing a public relations debacle with national implications, the USDA backed off, and trucked its infected sheep refuse to the Ames, Iowa, laboratory. "It became such a hassle, they decided not to do it," Acampora recalls with modesty; the truth is, she was the hassle. So many others—politicians, media, community groups—had failed over the years in trying to fight Plum Island. Finally, the task force had won a battle. It was a start.[5]

"Look at our space shuttle," Acampora says. "We spent billions of dollars on the space mission, hundreds of engineers, technicians, scientists, and safety experts to make sure the flight was fail-safe. We watched in awe. We clapped. And then it exploded in horror. That's what I'm worried about." Even to this day, while other local officials have accepted tour invitations, the feisty assemblywoman won't go to Plum Island. Why is that? "It's really a simple reason," she laughs. "They say that when you go there, you have to scrub down, and get all washed up. And no one's going to see my hair wet." One gets the impression her fears go far beyond a bad hair day. "There have been a lot of mishaps out there," she says, "and let's face it—Long Island has a *lot* of wind, and should something happen there, and the wind's blowing in the wrong direction, that could be pretty problematic."

"When you are a mile off the coast of a heavily populated area, an area for which we have no real evacuation, I have a concern." Indeed, there is only one arterial road off the North Fork and one out of the Hamptons, spilling into the heart of Long Island. From there, Long Island's 6.1 million people bottleneck into ten narrow bridges and tunnels that themselves empty into the cluttered congestion of New York City, where 11 million more people reside.[6] Indeed, Plum Island lies on the periphery of the largest population center in the United States.

"You know, a lot of people really don't know Plum Island even exists," Acampora muses as her smile fades to a smirk. "People move here from New York City, they come here, play here—and they have no idea what's going on one mile off the coast."

Whether the nation's exotic animal disease laboratory should remain on Plum Island remains the central question. "When the Manhattan

---

[5] Later, the USDA disclosed that tissue samples from the infected sheep had been sent previously to Plum Island and, unknown to the task force and the local community, were used in tests there.
[6] As of 2000, the island of Long Island holds 2.6 million people in Nassau and Suffolk counties, 2.2 million in Queens, and 1.3 million in Brooklyn. Over 8 million people live in Manhattan, 2.5 million in The Bronx, and 450,000 in Staten Island.

Project was developed during World War II," Karl Grossman points out, "they selected places that would be safe. The Manhattan Project laboratories were inland, at places like Oak Ridge in Tennessee, and Los Alamos. If it was a small island in the Pacific doing very sensitive research, through radar you could make sure no vessels came within a certain distance. But Plum Island . . . man." Grossman lets out a long sigh. "Plum Gut on a summer day is a main street for boaters and fishermen. If they aren't moving in, they are moving out. You can't screen [marine traffic] until they are onshore. There is a war consciousness today—is it appropriate to have [the laboratory] on an island that is jutting out there in the Atlantic, with lots of traffic?"

Can Plum Island's facilities be moved to the mainland, like the National Academy of Sciences urged back in 1983? Asked this question in his Plum Island office, Dr. Huxsoll says, "With the biocontainment technology today, you can probably put this anywhere. But the stakeholders are more comfortable with it being there because of the geographic location." Yes, the "stakeholders"—a fancy euphemism for the $100 billion agribusiness empire that claims this research is essential, but won't spend a dollar of its own money to do it itself. "The reason Plum Island survives today is money," says ex–union chief Ed Hollreiser. "The cattlemen and pig and poultry farmers want it continued."

"You could put it in the middle of New York City and obtain the right biosecurity," Huxsoll asserts. Sounds crazy, but he's right. While it sounds ironic, the work is safest not on an island but right next door, like the CDC in Atlanta, where tens of thousands of people mill past it each day. Why? Because that kind of closeness demands (and in the case of the CDC, enjoys) the most stringent protocols of safety possible. If past history has taught us any lessons, it's that Plum Island desperately needs a babysitter.

If Plum Island should go, there will be yet another accompanying consequence, though perhaps not as weighty as national and regional safety and security—the demise of a local economy. Dr. Breeze's ploy to split the workforce between New York and Connecticut crippled a good part of the east end tax base, and far fewer federal dollars were spent locally. With the departure of those skilled and professional positions remaining on Plum Island, the "brain drain" of educated folks will continue, leaving behind empty nesters and dwindling farm families.

The government closed Lab 257 for good in the spring of 1995. They cited the building's outdated biocontainment facilities and crumbling old age—eighty-four strenuous years of service under its belt, forty-one of them as a exotic germ laboratory—as reasons for the shutdown. All of Lab 257's laboratory operations were consolidated within the other research facility, Lab 101 on the island's northwest shore.

Although scheduled to be fully decontaminated and demolished in 1996, Lab 257 still stands today, rotting from weathered decay, harboring who knows what deep within. On one of my voyages to Plum Island, I had an opportunity to investigate it up close. It was an eerie, postapocalyptic scene. Corrugated metal plates clamored loudly against the lab's wall. Scores of rusty pipes sprouted out of the flat tar roof. A lonely two-story skeletal extension was partially connected to the near side of the building, exposing rusted beams. Pipes snaked around the insides, some twisting into air vents, others ending in midair with no connection. Two short, thick pipes rose along the side of the building, like turbines in a ship's boiler room.

The door to the gatehouse hung open, creaking away in the whistling wind. I carefully inched inside. On the right was a padlocked wooden box marked SUGGESTIONS. The guard's desk, straight ahead, was covered with a thick coat of soot. There was a flip calendar on the desk, one page half-turned over—paused on January 13, 1995. The clock overhead, stopped dead at 9:12 and some seconds. Leatherbound logbooks were piled haphazardly in the middle of the desk, just below the calendar. Sand blown in from the beach was piled in five-inch drifts at the floor's corners. It was a ghost lab.

Nailed to the door leading into the lab compound were rusted warning signs, stamped BIOHAZARD. They were signs reminiscent of Lab 257's shadowy past.

And a mile away, Lab 101 marched—and continues to march—on.

15

A Plum Island Prescription

For all of its warts, we still need Plum Island. Or a place like what Plum Island ought to be, where America can fight the emerging threat of biological terrorism against the nation's food supply—called agroterrorism in the new parlance. Back in the 1970s, a biological warfare expert told the *Newsday* investigative duo Cummings and Fetherston, "If you want to use biological warfare, you would be much better at striking livestock. If you use nerve gas on an army, you just kill the soldiers. If you destroy or damage a country's food supply, you strike at everyone." Those in power didn't listen then, but after the 2001 anthrax attacks that infected eleven and murdered five, bioterrorism became a new vocabulary word. Floyd Horn, shifted from USDA research czar to agroterror chief at the Department of Homeland Security, said this of the threat in May 2002:

> It is dead serious, and we really are at war. They want to get us, they want to get our economy, they want to kill us, they want to make us sick, get us out of the global marketplace and they want us to get out of their part of the world. And they have very few constraints about how they want to go about it. . . . When people lose their faith in government to protect the food supply, there's big trouble. This is ice cream for the average terrorist, and it's something we have to worry about every day here.

When Sultan Bashiruddin Mahmood was caught red-handed in Kabul, Afghanistan, with a dossier on Plum Island, Horn's startling warning became a grim reality.

Tom Ridge, secretary of the DHS, said in June 2003, "I understand completely that [Plum Island] is the first line of defense in agroterrorism. I mean, that's where we're going to see the contaminants. That's where we're going to identify the pathogens. . . ." And, said another DHS official, the aim of Plum Island "is not just science, it's protecting U.S. agriculture from agroterrorism." If this is what DHS chiefs expect from Plum Island, they may end up shortchanged.

Karl Grossman used to think it was a joke. Always the thorn in Plum Island's side, he playfully wrote in a 1995 column, "As to what nation has been interested in poisoning America's pigs and assassinating our chickens, that has been very unclear." But after September 11, 2001, and revelations of an Iraqi anti-agricultural and antipersonnel germ warfare program, Karl has had a change of heart. "After September 11th, it seems to me—finally—that with these people doing crazy things, that al Qaeda and the Islamic extremist movement attacking food and livestock is a real possibility. We are dealing with a challenging, inventive, and nutty enemy. So the defensive biological warfare mission on Plum Island suddenly has—and I don't know if it should be on Plum Island, that's another issue—but the mission itself suddenly has validity."

While calls to improve homeland defense are focused on airports and biological attacks on people, a ready-made terrorist opportunity lurks elsewhere. A terrorist could use germs targeted at animals, but like Rift Valley fever, the animal viruses could be zoonotic agents that infect animals and people simultaneously. Infecting domestic (and wild) animal populations that act as vectors would spread and amplify the disease in the very same manner that Lyme disease, West Nile virus, and Dutch duck plague spread from the New York area to the rest of the nation in a matter of years. Economically speaking, a nonzoonotic virus—one affecting only animals—directed against domestic livestock would be nearly as damaging.

Though we require it to live, food is a luxury taken for granted in America. For millions of urban, suburban, and exurban dwellers, food comes from the sparkling clean thirty-five-aisle supermarket, or the Friday's chain bistro in the shopping mall, or the McDonald's just off the highway exit. Rarely do we ever think about where our food comes from. It's that ignorance that terrorists will look to exploit. Recent events serve as a harbinger for things that may come.

t would be very easy to destroy the American food supply. In fact, if you wanted to, you could do it all by yourself. I sat down with a veterinarian-microbiologist named "Todd Barker" and discussed this possibility. "It's amazing what lab equipment you can buy on-line," says Todd. Websites like www.labjunk.com sell old equipment at bargain-basement prices, and from there you can start assembling your very own kitchen laboratory. Over the Internet, you can purchase anaerobic fermenters, a good centrifuge, a pressure cooker, bacterial media, and a milling device. A crude kitchen lab environment will be messy and things tend to spill, so you would want to go to a feedstore out west to pick up antibiotics and vaccines meant for cattle. "You'll get a sore arm and a reaction," says Todd of the crude animal prophylactics. But it will still protect you while you're playing mad scientist.

Then you will need a starter germ, and all that takes is some fake letterhead. "You could write someone and say you are a scientist from a phony lab, and you need it to test and make a diagnosis—and you'll get it. It's a common courtesy." If that doesn't come through, you can always go abroad and obtain viruses like foot-and-mouth disease virus and rinderpest in European and African countries where they are endemic, like Dr. Erich Traub did in Turkey during World War II for SS Reichsführer Himmler. If those two options fail, you can try to follow the Long Island researchers who recently created the first synthetic live virus (polio virus, to be exact): by using genetic recipes available over the Internet and mail-ordering the component gene snippets.

Then you set your target. Modern agribusiness presents the nation's most vulnerable bioterrorism target. In the last fifteen years, virtually all livestock farms have been vertically integrated. Take swine, for example. There is no national tag system. Pigs go into a sale barn where they mix with thousands of other pigs and then ship out. Every pig that is reared and fattened for slaughter has some one thousand miles of travel on it. They go from Arkansas to North Carolina to Indiana, then to market in Chicago. Over twelve thousand pigs move out of North Carolina a day, and they head out for contract barns in ten different states; if it has to, a pig producer will move pigs two thousand miles to a contract barn with an open slot. Two weeks later, it's off to twenty-odd swine-producing states. Using sophisticated computer inventory systems, it is no more difficult than Home Depot moving hardware inventory between store locations.

Cattle ranches orchestrate similar movements of "product" through sale barns to large stockyards. A million head will go through sale barns to market in a span of three months. Dairy "farms," as they are still called, are no different. Comparing today's dairy industry to his roots growing up on a small dairy farm, Dr. Huxsoll says, "You wouldn't recognize a single similarity—except maybe the presence of a cow." Several thousand cows are milked mechanically twenty-four hours a day, nonstop, on a mere forty acres of land. It's not really a farm anymore. Their milk goes directly to a six-thousand-gallon tanker truck; when the tanker's full, it pulls away and goes to the dairy processing plant to cool for a while. This is where the vulnerabilities arise. Sixty vehicles drive in and out of those dairy operations every day, bringing in feed, hauling off milk, driving workers to work, you name it. And germs travel extremely well when suspended in milk. Many of the sale barns are wide open places, and only a few have even token security, says Todd. "These operations claim to have real biosecurity, but it's superficial." Infiltrating these facilities, then, is not all that difficult.

But Todd has a far easier target for you: a cattle truck. There's not much biosecurity to a cattle truck sitting at a rest stop, on its way to a contract farm or a sale barn. Your compact rental car can pull right alongside the truck loaded full of animals after the driver steps into the Waffle House for a plate of eggs and cup of coffee. You can walk up to the side of the long open-corral trailer with your test tube of—insert virus of your choice here, say foot-and-mouth disease virus, held in liquid slurry form—uncork it, and swat its contents inside, like a priest anointing his congregation with holy water. On roughly the seventh day, results should materialize.

Your macabre blessing, unleashed on a truckload of cattle, just created a wide-scale food shortage and in so doing, proceeds to tank over one-tenth of America's total economic output.[1] You've sent America a brutal reminder of where its abundant food comes from, and created national mass hysteria in the process. Says Dr. Huxsoll, "If you had a major agricultural catastrophe, you'd see the disappearance of a lot of family farms and the disappearance of a lot of small rural communities."

Attacking the American food supply is the simplest form of biological warfare available to a terrorist. United States agriculture is vertically integrated, accessible, and concentrated. The fact that American livestock has been fortunate enough to be disease-free has a downside. Not being needed, vaccines aren't widely used, making herds extremely susceptible to infectious foreign germs. And because of the high level of animal intermingling and herd combination, recombination, and consolidation, there's no way an infection can be limited to an isolated occurrence, and there's no way it can be traced. "We're such a wide-open target, it's just unbelievable," Todd laments. "Because we're so damn incredibly efficient. Agribusinesses move pigs like Wal-Mart moves toilet paper and motor oil." There's signif-

---

[1]The American livestock industry makes up 13.3 percent of the entire U.S. gross domestic product, accounting for $100 billion in sales every year.

icantly less bioprotection in place for animals than there is for humans; and there's no expensive research and development, mass production, or bomb delivery device prerequisites either. A kitchen laboratory and chemistry textbook would suffice. If left unguarded, we won't know what hit us until that animal truck strewn with disease begets millions of hungry and panicked Americans scouring empty supermarket shelves coast to coast.

The worst part of this scenario is that it pales in comparison to what the al Qaeda terrorist network or a well-trained scientist-cum-terrorist already has in mind, waiting for the right moment to unleash upon an unsuspecting—and strikingly vulnerable—America.

In August 2002, CNN broadcast excerpts from al Qaeda videotapes showing three tests of a white oozing liquid being unleashed on Labradors, romping around in a closed test room. On film, the trapped dogs quickly turn dour. They begin to yelp and writhe in pain and go on to suffer agonizing deaths from these mysterious chemical or biological agents. Made available to a CNN reporter outside of Kabul, Afghanistan (home of Bashiruddin Mahmood's "charity"), the disturbing tapes show "meticulous planning and attention to the tradecraft of terror," in the words of terrorist expert Magnus Ranstorp. Recently, an Algerian terrorist cooperating with U.S. authorities testified he was taught by al Qaeda how to smear biological toxins on doorknobs. The techniques are similar to those used by German terrorists who smeared glanders bacteria on the gums of horses in Maryland, Pennsylvania, and New York during World War I.

If you don't think agro- or bioterrorism could be coming soon to an American town near you, you need to think again.

A former military intelligence commando, Gary Stubblefield runs a consulting firm that specializes in ultra-high security. Four years ago he was writing a report on the national security threats of radiological emission devices (REDs) and national food security, warning of possible terrorist threats. "All the national labs just looked at us and laughed," he recalls, when they read his report. "Said it was ridiculous." Today, no one is laughing, and the hulky Stubblefield is spending seventy-five hours a week flying all over the world for the Pentagon to stymie the threat of REDs, or "dirty bombs." Yet with all his time spent confronting the dirty bomb threat, he hasn't forgotten about the food security issues he also once raised. "They laughed us out of the room back then on that one, too," he recalls. But people are starting to realize that our food supply may be the next target.

"Here's one of the scary things," Stubblefield says. "Those guys watch CNN just like we do, they see what happened economically to us, and how we reacted to that small anthrax situation—and I only say small because it didn't spread widely. They recognize this, and it has been written in their intercepted literature that 'We can bring down the giant economically, and that too will kill him.' If they can find a way to stop our food supply, they believe they will bring America to a halt." Stubblefield says that since its inception, al Qaeda's modus operandi involves simultaneously attacking multiple targets, which can include planned attacks on our food.

The state of Montana decided to get serious. After September 11, 2001, officials asked Stubblefield to come up with a plan to protect the agriculturally rich state from agroterrorism. Among other defense options, Stubblefield proposed fiber-optic detection cables, run along fence lines. "If anything gets inside," he says, "an alarm will go off and tell you exactly where it was set off." Though Stubblefield's fiber-optic security solution is effective and easy to install, the cable is fairly expensive, and no one is stepping forward to pick up the tab. "The rancher doesn't feel he has to pay because he says it isn't his problem. His mission in life is to raise stock and sell it—they don't understand this. I don't think they have a clue." What about the USDA? "The feds are saying they can't do it because it's not their responsibility either. Everybody is sidestepping the issue." Montana hasn't implemented a full agroterrorism program, and no other state has either. Is the food supply close to being protected? "We are quite far away," he says. "I think we are still at the discussion level, and that is shortsighted. We should be more proactive."

One of the most insightful analyses of ways to defend against agroterrorism comes from Anne Kohnen, who designed specific recommendations in her October 2000 Harvard master's degree thesis. Kohnen proposed a four-level counterattack, from the organism level to the farm, sector, and national levels. The report requires Plum Island be upgraded to BSL-4 to combat all exotic animal diseases. When asked about the impact of her thesis—which is perhaps the most comprehensive report publicly available on the subject—Kohnen, who now works on nuclear proliferation, laments the inattention agroterrorism still receives. "Until now, it has been overlooked. Not of much interest to the policymakers—it's not a sexy topic like arms control or international trade. It hasn't received much press. But I don't think that's a bad thing—I'd just as soon have fewer articles out there about how easy it is." Gary Stubblefield takes the opposite approach. "I don't like putting these things out in the news. But sometimes it doesn't hurt to advertise that we are working on things." Speaking of our homeland security, Massachusetts Senator John Kerry said on *Meet the Press* that there are "enormous gaps and deficits in the preparedness level. . . ." Given the lack of preparedness, federal and state officials need to work much harder.

n the wake of the September 11 attacks on America, homeland security has become paramount to the nation's survival. Plum Island can play an important role in securing the home front. But not the Plum Island that exists today.

There are two worthy options for the island's future: closure or total rebirth. In 1983, the National Academy of Sciences implored the USDA to close Plum Island down, eight years before it was foolishly privatized. "Limitations of time and budgets, shortage or lack of primary containment equipment, and lack of a sufficient number of engineering and safety staff trained in biomedical containment facilities have made it difficult. . . ." Had the public known of this report, the shop would have long been closed. But they hadn't, and Plum Island still limps along, plagued by those same limitations and shortages. Plum Island management is asleep at the wheel, waiting to be jolted awake by an employee dying from contact with deadly exotic animal virus, or by a scientist burning in a laboratory fire the volunteer bucket brigade couldn't put out because the emergency exit door was locked, or by large numbers of swimmers in the Hamptons coming down with strange untraceable illnesses, or a boater in the Long Island Sound hooking a box full of test tubes.

The second possibility, a born-again Plum Island, must take strict measures in biological safety and biological security, truly separate matters that are too often confused. Science, security, and safety must go hand in hand and rate equally; never should one be sacrificed for another. Should there be an imbalance, it must tip in favor of security and safety, and rather than the pursuit of science.

In this end, a "Plum Island Prescription" follows:

Biological Safety—It begins with the transportation of virus and bacteria samples to Plum Island. When the USDA convinced Congress to let it transfer viruses over the mainland in the late 1950s, it duped the lawmakers by saying there would be only a few trips a year. That wasn't true then and it's not true today, as samples make their way anonymously along highways and local roads multiple times *each day*. Dr. Callis used a USDA courier to transport samples, saying, "We do not trust the mails." Today, Plum Island is far more trusting, entrusting germ samples to the U.S. Postal Service, airport limousines, and private shipping companies like Federal Express. Drivers don't know their packages' contents, nor are they trained on handling an accidental spill.

The USDA's private courier is marginally better. The precious cargo is placed in the trunk of a compact car or in the backseat of a Dodge Caravan minivan, and an unarmed driver takes it for a two-hour spin from John F. Kennedy International Airport to Plum Island. Nothing will break in the sturdy containers if dropped from heights of up to twenty-seven feet. Physics dictate, however, that a dropped box from that height will be traveling at 29 miles per hour upon ground impact. A 60 mph car crash will subject the same box to twice the force. That's far from unbreakable.

Virus shipments should arrive either by the original method of Navy ship transport or by armed escort over New York City and Long Island roads. In this case, notification of each trip must be given to emergency responders like fire and police departments, along with the nature of the samples and what to do in case of an accident (including a fatal one, like the deadly accident in 1970 involving the Plum Island germ courier).

It should go without saying that biological samples should brought immediately to Plum Island and locked away in storage and never be left on the side of Route 25 or in the Plum Island mailbox. At 7:05 p.m. on May 19, 2002, Southold town police received a worried call from the Orient Point private security guard. Someone had placed a suspicious-looking package inside the oversized black Plum Island mailbox along Route 25. The road was cordoned off, and 350 passengers on two Connecticut ferries were detained on the boats for more than three hours. The police called in the Suffolk County emergency services unit, who donned protective gear to rip open the mailbox and expose its contents. "After this examination," said the police report, "the items found in the mailbox were determined not to be of suspicious nature." The package was an envelope containing junk mail. Had it not been a false alarm, a three-hour response such as this would have posed tragic consequences.

The Plum Island safety office needs to be taken seriously and augmented with seasoned professionals, returning to the policy that the director of safety is a trained scientist. The Plum Island frying pan affair clearly demonstrates this. As part of a 1993 renovation of sixty thousand square feet in Lab 101, the windows needed to be replaced. The entire wing was opened to the atmosphere for the first time since 1956. The safety department hatched a jerry-rigged decontamination operation using two hundred electric frying pans, three hundred power extension cords, eighty rolls of duct tape, and a bunch of household fans. With frying pans placed on the floor every few feet, the safety office heated eighteen-gram blocks of solid formaldehyde to 450 degrees Fahrenheit for one hour, turning the solids into a poisonous decontaminating gas. When the cloud dissipated days later, they would collect 600 test strips of the *Bacillus subtilis* spore, a germ known for its hardiness (the same germ used in Fort Detrick's germ warfare tests in the New York City subway), to see if the formaldehyde had fully decontaminated the area. Climbing into the air lock and collecting the test strips, workers in space suits found that more than 120 of the strips turned murky with bacteria spores. The decontamination failed. On the second try, poisonous gaseous formaldehyde, in concentrations far greater than those that sickened Fran Demorest, leaked into the adjacent lab and forced a full evacuation to the Army chapel. Again, Plum Island escaped the reaper—no one was seriously injured.

What caused this potentially deadly leak? It was something that plagued Labs 257 and 101 often in the past, particularly during Hurricane Bob. On the day of the second try, a power outage just after 7:30 a.m. shorted out the negative airflow system and stopped the forward airflow. Soon after, thirty unsuspecting people working in the laboratory smelled a pungent odor, and their eyes began to tear. No viruses escaped, the USDA said. Just formaldehyde gas.[2] The USDA blamed it on the Long Island Power Company, which in turn blamed it on an osprey that crashed into a power line, disrupting the power running through the undersea cable to Plum Island. Neither mentioned why the emergency generators at the power plant failed to kick in. The same people that cooked up the frying-pan fiasco still run the skeletal Plum Island safety office. This critical department, arguably the most important on the island, must be staffed with a multitude of professional experts, rather than a handful of *McGyver*-like handymen solving biological problems with pots and pans.

Fire safety remains an important issue. With a single dedicated fireman and a bucket brigade of employees (performing other responsibilities) who are supposed to sprint to the firehouse at the sound of an alarm, the ability of Plum Island to control laboratory fires seems dubious at best. According to workers now on strike, the contractor was essentially forcing workers to be part of the fire brigade. "The sad part about it," says one worker, "is that we don't want to—and we never went to school to—fight a fire! And fire training today is a fifteen-minute confined space rescue drill." On top of that, it appears that crucial mutual aid from nearby Long Island in the event of an out-of-control blaze is a nonstarter. Many firefighters, all community volunteers, privately admit they will not go to Plum Island to fight a fire, fearing biological infection, regardless of any mutual aid agreement. Trained to chop holes in roofs to put out fires, the Long Island fireman would have a far different task inside a biological containment laboratory that must remain sealed to the outside world. There was to be comprehensive fire training on Plum Island—possibly the first of its kind—for the Long Island fire departments. But the training was canceled after the strike.

---

[2] In an amusing aside, this only-on-Plum Island disaster occurred the same day the battered Dr. Jerry Crawford announced his departure. One can picture Crawford, packing up his desk, breathing a long sigh of relief.

"Even if they went to the fire," says a former Plum Island fire brigade member of the local fire departments, "how long would it take to get to the island? At full steam that's still about forty minutes in response time." At the very least, then, the bucket brigade should be disbanded and the professional fire department reinstated.

My interview with Dr. Huxsoll in his Plum Island office—before he departed, leaving no one on site in charge of the Plum Island helm for months—adds further weight to the vision of chaos that surrounds the island. Our meeting was conducted in stifling heat because, as luck would have it, there was a power outage that morning, forcing the evacuation of all nonessential personnel. I entered Lab 101 to a commotion: workers arguing about how to safely store the arriving boxes, loaded with live biological samples, that had traveled across Plum Gut on the same ferry as me. Because of my meeting with Huxsoll, I was spared from the evacuation and rushed up to his office. We sat there speaking while sweating profusely, both pretending nothing was wrong. We did our best to ignore the loud piercing beeps that periodically blasted from the hallway, reminding us that something was going very wrong, again, on Plum Island.

The safety office should also check employees regularly for exposure to pathogens and provide assistance with medical diagnosis if exposures occur. Plum Island must adhere to the 1982 outside safety review committee report issued in the wake of the virus outbreak: "[I]n the event of active suspected laboratory-associated infections, however, it is recommended that diagnostic examinations be conducted at a facility other than [Plum Island]. . . . Examinations of active cases should be conducted with the knowledge and concurrence of the USDA Medical Officer and the employee's private physician." Two decades later, the edict remains ignored. Instead, the Plum Island safety office turned its back on Phillip Piegari, Pete Swenson, Shine Mickaliger, and others.

Biological Security—Plum Island needs to be guarded around the clock by a platoon of trained, armed guards, like the thirty-four-man department it used to have in the 1950s. The current force of retired NYPD cops guarding Orient Point and Plum Island isn't even a start. Since his first visit in 1971, reporter Karl Grossman has seen more of Plum Island than anyone not connected the government. "In my earlier visits," recalls Grossman, "it looked like what you would expect out of a James Bond movie, with armed federal agents who patrolled the beach constantly. Later they had privatized the security, and you were walking past the kind of people you would see in front of a convenience store in a bad neighborhood." Regular helicopter and marine patrols, proposed by Senator Schumer and Assemblywoman Acampora, are also imperative.

Plum Island is more vulnerable to enemy and terrorist attack than it was prior to September 11, 2001, though the USDA strongly disagrees. "It's really no more a terrorist target than the local federal courthouse," says spokeswoman Sandy Miller Hays. "It's just not that attractive a target." I suggest that Hays relay the USDA's viewpoint to Bashiruddin Mahmood, c/o Islamabad, Pakistan, and to his unsavory cohorts. Security needs to be on par with Steve Nostrum's armed paramilitary unit that protected Shoreham, the nuclear power plant. Thanks to the discovery of Mahmood's cache, there is proof that Plum Island is as vulnerable a target as a nuclear power plant, and in many ways a far greater one.

After the terrorist attacks, the American public called for increased security at airports, focusing particularly on private companies that hire minimum-wage workers to X-ray baggage and run the passenger metal detectors. By federalizing airport security—placing guards under federal control—better trained and equipped people will vastly improve lax safety measures, say supporters. Plum Island is an installation as vulnerable as any airport and arguably far more exposed. The federal A-76 privatization program stated that the private enterprise system, "characterized by individual freedom and initiative, is the primary source of economic strength." But wherein lies the primary source of national security strength? Or the strength required to protect the public from biological and environmental harm? That lies with the people's government, not in the private sector, where the "governing" is motivated by profit. Still, Plum Island's support functions and its paucity of island security remain in private hands. If there is strength in numbers, then Plum Island is exceptionally weak. From a support staff high of 156 employees before privatization, the workforce was trimmed from 156 to 100, and then pared to 80—*half* of its original size. In some instances, workers are assigned job tasks once performed by three separate individuals. This is a recipe for disaster, evidenced by this one sewage plant worker who kept a private log:

> [We] proved the problem can be kept under control by concentrating more operator time running the plant. Due to the lack of manpower, we could only do this for a short time by ignoring other island operations, which contributed to the old plant being out of compliance this month. . . . Requested covers be installed to avoid freezing problems during winter. Request denied due to "Who will pay?" syndrome. Systems around the island that we have been ignoring are beginning to fail. Still no extra manpower as promised. High temperatures shutting down pumps. But EQ tank full. Must let go through plant HOT (100 degrees F[ahrenheit]). . . . Bad storm blew poly pumps and plumbing apart. . . . Some days with no operator at all! During this time, a major plant spill occurred due to this lack of man-

power. Plant is undersize for current and future needs of island. . . .
Plant requires much more operator manpower to overcome design
problems. . . . Budget cutbacks have wiped out availability of normal
maintenance.

During my personal visits to Plum Island, I noticed that the gates leading into the Lab 101 compound were left open all day, and utility vehicles
and box trucks (presumably carrying biological samples and animals) freely
moved in and out. The laboratory building itself needs to be better guarded.
"Take the new lobby," says Dr. Carol House of Lab 101, who's been there a
few times since her retirement. "They built that nice brand-new lobby, but
the staircase blocks the doors into the lab, and there was no pass card to get
into the lab for years. There is now, maybe installed a month ago. Anybody
in that building can walk into those laboratories at will. Just look at the
physical setup—the [security] guy sitting at reception is flush with the wall,
and if he's sitting down, he can't see the entrance to the lab because the
staircase is in the way. All you have to do is slip in—I mean, c'mon!"

Maintaining proper biological security also means screening scientists
and workers. In the early days, Dr. Callis defended the security checks and
taking oaths before researchers were hired, telling a panel, "We wouldn't
want anybody working here who would maliciously sabotage our work."
But over the years Plum Island has welcomed many foreign scientists in
residence with little or no background checks, from nations like China,
Ethiopia, Nigeria, Uganda, and—of all places—Iraq. While doctors from
staunch American allies like Japan and Taiwan should be, and are, welcomed to Plum Island, those hailing from nations with suspect political
affiliations ought to be barred. "A highly skilled individual could take
advantage of the opportunity afforded by a position of trust in a laboratory . . . in human or veterinary medicine. Such persons would use to full
advantage the appearance of loyalty and transparent honesty," Canadian
officials wrote this in December 1941 at a time of world turmoil. It rings
just as meaningfully today.

Background checks should be run, like they were in the 1950s, on all
nonprofessional personnel as well—yet another discarded procedure.
Recently, a worker in charge of preparing Plum Island identification badges
was caught making fake badges and was quietly dismissed. Dr. Carol
House speaks of a USDA employee who was accused of taking large bribes
at the Port of Miami quarantine facility, and was previously convicted for
assault and battery. The employee was transferred to work at Plum Island.
House noticed the man, an inventory keeper, on Plum Island on a Saturday
morning, for no apparent reason. "I'm thinking to myself, 'This is weird—
what's *he* doing here on a Saturday?' He knows where all the keys are, and

knows all the combinations to the locks." On Monday morning, House privately reported her observation to security, who never properly investigated the claim. What had he been doing there that day? "I don't know," says House. "I wasn't going to trail him. I went home earlier in the day and he stayed on after that." The man mysteriously left Plum Island about two months later. "This was the mentality of security," House recalls. Bungle it, or honor it in the breach.

A final concern involves Plum Island's airspace. At some point during the gradual decline of the island's security, the restricted "no-fly zone" designation over Plum Island's airspace was suspended. Its airspace should be redesignated RESTRICTED on navigational flight charts and enforced by the U.S. Air Force or the National Guard.

Privatization—A democratic government should perform essential functions that are economically and practically infeasible for private citizens and industry to perform themselves. For example, maintaining a private armed security force to guard one's home is economically beyond most people's reach, so everyone pays taxes to support a municipal police force that guards one's person and property. Applied to Plum Island, research on exotic animal disease has been deemed inefficient and impractical for the private sector to finance, so it has been delegated to government. Because of the ultrasensitive nature of operations, its support workers should be of the highest caliber. Skill—not a willingness to work for low wages and benefits—should drive recruitment. Safety and security—not year-end profits realized on a balance sheet—should be the primary motivator. This is not an indictment of the private sector; it is a realization that government takes on inefficient roles to provide a necessary benefit and protect its citizens. Plum Island, like airport security, *must* be federalized.

A stark reminder of this occurred at midnight on August 13, 2002, when seventy-six workers, with a combined 758 years of professional experience running Plum Island, went on strike for the first time in the island's history. Citing living wages and medical benefits that lag behind similar operations elsewhere, Local 30, the Plum Island union, took action after going without a contract for eleven months.[3] The union asked for a forty cents per hour per worker increase, or $850 per worker. The private contractor rejected their modest demands, and instead airlifted some forty replacement workers ("scabs" in union parlance) from as far away as Colorado to Connecticut, and secretly ferried them to Plum Island under cover of night. The replacements circumvented the twenty-four-hour picket line

---

[3]The earlier 1998 contract that expired was only agreed to after a federal mediator was brought in to resolve a threatened strike. The contractor agreed to 3-percent-per-year wage increases, slightly lower than U.S. inflation increases over the same period.

at the Orient Point dock—christened with a camping tent, a twenty-foot-high inflatable balloon rat, and an American flag. The scabs were taking turns working the island's critical systems and sleeping on cots huddled together in the big warehouse at Plum Island harbor. Food and toiletries were ferried in for them.

Responding to concerns over whether Plum Island could be safely run during a walkout by the workers trained to sustain it, a USDA spokeswoman said simply, "We do not feel it is necessary to shut down the island." But the replacement workers' mishaps contributed to Plum Island's circus-gone-wrong atmosphere. Scabs caused three separate ferryboat incidents: one in which the ferry ran over a buoy in Plum Gut, causing $10,000 in damage; another where the boat crashed into the Old Saybrook, Connecticut, dock; and yet another in which a six-hundred-gallon tank of liquid nitrogen fell off the back of the boat into Plum Island Harbor. Thankfully the tank did not rupture, but the parade of horribles marched on. One worker, placed in charge of the laboratory's critical negative air containment system, had been previously charged with malicious assault on three occasions. The man went missing for three days, taking with him a laptop computer which remotely controlled the laboratory's biological containment system by dial-up access. Seven hundred gallons of oil spilled near the power plant (on the same site of a 1,500-gallon spill in 2000), 200 of which seeped into the ground. Two replacement workers drove off with a 2002 Dodge Caravan assigned for local errands and never returned. Later found in New York City, the minivan's cargo was unknown. OSHA returned, again, to Plum Island and cited the contractor for at least six workplace safety violations, including repeat infractions all too familiar now: poor handling of hazardous materials, inadequate training in blood-borne pathogens, and no radiation hazard training. At this writing, a second laptop computer is missing from inside the biological containment area, one that according to sources contains laboratory experiment data. The DHS is investigating yet another sensitive computer believed to be stolen. And in yet another instance of history repeating itself on Plum Island, there were two power outages, each multiple hours long, in December 2002. The emergency power generators failed, and the ever-vigilant Plum Island safety office attempted to seal containment doors with ordinary duct tape to keep germs from escaping. Although three employees were marooned inside biological containment areas until power could be restored, one scientist referred to the tape-up job as "standard operating procedure."

On Plum Island, speaking out was forbidden after the strike. Worker Jim McKoy came forward and told USDA and DHS officials about lax safety and security procedures. He told them that he saw employees working with asbestos without proper protections; that he saw visitors walking around and about Plum Island unescorted; and that he exchanged ID badges with another worker as a test of the security system. The ID switch, he said, went unnoticed for a whole day. "I really felt I had to do something," McKoy told a local newspaper. "I could go anywhere in the lab. . . . That place is a powder keg." USDA and DHS told McKoy there would be no retaliation by his employer. Then they changed their tune and said government officials couldn't get involved in the affairs of a private contractor. The Plum Island contractor fired McKoy. The contractor said that the whistleblower had left his post without permission.

Chaos reined on both sides of the coin. Mark DePonte, in charge of Plum Island's water treatment plant during the strike, left a parting gift when he walked off the job: he shut down the complex water supply system, crucial to both fire suppression and biological decontamination. Then he sent DHS Secretary Tom Ridge an e-mail warning of "catastrophe" if the union workers did not return to their posts. Asked by a local reporter about the water crisis, DePonte said everything was fine when he left his post. "I can't help it if the new workers don't know which switches to turn on." After management spent hours figuring out the intricate maneuvers required to restore water, they called in the FBI, which began to investigate. Soon after, DePonte pleaded guilty to tampering with federal government property.[4]

Though the whole mess could have been averted with a forty-cent-per-hour wage increase for a total of $62,000 a year, the strike cost an additional $12.82 per worker per hour, or $45,000 *per week*—outrageous expenses passed along to, and essentially a strike shouldered by, the U.S. government and the taxpayers. "I am bewildered at why it had to get to this point," exclaimed Senator Hillary Clinton to the *New York Times* when asked about the strike. At the end of the year, Clinton demanded that the USDA

---

[4]A union source says DePonte shut down the system because the line pump wasn't working properly, and that following procedure, he entered the shutdown into the logbook and recorded it on a dry-erase system status board in the power plant. Another worker familiar with the situation says it was shut down because from time to time the antiquated system—used in lieu of a recently installed replacement system—did not properly maintain the correct pH level. "This was standard procedure. The new system never worked, and management had it redesigned twice and it still didn't work right, so the old system is used." Failing to shut off the system, says the worker, would have caused the pH level to move in the opposite direction. "If [DePonte] left the water system running, they would have said he was trying to poison people." Why, then, did he plead guilty? "Dollars and cents, plain and simple," says the union source of DePonte's mounting legal fees, and calling the affair a "red herring," points to the fact that despite the extensive FBI investigation, DePonte retains his regulatory operator's license and is working at a municipal plant in Connecticut. "They could not troubleshoot it [after the strike] because they had no one capable of doing it, so they blamed someone. And they were looking for something they thought would turn the media against us—and it didn't work."

cease operations at Plum Island until labor and biological safety lapses were rectified. And along with Republican congressman Rob Simmons from Connecticut, she formally requested that the Plum Island workforce be completely un-privatized and re-federalized. Unfortunately, their urgings fell upon deaf ears.

This is the state of the Plum Island Animal Disease Center.

Placing all employees at this germ-ridden utopia—if it is to continue its mission there—on the rolls of a well-trained Department of Homeland Security might prevent it from reaching this point again. It might ensure the safety and biological security of this troubled island laboratory. But this hasn't been done.

It goes without saying that Plum Island's transfer to DHS should be far more than a "paper transfer," as assistant director Santoyo put it. Above all, Plum Island needs something that's completely new in its long history: transparency. The past has proven, time and again, that the island's stewards cannot pilot the ship safely without close supervision. This secretive island has been exploited, both as a monarchical fiefdom and as a resumé-building career stepping-stone. It should be neither. If Jerry Callis can be compared to a fallen angel, Roger Breeze can be likened to a vainglorious "doomsday machine," and privatization to an absentee slumlord.

Former Plum islander Phillip Piegari remembers a saying among the workers: "Plum Island is always taking calculated risks, and somehow they get away with it." The public must not allow the government to continue its research without being fully supervised, because it's running Plum Island like it's a game of Russian roulette—where its luck can only last so long.

The true story of Plum Island reaches a most disturbing conclusion.

The laboratory once glorified as the "World's Safest Lab" is today the world's most dangerous.

In light of everything presented in these pages, consider these queries:

- Is such a facility necessary and important to the United States and its national security interests?
- Do the past catastrophes, the present level of safeguards, and its future plans justify the laboratory's existence on Plum Island?

Plum Island is a ticking biological time bomb. The U.S. Department of Agriculture set the timer, and the clock has been ticking away for years. Today the island is more vulnerable than ever before to a germ outbreak, and it remains wide open to sabotage and terrorist attack. Now is the time to reconsider its existence or take meaningful corrective measures. In other words, there's still some time to dismantle the bomb.

Or will the people of America have to wait until Plum Island reaches the point of no return?

The choice is ours.

Source Notes

Although Plum Island has been on my mind as long as I can remember, I began working on this book in 1997. It is based upon many hours of in-person and telephone interviews, and extensive research at local and national libraries, historical societies, local government archives, the National Academy of Sciences, and the National Archives. For a time, the USDA granted me access to Plum Island. After six visits to the island, the department denied me further access, citing national security concerns in the wake of September 11, 2001.

Part of the material presented in the book is based upon documentation obtained under the Freedom of Information Act. Although the USDA evaded the letter—and most certainly the spirit—of the law by failing to produce any meaningful information about Plum Island after I filed a two-page detailed FOIA request with that agency in early 2002, other government agencies, including the Environmental Protection Agency, the Occupational and Safety Health Administration, the Nuclear Regulatory Agency, and even the Department of Defense were all very cooperative and sent me reams of material that proved important to the book.

For the interested reader, a compilation of selected primary and secondary source material that I found useful follows, roughly in chronological order, followed by a listing of books, newspapers, and periodicals consulted.

PLUM ISLAND EARLY HISTORY

"Parchment Deed of Plumm [sic] Island-Gull Island: Samuel Willis [sic] of Hartford, Conn. to Joseph Dudley of Roxbury, Mass., June 7, 1688." New London, Connecticut, Historical Society.

Correspondence from Captain Charles Paget, *H.M.S. Superb*, anchored off Plum Island, New York, to Joseph Terry, Justice of the Peace, Oyster Ponds, 4 August 1812. Oyster Ponds Historical Society.

History of the Beebe Family (circa 1850). Oyster Ponds Historical Society.

"The Early History of Plum Island," *Republican Watchman*, 26 April 1884.

"Plum Island: Something About It—Why There is No Hotel," *The Day*, 23 November 1889.

"A Trip to Plum Island," *The Sun*, Sunday, 2 November (circa 1890).

Deed "Hewitt to United States of America—150 acres, February 2 1897," Book 456, Page 31, Suffolk County Records, Riverhead, New York.

"Plum Island Fifty Years Ago," Essex, Connecticut: 3 September 1908 (unpublished manuscript).

Fair Value Appraisal Report: Fort Terry, Plum Island, Town of Southold, County of Suffolk, State of New York. National Archives and Records Administration, New York Regional Branch (circa 1945).

Correspondence from Maurice S. Shahan, Director, Plum Island Animal Disease Laboratory, to Dr. C. N. Dale, regarding early Plum Island history, 11 September 1953.

Abram S. Hewitt, former mayor of New York City, entry in *The Dictionary of American National Biography*. Cambridge: Oxford University Press, 2000.

60 Minutes. 30 July 2000. "Lord of the Manor: Gardiner's Island." CBS News transcripts.

USDA Internet website, <www.usda.gov>.

Newsday history series "Long Island: Our Story," <www.lihistory.com>.

PREFACE

"Pakistan Detains Nuclear Scientists in Probe Over Taliban Links," *Bloomberg News*, 24 October 2001.

"U.S. Worries about Pakistan Nuclear Arms," *Washington Post*, 4 November 2001.

"Fighting Terror: A Disenchanted Researcher's Bin Laden Tie," *Boston Globe*, 16 December 2001.

"A Nation Challenged: Nuclear Secrets," *New York Times*, 17 December 2001.

"Al Qaeda Runs for the Hills," *Newsweek*, 17 December 2001.

"Pakistan to Forgo Charges Against 2 Nuclear Scientists; Ties to Bin Laden Suspected," *Washington Post*, 30 January 2002.

"Pakistani Scientist Who Met Bin Laden Failed Polygraphs, Renewing Suspicions," *Washington Post*, 3 March 2002.

1975: THE LYME CONNECTION

"Interrogation of General Gerhard Rose, Vice-President of the Robert Koch Institute, Berlin, Chief Consultant in Tropical Medicine to the German Air Force," Combined Intelligence Objectives Sub-Committee, 25–26 June 1945.

Memorandum "Families of Detainees of DUSTBIN," G. W. Staenpfli, 1 July 1945.

"Interrogation of Kurt Blome," E.W.B. Gill, Major, 1 July 1945.

"Preliminary Interrogation Report (PIR) No. 1: Prisoner Dr. Blome, Kurt," Headquarters, U.S. Forces, European Theater, Interrogation Center, 2 July 1945.

Memorandum "Interrogation of Dr. Blome, Kurt," from Howard Hansen, Colonel, Executive, Headquarters, U.S. Group C-C, Public Health and Welfare Division to Interrogation Center, HQ USFET, 12 July 1945.

"Interrogation of Blome, Director of German B.W. Activities," Headquarters, United States Forces, European Theater, Alsos Mission, 30 July 1945.

Memorandum "Commandant DUSTBIN," from Lieutenant Colonel G. S. to Major Wilson, 1 August 1945.

Memorandum "The Current Movement of German Scientists into Russian Territory," from the Field Information Agency, Office of the Director of Intelligence, Office of Military Government for Germany (U.S.) to William S. Culbertson, 17 October 1945.

"Miscellaneous Interviews on Medical Practice and Research in Germany," Combined Intelligence Objectives Sub-Committee, 22 October 1945.

"Aviation Medicine, General Medicine, Veterinary Medicine, Chemical Warfare," Hans G. Schlumberger, Combined Intelligence Objectives Sub-Committee, 12 December 1945.

Memorandum "German Specialists in the United States—Procedure for clearance of," from J. M. Horan, Commander, USNR for the Director, Joint Intelligence Objectives Agency, 15 May 1946.

Memorandum for Information "Exploitation of German and Austrian Scientists and Important Technicians," Bosquet N. Wev, Captain, USN, Director, Joint Intelligence Objectives Agency, 3 July 1946.

Final statement of Defendant Blome, and "Description of the official functions of the REICHSARZTEKAMMER," Kurt Blome, Nuremberg Doctors' Trial, 1947.

Memorandum "Conditions in the Russian Zone: Project 113/30," from Bruno Deussing, Headquarters, Sub-region Darmstadt, Counter Intelligence Corps Region III, to the Officer in Charge, 25 January 1947.

"Interview of German Scientist, German Research on Biological Warfare," Office of the Chief of Chemical Corps, HQ EUCOM, 10 November 1947.

Memorandum "German Specialist, Dr. Erich Traub: Request for services of," from Chief of Naval Operations to Director, Joint Intelligence Objectives Agency, 2 February 1949.

Memorandum "Professional Status of Former Paperclip Specialists Following Their Immigration," H. E. Michelet, Lt. Colonel, GSC, Acting Chief, Administrative and Liaison Group, 6 March 1950.

Memorandum from Charles M. McPherson, EUCOM Special Projects Team, Office of the Land Commissioner for Hesse, to Major Barrick, Intelligence Division, EUCOM, Project PAPERCLIP, 27 March 1951; 27 November 1951.

Memorandum for the Governing Committee Joint Intelligence Objectives Agency, "Procurement of PAPERCLIP Scientists," 4 February 1949.

File A-7183623 Re: Eric Traub, Immigration and Naturalization Service, U.S. Department of Justice, 23 February 1951.

Memorandum "Repatriation of PAPERCLIP Scientist, Eric Traub, and Dependents," from Joint Intelligence Objectives Agency to Major Lewis W. Saxby, Chief, Collection and Dissemination Division, Intelligence, Department of the Army, circa 1951.

"African Swine Fever," Plum Island Animal Disease Laboratory Project Abstract, 22 February 1978.

"Ticks of Veterinary Importance," Animal and Plant Health Inspection Service, USDA, Agriculture Handbook No. 485, circa 1980.

Endris, R. G. et al. "Techniques for Mass Rearing Soft Ticks (Acari: Argasidae)," *Journal of Medical Entomology*, May 1986: 23(3).

Endris, R.G. et. al. "A Hemolymph Test for the Detection of African Swine Fever Virus in *Ornithodoros coriaceus*," *Journal of Medical Entomology*, March 1987: 24(2).

Endris, R.G. et. al. "Experimental Transmission of African Swine Fever Virus by the Tick *Ornithodoros puertoricensis*," *Journal of Medical Entomology*, November 1991: 28(6).

Beyerchen, Alan. "Secret Agenda: The United States Government, Nazi Scientists, and Project PAPERCLIP," *Science*, 24 January 1992, 255 (5043): 481.

"Nazis Planned to Use Virus Against Britain," *The Times*, 12 March 2001.

"Biological Warfare Operations," Research and Development Annual Technical Progress Report, Department of the Army, 1951.

Memorandum "Repatriation of PAPERCLIP Scientists, Eric Traub and dependents," from Commanding Officer, Naval Medical Research Institute to Joint Intelligence Objectives Agency, 31 October 1952.

Memorandum "Conversation Between Doctors H. W. Schoening, M. S. Shahan and Erich Traub," M. S. Shahan, 10 December 1952.

Memorandum "Proposed Visit to Plum Island by Dr. Erich Traub," from R. A. Carlson, Administrative Officer, ARS, to M. S. Shahan, Director, Plum Island Animal Disease Laboratory, undated.

Correspondence between Dr. A. D. MacDonald, Animal Disease and Parasite Research Branch, ARS, USDA, and Dr. Eric Traub, Bundesforschungsanstalt Fur Viruskrankheiten der Tiere, 14 June 1954; 20 October 1954.

Correspondence between Edwin R. Goode, Jr., Acting Director, Animal Disease and Parasite Research Division, ARS, USDA, and Dr. Eric Traub, Bundesforschungsanstalt Fur Viruskrankheiten der Tiere, 23 July 1957; 25 September 1957.

Memorandum "Justification for Employment of Dr. Erich Traub," from Howard V. Johnson, Director, Animal Disease and Parasite Research Division, USDA, to J. H. Starkey, Director, Personnel Division, ARS, 22 October 1958.

Letters of reference regarding Erich Traub from William A. Hagan, Dean, New York State Veterinary College, Cornell University; Richard E. Shope, Rockefeller Institute for Medical Research; and C. A. Brandley, Dean, College of Veterinary Medicine, University of Illinois, to Dr. Howard V. Johnson, Director, Animal Disease and Parasite Research Division, USDA, 10, 12 March 1958.

Correspondence from H. W. Schoening, Consultant, Animal Disease and Parasite Research Division, USDA, to Dr. Howard V. Johnson, Director, Animal Disease and Parasite Research Division, USDA, 13 March 1958.

1999: EAST END MEETS WEST NILE

Baseman, Joel B., and Joseph G. Tully. "Mycoplasmas: Sophisticated, Reemerging, and Burdened by Their Notoriety," *Emerging Infectious Diseases*, January–March 1997, 3 (1).

"Summary of West Nile Virus in the United States, 1999." Animal and Plant Health Inspection Service, U.S. Department of Agriculture.

Memorandum for the Record "Investigation of a Cluster of Equine Neurologic Illness—Jamesport, NY," from Eastern Regional Early Response Team, USDA, 21 October 1999.

"Prevention and Control of West Nile Virus Infection in Equine and Other Livestock or Poultry." Veterinary Services, Animal and Plant Health Inspection Service, USDA 31 May 2000.

"Update on the Current Status of West Nile Virus: Period from 8 October through 17 October, 2000." Animal and Plant Health Inspection Service, USDA.

"West Nile Virus Kills Three Birds Belonging to Zoo," *Washington Post*, 18 July 2002.

"West Nile Virus Kills Four in Louisiana," *Washington Times*, 3 August 2002.

"Anxiety Felt in Louisiana After 4 West Nile Deaths," *New York Times*, 4 August 2002.

"Expect the Unexpected: The West Nile Virus Wake Up Call," Senate Committee on Governmental Affairs, 24 July 2000.

1967: THE DEMISE OF THE DUCKS

Leibovitz, Louis and Jen Hwang. "Duck Plague on the American Continent," Duck Research Laboratory, Cornell University, 7 August 1967.

"Fatal Virus Found in Wild Ducks on L.I.," *New York Times*, 24 December 1967.

"Wood, Field and Stream: Biologists Winning War in So. Dakota Against Ravages of 'Dutch Duck Plague,'" *New York Times*, 1 April 1973.

GENESIS

U.S. BIOLOGICAL WARFARE PROGRAM

Correspondence from Edwin B. Fred, Dean of Graduate School, University of Wisconsin, and Chairman, War Bureau of Consultants Committee, to Frank B. Jewett, President, National Academy of Sciences, 28 December 1941.

Correspondence from Frank B. Jewett to Harvey H. Bundy, Special Assistant to the Secretary of War, 31 December 1941; 31 July 1942; 28 September 1942.

Correspondence between Edwin B. Fred and William A. Hagan, Dean, New York State Veterinary College, Cornell University, 13 January 1942; 19 January 1942; 26 January 1942; 31 July 1942; 2 September 1942; 20 September 1942; 13 November 1942; 16 November 1942; 26 November 1942; 6 December 1942; 15 December 1942; 9 March 1943; 7 May 1943; 12 May 1943; 18 May 1943; 30 April 1943; 25 May 1943; 15 June 1943; 15 June 1943; 21 August 1943; 20 October 1943.

Correspondence from E.G.D. Murray, Department of Bacteriology and Immunity, National Research Council, Canada, to Edwin B. Fred, 26 January 1942.

Hagan, William A. "Report of the W.B.C. Committee, February 17, 1942: Foot and Mouth Disease," National Academy of Sciences.

War Bureau of Consultants Committee. *Report of the W.B.C. Committee*. Washington, D.C.: National Academy of Sciences, 19 February 1942.

Correspondence from Frank B. Jewett to Henry L. Stimson, Secretary of War, 27 February 1942.

Correspondence from Edwin B. Fred, Chairman, W.B.C. Committee, to Dr. Eugene C. Auchter, Chief of Bureau of Plant Industry, USDA, 26 May 1942.

Memorandum "The Next Attack on Oahu—Bullets—or Bacteria!" from (illegible) Honolulu doctor to Henry L. Stimson, Secretary of War, 12 June 1942.

Excerpt from W.B.C. Committee Report of June 18, 1942, "Grosse Ile Project."

Memorandum "Grosse Ile Project," from W.D. Styer, Brigadier General, G.S.C., Chief of Staff to Chief, Chemical Warfare Service, 29 July 1942.

Correspondence from Herbert S. Gasser, Director, The Rockefeller Institute for Medical Research, to Henry L. Stimson, Secretary of War, 25 August 1942.

Memorandum "E.B. Fred's visit with George W. Merck in New York City on Monday, August 31, 1942," Edwin B. Fred, 1 September 1942.

Correspondence from Claude R. Wickard, Secretary of Agriculture, to Henry L. Stimson, Secretary of War, 18 September 1942.

Expense report "War Research Service Project No. 1: Anthrax (Code "N")," National Academy of Sciences, January 1, 1943, through June 30, 1944.

Memorandum "Digest of Information Regarding Axis Activities in the Field of Bacteriological Warfare," War Research Service, National Academy of Sciences, 8 January 1943.

Memorandum "Concerning Fowl Plague and Newcastle Disease," Edwin B. Fred, 9 January 1943.

Memorandum "Concerning Fowl Plague and New Castle Diseases," Harry W. Schoening, U.S. Department of Agriculture, 11 January 1943.

Schoening, Harry W. "Report on W.R.S. Project 6: Foot and Mouth Disease," National Academy of Sciences, 10 April 1943.

Correspondence from Claude R. Wickard, Secretary of Agriculture, to George W. Merck, Director, War Research Service, National Academy of Sciences, 24 April 1943.

Expense report, "War Research Service Project No. 13 (a): Anthrax Immunization (Code 'MN')," National Academy of Sciences, March 1, 1943, through June 30, 1944.

Expense report "War Research Service Project No. 12: Fowl Plague & Newcastle Disease (Code 'OE')," National Academy of Sciences, March 8, 1943, through August 31, 1944.

Correspondence from George W. Merck, Director, War Research Service, National Academy of Sciences, to Lt. General Delos G. Emmons, Commanding General, U.S. Army Forces, Hawaiian Department, 19 March 1943. Attachment: "Probability of a Bacteriological Attack on the Hawaiian Islands."

Correspondence from Edwin B. Fred to Richard E. Shope, Commander, Grosse Ile Project, 7 May 1943.

Memorandum "Protection Against Bacteriological Warfare," from Brehon Somervell, Lieutenant General, U.S.A., Commanding to the Chief of Staff, United States Army, 6 January 1944.

Hagan, William A., W. H. Feldman, Raymond A. Kelser, A. W. Miller, H. Snyder, and Harry W. Schoening. "Report of the Joint Committee of W.R.S. and the United States Department of Agriculture on Foot-and-Mouth Disease," National Academy of Sciences, 8 January 1944.

Minutes of panel discussion entitled "Use of Glanders and Melioidosis organisms as Biological Warfare agents," held in Edwin Fred's office, 22 March 1944.

Memorandum "Research on Glanders and Melioidosis," from George Merck, Director, War Research Service, National Academy of Sciences, to Chief, Chemical Warfare Service, 5 May 1944.

Correspondence from Henry L. Stimson, Secretary of War, to President Franklin D. Roosevelt, 12 May 1944.

Correspondence from Franklin D. Roosevelt to Henry L. Stimson, 8 June 1944.

Weiss, Freeman A., W. W. Diehl, and E. B. Lambert. *Biological Warfare Report*. National Academy of Sciences Archives, November 1944.

Merck, George W. Biological Warfare: "Report to the Secretary of War by Mr. George W. Merck, Special Consultant for Biological Warfare." National Academy of Sciences, 3 January 1945.

SPECIFIC TO PLUM ISLAND

"Certificate of Decontamination," Leander H. Anske, Capt., Post Engineer, Headquarters, Harbor Defense of Long Island Sound, 2 November 1948.

"For Sale: Plum Island—Ideal for Resort Development," advertisement submitted to the *New York Times*, the *New York Herald Tribune*, and the *Wall Street Journal*, General Services Administration, 1949.

Correspondence from F. W. Salfingere, Major, Corps of Engineers, Executive Officer, to A. J. Intermont, Chief, Property Management Division, War Assets Administration Region II, 9 May 1949.

Correspondence from Carl P. Malmstrom, Deputy Regional Director, Office of Real Property Disposal, to Milton T. Smith re: Fort Terry, Plum Island, 17 May 1949.

"Final Certificate of Decontamination," Nicholas J. Culhane, Post Engineer, War Department, 16 June 1949.

Correspondence from A. J. Intermont, Chief, Property Management Division, to Commander of the Third Coast Guard District re: Fort Terry, Plum Island, Long Island Sound, New York, 4 November 1949.

Correspondence from H. S. Berdine, Captain, U.S. Coast Guard, Chief, Operations Division, to A. J. Intermont, Chief, Property Management Division, General Services Administration, War Assets, 17 November 1949.

Memorandum "Fort Terry, Plum Island, Long Island Sound, New York WD-1300 W-NY-141," from W. A. Boyd, Chief, Rehabilitation and Maintenance Branch, War Assets, to A. J. Intermont, Chief, Property Management Division, General Services Administration, War Assets, 25 November 1949.

Correspondence from Charles H. Duryea, Chairman, Board of Supervisors, Suffolk County, to General Services Administration, War Assets, 9 December 1949.

Notes "Meeting at Timber Point, Saturday, February 25, 1950, on a Potential Application of the County of Suffolk, State of New York for Fort Terry, Plum Island for Recreational/Educational Use," Dr. A. L. Harris and Mr. David H. McBuen, 3 March 1950.

Memorandum "Interim March Report for New York," from T. L. Peyton, Director, Real Property Disposal Division, Liquidation Service, General Services Administration, to John H. Joss, Commissioner, Liquidation Service, General Services Administration, 21 March 1950.

Memorandum "Proposed sale of approximately 792 acres on Plum Island known and identified as Fort Terry," from Carl P. Malmstrom, Deputy Regional Director, Office of Real Property Operations, Liquidation Service, General Services Administration, to Regional Director, Office of Real Property Operations, Liquidation Service, General Services Administration, 5 April 1950.

Correspondence from Edgar F. Hazleton, County Attorney, Suffolk County, to Mr. Sirillo, Liquidation Service, General Services Administration re: sale of Plum Island, 19 June 1950.

Correspondence from H. L. Mathews, Captain, U.S.N., Chief, Office of Construction, Munitions Board, to Mr. Peyton, 14 November 1950.

"Fort Terry Historical Report," 1 January 1951–30 June 1952; 1 July–30 September 1952; 1 October–31 December 1952; 1 January–31 March 1953; 1 April–30 June 1953; 1 July–30 September 1953; 1 October–31 December 1953; 1 January–30 June 1954, U.S. Army Chemical Corps, 1951–1954.

Correspondence from A. M. Martinson, Captain, U.S. Coast Guard, Chief of Staff, 3rd CG District, to John D. Halloran, Chief, Real Property Disposal Division, General Services Administration, Public Buildings Service, 16 April 1951.

Quarterly Technical Summaries, Biological Department, U.S. Army Chemical Corps, Medical-Veterinary Division, 1 April through 30 June 1949;

30 September 1949; 31 December 1949; 31 March 1950; 30 June 1950; 30 September 1950; 31 December 1950.

"Report of a Trip to Boston to Investigate Possible Island Sites," Dr. William R. Hinshaw, MVM Division, Biol. Labs, Camp Detrick, 18–19 April 1951.

Telegram from P.B. Finegan, Chief, RPA and UBRR, to W.F. Downey, Regional Director, G.S.A. N.Y. re: Defense Need for Plum Island, 25 May 1951.

Memorandum "Reassignment of Plum Island (Fort Terry) to the Chemical Corps for Responsibility," from Brigadier General Charles E. Loucks, Deputy Chief Chemical Officer, to Lieutenant Colonel R.O. Ball, Chief Chemical Officer, Department of the Army, 24 September 1951.

Correspondence from Walter F. Downey, Regional Director, to County Attorney, Suffolk County re: Fort Terry, Plum Island, New York, 8 January 1952.

Correspondence from Harvey F. Gibson, Chief, Real Property Acquisition, Utilization Division, Public Buildings Service, to Mr. Lawson B. Knott, Jr., Acting Chief, Management & Disposal Division, Department of the Army re: WD-1300-W-NY-14 1, Fort Terry, Plum Island, New York, 10 January 1952.

Correspondence from Fred Munder, County Attorney, Suffolk County, to Walter F. Downey, Regional Director, General Services Administration, re: Fort Terry, Plum Island, New York, 11 January 1952.

Memorandum "Acquisition of Plum Island (Fort Terry), Long Island Sound, New York," from H. O'Neill, Acting Chief, Management and Disposal Division Real Estate to Assistant Chief of Staff, Department of the Army, 18 January 1952.

Correspondence between Richard E. Shope, Rockefeller Institute for Medical Research, and Maurice S. Shahan, 26 June 1952; 3 July 1952; 1 August 1952; 13 August 1952.

W.S.E.G. Report No. 8, "An Evaluation of Offensive Biological Warfare Weapons Systems Employing Manned Aircraft," Weapons Systems Evaluation Group, Office of the Secretary of Defense, 15 July 1952.

Memorandum "Activation of Work on Technical Projects at Fort Terry," from Headquarters Fort Terry through Camp Detrick to Commanding General, Army Chemical Center, 19 November 1952.

Memorandum, "Contract DA-30-075-Eng-5509," from William C. Barnholt, Vitro Corporation of America, to C.K. Panish, Authorized Representative, Corps of Engineers, New York District Office, 26 December 1952.

Correspondence from F.C. Morris, Architect, Plum Island Animal Disease Research Institute, to Charles K. Panish, District Engineer, Corps of Engineers, New York District Office, 29 December 1952.

Correspondence from William B. Hinshaw, Chief, VM Division, Chemical Corps Biological Laboratories, Camp Detrick, to Dr. Stewart H. Madin, Department of Bacteriology, Naval Biological Laboratory, 5 February 1953.

Correspondence between Maurice S. Shahan and Stewart H. Madin, 4 February 1953; 10 February 1953.

Memorandum "Proposed Agreement—Department of Agriculture—Re: Fort Terry, N.Y.," from Deputy Commander, Colonel W. R. Currie to Commanding Officer, Chemical Corps Procurement Agency, Army Chemical Center, Md., 20 February 1953.

Correspondence from Maurice S. Shahan to Joseph L. Melnick, Section of Preventative Medicine, Yale University School of Medicine, 13 March 1953.

Correspondence between Jerry J. Callis, Acting Official in Charge, Bureau of Animal Industry, and Maurice S. Shahan, 5 February 1953; 11 March 1953; 16 March 1953.

Memorandum for the Record "Report of Conference on Agent Transportation," Headquarters Fort Terry, 17 March 1953.

Correspondence from Maurice S. Shahan to Lt. Col. Don L. Mace, Commanding Officer, Fort Terry, Plum Island, 19 March 1953.

Correspondence from Maurice S. Shahan to Dr. Howard L. Bachrach, Virus Laboratory, University of California, 21 January 1953; 5 May 1953; 17 June 1953.

Memorandum "Biological Warfare: Support of Research and Development in Anti-Animal and Anti-Crop Agents," from Ralph O. Moore, Secretary, Armed Forces Policy, Department of Defense, to Deputy Secretary of Defense, Secretary to the Army, Special Assistant to the Secretary of Defense (Research and Dev.), 11 August 1953.

"Report by the Joint Strategic Plans Committee to the Joint Chiefs of Staff on Chemical (Toxic) and Biological Warfare Readiness," 13 August 1953.

"Inventory of Animal Viruses and Antisera Procured by the Cooperation Between the Chemical Corps and the U.S. Dept. of Agriculture and Stored on Plum Island," U.S. Army Chemical Corps, 1954.

"Report of visit to Plum Island to review general supply activities," John B. Holden, Chief, Division of Procurement and Property Management, USDA, 28 May 1954.

Memorandum "Plum Island," from S. P. Williams, Director, Administrative Services Division, ARS, to F. H. Spencer, Assistant Administrator for Management, ARS, 22 June 1954.

"Plum Island Animal Disease Laboratory," Animal Disease and Parasite Research Branch, Agricultural Research Service, USDA, 8 July 1954.

Memorandum "Meeting at Plum Island: July 7 and 8, 1954," from H. W. Shoening to B. T. Simms, 12 July 1954.

Correspondence from A. W. Linquist, Head, Insects Affecting Man and Animals Section, Animal Disease and Parasite Research Branch, ARS, to George E. Cottral, Plum Island Animal Disease Laboratory, 7 September 1954.

Memorandum "Reassignment from Department of Defense to Department of Agriculture of Responsibility for Protection of Crops and Domestic Animals against Biological Warfare (N.S.C. Action No. 1120)," Edwin H. J. Carns and Richard H. Phillips, Joint Secretariat, Joint Chiefs of Staff, 9 July 1954.

Memorandum "Telephone Conversation with Murray Davis, New York World-Telegram and Sun," Maurice S. Shahan, Plum Island Director, 14 July 1954.

"Notes of Telephone Conversation with Dr. Schoening," Maurice S. Shahan, 14 July 1954.

"Notes of Telephone Conversation with Mr. McPhee," Maurice S. Shahan, 14 July 1954.

"Research Begins on Plum Island," Agricultural Research Service, USDA, August 1954.

Fellowes, O. N., G. T. Dimopoullos, and J. J. Callis, "Isolation of Vesicular Stomatitis Virus from the Blood of an Infected Worker," *Biological Proceedings,* 1955.

Press release, "USDA Reports First Results of Foot-and-Mouth Disease Research in U.S.," USDA, 10 October 1955.

Correspondence between William B. Hinshaw, Chief, Virus and Rickettsial Division, Chemical Corps Research and Development Command, U.S. Biological Warfare Laboratories, Fort Detrick, and Maurice S. Shahan, 5 May 1952; 18 June 1952; 28 August 1956.

"Interim Report of the PIADL Committee on Foot-and-Mouth Disease Vaccination," Plum Island Animal Disease Laboratory, August 1960.

GENESIS–PERIODICALS

"The New Dean," *The Cornell Veterinarian,* July 1932, 22 (3): 211.

Fox, Major Leon A. "Bacterial Warfare: The Use of Biologic Agents in Warfare," *The Military Surgeon,* May 1942, 90 (5).

"U.S. Gets Suffolk Plan for Plum Is. Today," *Newsday,* 8 December 1949.

"Plum Island Bids Thrown Out by Govt. as Too Low," *East Hampton Star,* 15 December 1949.

"County Gets Plum Island at Bargain Price; Could Already Sell at a Profit," *Long Island Traveler-Watchman,* 30 March 1950.

"Plum Island Adds Another Blessing to Suffolk," *Long Island Traveler-Watchman*, 13 April 1950.

"Suffolk Holding Plum Island but Profit Considered," *East Hampton Star*, 1 June 1950.

"Suffolk County to Buy Plum Island on Cash Basis," *East Hampton Star*, 3 August 1950.

"Plum Island 'Frozen' by Government Order During Emergency," *East Hampton Star*, 21 August 1950.

"Plum Island Not 'Frozen' by Federal Government," *East Hampton Star*, 7 September 1950.

"Plum Island Sale to Suffolk Is Off," *East Hampton Star*, 31 January 1952.

"Lab for Hoof-and-Mouth," *The News*, 23 July 1952.

"Report on 'Plague Island,' " *Saturday Evening Post*, 1959.

"Dr. William A. Hagan, 1893–1963," *Journal of the American Veterinary Medical Association*, 15 February 1963, 142: 423.

"Dr. W. A. Hagan, 1893–1969," *Veterinary News*, March–April 1963, 27 (3): 16–18.

Bernstein, Barton J. "The Birth of the US Biological-Warfare Program," *Scientific American*, June 1987, 256: 116–121.

Piller, Charles. "Lethal Lies about Fatal Diseases," *The Nation*, 3 October 1988.

AGE OF SCIENCE–DOCUMENTS

Memorandum "Material on Arrests," from Harry H. Ramm, Chief, Procurement Branch, ARS, to Lyle B. Shanks, Asst. Director for Management, Plum Island Animal Disease Laboratory, 24 November 1954.

Memorandum "Safety Regulations," from R. A. Carlson, Administrative Officer, ADP Branch, ARS, to Maurice S. Shahan, 15 December 1954.

Memorandum "Sensitive Clearance for Isaac Gaston," from Lyle B. Shanks to C. E. Schoenhals, Agency Security Officer, 22 December 1954.

Correspondence from Maurice S. Shahan to E. G. Moore, Director, Information Division, ARS, USDA, 20 October 1955.

Correspondence from Robert R. Hurt, Assistant Chief, Reports, Training and Orientation Division, Chemical Corps Research and Development Command, Biological Warfare Laboratories, Fort Detrick, to Maurice S. Shahan, 26 July 1956.

Correspondence from Herald R. Cox, Director, Viral and Rickettsial Research, Lederle Laboratories Division, American Cyanamid Company, to Maurice S. Shahan, 1 August 1956; 31 August 1956; 5 September 1956.

Memorandum "Invitations to PIADL Symposium, September 27 and 28," from Maurice S. Shahan to H. W. Johnson, Chief, Animal Disease and Parasite Research Branch, ARS, USDA, 8 August 1956.

Correspondence between Maurice S. Shahan and Major General William M. Creasy, Chief Chemical Officer, Chemical Corps, U.S. Army, 20 August 1956; 23 August 1956; 30 August 1956; 5 September 1956.

Press release "USDA Plans Dedication of New Plum Island Animal Disease Laboratory," USDA, 22 August 1956.

"Attendees of 1956 Dedication and Symposium," Plum Island Animal Disease Laboratory, September 1956.

"Plum Island Animal Disease Laboratory Dedication Week Ceremonies: Final Report—Committee X," September 19, 1956.

"Research—Defense Against Diseases," address by Secretary of Agriculture Ezra Taft Benson at the dedication of the Plum Island Animal Disease Research Laboratory, Plum Island, New York, September 26, 1956.

Correspondence from Claude W. Gifford, Associate Editor, *Farm Journal*, to Maurice S. Shahan, Director, Plum Island Animal Disease Laboratory, 3 October 1956.

Correspondence from Ezra T. Benson, Secretary of Agriculture, to Maurice S. Shahan, 18 October 1956.

Callis, Jerry J. "Responsibilities and Scope of Research at the Plum Island Animal Disease Laboratory," Second Annual ADP Scientific Workshop, 8 June 1964.

Memorandum "Purpose, objective and goal—PIADL," from Jerry J. Callis, Director to Plum Island Animal Disease Laboratory employees, 14 August 1964.

Callis, J. J. "Plum Island Laboratory: Its Role in Foreign Animal Disease Research," *Agricultural Science Review*, 9(3), 1971.

Booklet, The Plum Island Animal Disease Center, USDA, August 1982.

AGE OF SCIENCE–PERIODICALS

"Animal Disease Lab on Plum Island to Get New Building," *Sunday Mirror*, 26 February 1956.

"Plum Island Laboratory to Be Dedicated During Week of Sept. 24–28th," *Long Island Traveler*, 30 August 1956.

"Benson to Dedicate Animal Disease Lab," *Newsday*, 30 August 1956.

"Animal Disease Lab on Plum Island Will Be Dedicated Week Sept. 24," *Suffolk Times*, 31 August 1956.

"Plum Island Laboratory Dedication Sept. 24–28," *Long Island Traveler*, 6 September 1956.

"Plum Island Laboratory Ready for Dedication This Month," *Bridgehampton News*, 7 September 1956.

"Dedication at Plum Island," *Riverhead, NY News Review*, 13 September 1956.

"Secretary Benson to Be Speaker at Plum Island Dedication Ceremonies," *Long Island Traveler*, 13 September 1956.

"Dept. Agriculture Plans Plum Isl. Lab Dedication," *Suffolk Times*, 14 September 1956.

"Secretary Benson to Speak at Plum Island," *Long Island Traveler*, 20 September 1956.

"The Plum Island Laboratory," *Suffolk Times*, 25 September 1956.

"Laboratory Set for Plum Island," *New York Times*, 26 September 1956.

"U.S. to Open Farm Ills Laboratory: Building Isolated on Plum Island," *New York Herald Tribune*, 26 September 1956.

"Animal Disease Lab on Plum Island Opened by Secretary Ezra Benson," *Long Island Traveler-Watchman*, 27 September 1956.

"Benson Cites Rise in Farm Research," *New York Times*, 27 September 1956.

"Benson Opens $10 Million Lab for Hoof-and-Mouth at Plum," *Bridgehampton News*, 28 September 1956.

"Dedication of Plum Island Animal Disease Laboratory," *The Federal Veterinarian*, November 1956, 13 (6).

"Vaccine May End Animal Sickness," *New York Times*, 26 March 1967.

SYMPTOMS

Press Release "Laboratory Is World's Safest for Virus Research," USDA, 21 October 1971.

Press Release "Laboratory Prepared to Diagnose Foreign Animal Disease," USDA, 21 October 1971.

"Out of 'Andromeda Strain' Right Here on Long Island," *Long Island Press*, 22 October 1971.

"Press Visited Plum Island for First Time," *Long Island Traveler-Watchman*, 28 October 1971.

Correspondence from Downey to Honorable Earl L. Butz, Secretary of Agriculture, 12 November 1975.

Confidential notes, Downey Investigation Interview of "Bruce Becker" and "Charlie," 30 July 1976.

Letter, Information in Response to Inquiry of the Honorable Thomas J. Downey, U.S. Congressman, Regarding Research at Plum Island, New York, Plum Island Animal Disease Center, August 1976.

Correspondence from W. F. Davids to Honorable Thomas J. Downey, circa August 1976.

Memorandum "Infectious Animal Disease," from Janet E. Lemke, Science Analyst, Congressional Research Service, to Downey, circa August 1976.

Correspondence from Downey to Dr. Jerry Callis, Director, Plum Island Animal Disease Center, 10 August 1976.

Correspondence from John L. Naler, Chief, Investigations and Legislative Division, Department of the Army, to Downey, 19 December 1975; 22 July 1976; 17 August 1976.

Downey Investigation Interviews of J. Pugsley, 17 August 1976; Steve Lucas, 16 July 1976; John Cummings, 16 July 1976; Paul Rose, 10, 18 August 1976; Louis Grilli, 10 August 1976; Edward Kramer, 18, 23 August 1976.

"Cuban Outbreak of Swine Fever Linked to CIA," *Newsday,* 9 January 1977.

"A Call for Probe of Swine Report," *Newsday,* 10 January 1977.

Zilinskas, Raymond A. "Cuban Allegations of Biological Warfare by the United States: Assessing the Evidence," *Critical Reviews in Microbiology,* 25(3), 1999.

"Washington Accuses Cuba of Germ-Warfare Research," *New York Times,* 7 May 2002.

"THE DISASTROUS INCIDENT"–DOCUMENTS

Press release "Foot-and-Mouth Disease Incident," PIADC, September 1978.

Memorandum "Final Report Accession 844–78 FMD Suspect, Plum Island Animal Disease Center," from A. H. Dardiri, Laboratory Chief, Diagnostic Research, to J. J. Callis, Director, 15 September 1978.

Press release "Foot-and-Mouth Disease Diagnosed," PIADC, 15 September 1978.

Memorandum "Exploratory analysis—FMD outbreak in Animal Supply," from J. J. Callis, Director, to C. H. Campbell, C. A. Mebus, P. D. McKercher, J. S. Walker, M. E. Wiggin, W. E. Moulton, 19 September 1978.

Newcomb, Stanley S., "Report of Operations Related to F.M.D. Alert," PIADC, 15–21 September 1978.

Draft memorandum "The Committee Dilemma," from Merlon E. Wiggin, Chief, Engineering and Plant Management, 27 September 1978.

Memorandum for the Record "Inspection of Incinerator and Filter Changing of This Area," from Merlon E. Wiggin, Chief, Engineering and Plant Management, 28 September 1978.

Memorandum "Inspection and Testing of Filter System—Lab 101," from Merlon E. Wiggin, Chief, Engineering and Plant Management, to Investigation Committee, 11 October 1978.

Memorandum "Foot-and-mouth Disease, Type O, Incident, PIADC, September 15, 1978," from J. J. Callis, Director, to S. C. King, Acting Regional Administrator for Agricultural Research Service, USDA, 29 November 1978.

Memorandum "Plum Island FMD Break—Interim Report," from William M. Moulton, Chief, International Operations, Animal Health Programs, Veterinary Services, to J. K. Atwell, Assistant Deputy Administrator, Animal Health Programs, 30 November 1978.

Memorandum "FMD Incident—September 15, 1978," from J. S. Walker, Chairman, Investigative Committee, to J. J. Callis, Director, 8 December 1978.

Memorandum "Report of Committee Investigating FMD Outbreak at PIADC," from T. W. Edminster, Deputy Director for Agricultural Research, Science and Education Administration, USDA, to J. J. Callis, Director, 29 December 1978.

Memorandum "Ad hoc operations review committee," from J. J. Callis, Director, to J. H. Graves, J. S. Walker, M. E. Wiggin, C. H. Campbell, and P. D. McKercher, 9 January 1979.

Memorandum "Final Committee Report: Exploratory Analysis—FMD Outbreak in Animal Supply," from C. H. Campbell, W. Moulton, P. D. McKercher, J. Walker, C. Mebus, and M. E. Wiggin to J. J. Callis, Director, 9 January 1979.

Memorandum "Follow-up: September 15, 1978 FMD Incident," from J. J. Callis, Director, to T. W. Edminster, Deputy Director for Agricultural Research, Science and Education Administration, USDA, 12 January 1979.

Memorandum "Findings of Plum Island Safety Investigation Committee," from Jerry S. Walker, John W. McVicar, and Edward M. Wolfe, to J. J. Callis, Director, 14 February 1979.

Memorandum "Realignment of E&PM Supervisory Responsibilities," from John Graves, Acting Director, to M. E. Wiggin, H. J. Townsend, E. C. Hassildine, T. J. Roslak, A. F. Bohlke, J. R. Heaney, and T. H. Cook, 16 February 1979.

Memorandum "Ad Hoc Committee Report on Status of Operation of the Plum Island Animal Disease Center," 26 February 1979.

Memorandum "Ad Hoc Committee Report," from Merlon E. Wiggin to J. J. Callis, 5 March 1979.

"THE DISASTROUS INCIDENT"–PERIODICALS

"$10,049,000 Project Starts at Plum Island," *Suffolk Times*, 21 October 1976.

"$10 Million Plum Island Expansion," *Long Island Traveler-Watchman*, 28 October 1976.

"Hoof-and-Mouth Disease on Plum I.," *Newsday*, 16 September 1978.

"US Tracking Plum I. Visitors," *Newsday*, 18 September 1978.

"Source of Plum Island Germ Leak Sought," *Newsday*, 19 September 1978.

"Experts Probe Animal Disease," *Newsday*, 20 September 1978.

"Hoof-and-Mouth Disease Outbreak on Plum Island," *Long Island Traveler-Watchman*, 21 September 1979.

"Hoof-and-Mouth Stays on Plum Island," *Long Island Traveler-Watchman*, 28 September 1978.

"Outbreak Still Probed," *Suffolk Times*, 5 October 1978.

"P.I. Investigation," *Suffolk Times*, 12 October 1978.

"Animal Research Contract Probed," *Newsday*, 18 February 1979.

"The Plum Island Probe," *Suffolk Times*, 22 February 1979.

"Plum Is. Kickback Counts Probed," *Suffolk Times*, 22 February 1979.

"Plum Island Aide Reinstated in Job," *Newsday*, March 1979.

"USDA Still Silent on Inquiry at Plum Island," *Suffolk Times*, 1 March 1979.

"2 Plum I. Officials Deny Blame in Bacteria Leak," *Newsday*, 3 March 1979.

"US Halts Plum Is. Lab Construction Job," *Suffolk Times*, 8 March 1979.

"Builder Loses Plum Island Job," *Newsday*, 10 March 1979.

"Plum Island Breaks off with Contractor," *Suffolk Times*, 22 March 1979.

"Plum Island Mismanagement Charged," *Suffolk Times*, 31 May 1979.

"Plum Is. Project is Restarting," *Suffolk Times*, 28 June 1979.

"P.I. Construction Postponed Again," *Suffolk Times*, 12 July 1979.

"Worker Says Plum Island Flaws Ignored," *Newsday*, 14 August 1979.

"More Plum Island Changes," *Long-Island Traveler Watchman*, 16 August 1979.

"P.I. Official Charges Slander, Lax Security," *Suffolk Times*, 16 August 1979.

"Merlon Wiggin Calls It Quits at Plum Island," *Suffolk Times*, 30 August 1979.

"P.I. Construction Is Revived," *Suffolk Times*, 11 October 1979.

"Plum I. Project Called 'Horror'," *Newsday*, 21 February 1980.

"Plum Island Contractor Comes Under Fire," *Suffolk Times*, 13 March 1980.

"Charges in Plum Island Project," *Newsday*, 16 August 1980.

"Contracting-out at USDA Found 'Shoddy' by Congress," *The Government Standard*, February 1981.

"P.I. Aims for Construction Project," *Suffolk Times*, 26 February 1981.

"US Sues Over Plum Island Construction," *Newsday*, 2 March 1982.

"US Sues Over Work at Plum Island Lab," *Long Island Traveler-Watchman*, 4 March 1982.

"US Gets $10.6M Check from Jury Award," *Newsday*, 23 October 1987.

"THE DISASTROUS INCIDENT"–LEGAL DOCUMENTS

Grand Jury Indictment. United States of America v. Joseph Morton Company, Inc., Joseph Battaglia, and Samuel Semble, United States District

Court, Eastern District of New York (T. 18, U.S.C. §§371, 1001, 1503, 3237 and 2).

Joseph Morton Company, Inc., Joseph Battaglia, Samuel Semble, Defendants. Judgment and Probation/Commitment Order, United States District Court, Eastern District of New York.

Appropriation Bill 1981. Agricultural Research and Development and Related Agencies. Report No. 961030. Calendar No. 1162. 96th Congress, 2nd session, 24 November 1980.

RIFT VALLEY FEVER

"The Mystique That Is Plum Island," *Suffolk Times,* 31 August 1978.

"Let's Take No Chances on Plum Island Research," *Newsday,* 11 September 1978.

"On Rift Valley Fever," *Suffolk Times,* 21 September 1978.

Correspondence from John V. N. Klein, Suffolk County Executive, to Honorable Bob Bergland, U.S. Secretary of Agriculture, 28 September 1978.

Correspondence from T. B. Kinney, Acting Associate Deputy Director, Federal Research, USDA, to John Klein, 20 October 1978.

Correspondence from John Klein to Bob Bergland, 26 October 1978.

"NFEC Replies to Letter on Rift Fever," *Suffolk Times,* 26 October 1978.

Correspondence from John Klein to Honorable Thomas J. Downey, United States Congressman, 14 November 1978.

"Worried Klein Tours Plum Island," *Newsday,* 5 December 1978.

"Rift Research," *Suffolk Times,* 22 February 1979.

"Rift Valley Study Ended," *Suffolk Times,* 29 March 1979.

"Seals Are Used as Guinea Pigs in Viral Disease Experiments," *Newsday,* 3 April 1979.

CROSSING THE RUBICON

"The Plum Island Animal Disease Center: Research on Foreign Diseases of Animals," Agricultural Research Service, USDA, May 1975.

"Plum Island Lab Seeks OK for Virus Test," *Newsday,* 8 November 1979.

"Plum Island Begins Gene Cloning Project," *Long Island Traveler-Watchman,* 15 November 1979.

"Research of Disease OKd for Suffolk Lab," *Newsday,* 7 December 1979.

Plum Island Animal Disease Center Consultants' Reports, August 1980; June 1982; June 1983; June 1984.

"Plum Island Concedes Cat 'Overkill,'" *Newsday,* 25 March 1981.

"Vaccine Developed by Genetic Splicing," *New York Times,* 19 June 1981.

"Gene-Splicing Creates Foot-and-Mouth Vaccine," *Washington Post,* 20 June 1981.

"Director Suspended," *Suffolk Times,* 25 November 1982.

"Plum I. Chief Erred, Was Suspended," *Newsday*, 30 November 1982.

Correspondence from Jane Teas, Ph.D., Research Fellow, Harvard School of Public Health, to Dr. J. J. Callis, Director, PIADC, 15 April 1983.

Letter submitted to *The Lancet* for publication, "Could AIDS Be a New Variant of ASFV?," Jane Teas, Ph.D., Research Fellow, Harvard School of Public Health, 1983.

"Long-Term Planning for Research and Diagnosis to Protect U.S. Agriculture from Foreign Animal Diseases and Ectoparasites," Subcommittee on Research and Diagnosis on Foreign Animal Disease, National Research Council, 1983.

Beldekas, John, Jane Teas, and James R. Herbert. "African Swine Fever and AIDS," *The Lancet*, 8 March 1986.

"Will Plum Island Study AIDS?," *Suffolk Times*, 9 February 1984.

"Agriculture Dept. Ponders Swine Fever, AIDS Study," *American Medical News*, 9 March 1984.

"P.I. 'Probably' Will Research AIDS," *Suffolk Times*, 16 February 1984.

"P.I. Probe Confirmed," *Suffolk Times*, 8 November 1984.

"Test to Seek Swine Fever-AIDS Link," *Newsday*, 26 June 1985.

"New Look at AIDS Suspect," *New York Times*, 2 July 1985.

"A Disputed AIDS Link," *Newsday*, 24 September 1985.

"No Swine Fever Link to AIDS Seen," *New York Times*, 23 September 1986.

Leishman, Katie. "AIDS and Insects," *Atlantic Monthly*, September 1987.

Callis, Jerry J. "PIADC Notes—January 1986," 15 January 1986.

Attached Memorandum "ASF/AIDS Workshop Sept. 11, 1986," from R. E. Shope et al. to Dr. Terry Kinney, ARS, and Mr. Bert Hawkins, APHIS.

Correspondence from Frances K. Demerest to Jerry J. Callis, 24 November 1988.

Facsimile from R. E. Shope, Yale Arbovirus Research Unit to Jerry Crawford, Director, PIADC, 8 January 1993.

"Report of the USAHA Committee on Foreign Animal Diseases: Foreign Animal Disease and Pest Diagnosis and Research," 1998 Committee Reports, U.S. Animal Health Association.

"Report of the USAHA Committee on Foreign Animal Diseases: Biocontainment," 1998 Committee Reports, U.S. Animal Health Association.

THE KINGDOM AND THE GLORY

"Report of Commercial/Industrial Program Options and Opportunities at Plum Island Animal Disease Center and National Animal Disease Center," Science and Education Administration, USDA, 25 January 1980.

"U.S. Views Cuts in Plum Island Payroll," *Newsday*, 21 February 1980.

"Plum Is. Support Work by Contract Eyed," *Suffolk Times*, 21 February 1980.

Correspondence from Frank Bush, Vice President—America Federation of Government Employees Local 1940 to Honorable William Carney, United States Congressman, 1 April 1981.

"Plum Island Plan Dropped," *Newsday,* 1 November 1983.

Plum Island Diary, July 1987.

Plum Island Diary, September 1987.

"Changes in the Wind at Plum Island," *Suffolk Times,* February 1988.

"Plum Island Researchers Get the Axe," *Suffolk Times,* February 1988.

"USDA Honchos Visit Plum Island," *Suffolk Times,* February 1988.

BOOMERANG-DOCUMENTS

Aerometric Informational Retrieval System Facility Subsystem Updates, U.S. Environmental Protection Agency, 1975–1988.

Certificate to Operate an Air Contamination Source, Incinerator Renewal Application, 6 April 1987.

"A-76 Implementation Booklet," Agricultural Research Service, circa 1990.

Memorandum "Plum Island Animal Disease Center In-house Bid Evaluation Panel Report," from Louise S. Welker, Chairperson IBEP, to T.J. Clark, Deputy Administrator, 20 July 1990.

Memorandum "In-House Bid Evaluation Panel Report for PIADC," from T.J. Clark to Roger Breeze, Center Director, PIADC, 13 August 1990.

Memorandum "American Federation of Government Employees, Local 1940, Appeal of Cost Comparison," from Edward L. Hollreiser to A.M. Campbell, ARS Special Projects Division, circa August 1990.

Potential Hazardous Waste Site Preliminary Assessment of Plum Island, E.P.A., 9 April 1992.

Memorandum "Inspection of the PIADC April 30, 1992," from Eileen Governale, Department of Health Services, Suffolk County, to Claude Crawford, 5 May 1992.

Memorandum "Biosafety Practices," from R.W. Powitz, Safety Officer, to Claude Crawford and Roger Breeze, 20 July 1992.

Summary of Safety and Security Regulations for Contractors, 20 July 1992.

Log Book Entries, Lab 257, 29 October 1992.

Notices of Unsafe or Unhealthful Working Conditions, Occupational Safety and Health Administration, 1988–1993.

Citations and Notification of Penalty, Occupational Safety and Health Administration, 1988–1993.

Letter "Notice of Violation, Plum Island Animal Disease Center," from George C. Meyer, Chief Hazardous Waste Compliance Branch, New York State Department of Environmental Conservation, to Dr. Robert Powitz, Environmental Director, PIADC, circa 1993.

Memorandum "Suffolk County Health Department Review of Sewage and

Waste Treatment Facility," from Jack Rief, PIADC, to Charles Wenderoth, EPS Manager, 6 January 1993.

Letter "Notice of Violation and Request of Information Pursuant to Sections 308 of Clean Water Act," from Richard L. Caspe, Director, Water Management Division, EPA, to Mr. Claude G. Crawford, Deputy Area Director, PIADC, 4 February 1993.

Memorandum "TSCA Section 6(e) PCB Inspection of PIADC," from J. R. Olechowski, Chemist, E.P.A., to Daniel J. Kraft, Chief of the Toxic Substances Section, 3 March 1993.

Memorandum "Asbestos Information Requested by EPA," from Carlos Santoyo, General Engineer, to R. W. Powitz, Safety Officer, 24 March 1993.

Letter, "Plum Island Animal Disease Center Inquiry," from C. G. Crawford to Conrad Simon, 16 March 1993.

Press release "E.PA. Seeks Fines Totaling $190,700 for Federal Hazardous Waste Rule Violations at the Plum Island Animal Disease Facility," 23 January 1995.

ARS Plum Island Research News, Issue 2, February 1995.

Federal Facilities Enforcement and Compliance Accomplishments Report F.Y. 1999, E.P.A., December 2000.

BOOMERANG–PERIODICALS

"Plum Island Director Moved to Research," *Newsday,* 1 February 1991.

"US Contract Switch Places Plum Island Jobs in Jeopardy," *Long Island Traveler-Watchman,* 28 February 1991.

"Plum Island Hiring Plan Disputed," *Newsday,* 7 March 1991.

"Feds Lose Plum Island Appeal," *Suffolk Times,* 9 May 1991.

"Job Havoc on Plum Island," *Newsday,* 10 November 1991.

"Off North Fork, Two Islands in Disputes," *New York Times,* 24 November 1991.

"Plum Island Tug of War Gets Tugging," *Suffolk Times,* 13 February 1992.

"U.S. Bulwark Against Animal Disease Cuts Back," *New York Times,* 26 November 1992.

"Island's Shroud of Secrecy Lifted," *Newsday,* 15 December 1992.

"Research Unit Is Postponing Major Projects," *New York Times,* 24 December 1992.

"Plum Island to Be Un-privatized?" *Suffolk Times,* 3 June 1993.

"Plum Is. Administrator Moving On," *Suffolk Times,* 10 June 1993.

"Savings Questioned at Plum Island," *Newsday,* 11 March 1995.

"Privatizing Plum Island," *Southampton Press,* 3 August 1995.

"Plum Island Bird Kill Probed," *Newsday,* 2 July 1993.

"Plum I. Cited for Lax Safety," *Newsday,* 23 March 1993.
"Plum Island Cited for 61 Violations," *Suffolk Times,* 27 October 1988.
"More Violations Cited on Plum Is.," *Suffolk Times,* 25 March 1993.

BOOMERANG–LEGAL DOCUMENTS

Letter, "Plum Island Animal Disease Center," from Conrad Simon, Director, Air and Waste Management Division, U.S. Environmental Protection Agency, to Claude G. Crawford, 17 February 1993.
Letter, "Plum Island Animal Disease Center," from James H. Pim, Office of Pollution Control, Suffolk County, to Robert J. Wing, Chief Federal Facilities Section, E.P.A., 18 February 1993.
E.P.A. Complaint Compliance Order and Notice of Opportunity for Hearing Pursuant to 3008 of the Solid Waste Disposal Act, In the Matter of U.S.D.A. Plum Island Animal Disease Center, 20 October 1994.
E.P.A. Federal Facility Compliance Agreement, Plum Island Animal Disease, 20 January 1994.
E.P.A. Complaint Compliance Order and Notice of Opportunity for Hearing Pursuant to 3008 of the Solid Waste Disposal Act, In the Matter of U.S.D.A. Plum Island Animal Disease Center, 22 December 1994.
E.P.A. Consent Agreement and Consent Order Pursuant to 3008 of the Solid Waste Disposal Act, In the Matter of Burns & Roe Services Corporation, 15 March 1996.
E.P.A. Consent Agreement and Consent Order Pursuant to 3008 of the Solid Waste Disposal Act, In the Matter of U.S.D.A. Plum Island Animal Disease Center, 16 February 1999.

MELTDOWN

Maximum Hurricane Surge Elevation Table for Suffolk County, Plum Island Quadrangle, Drawing No. P1 (date unknown).
Memorandum for the Record "Contamination of Building 257 Equipment Room," from Michael Spillane, Safety Office, PIADC, 4 March 1983.
Memorandum for the Record "Exposure from Contaminated Sewer Line Drainage from VSV Room," from J. S. Walker, Safety Officer, PIADC, 2 March 1983.
Correspondence "Re: Sinowski Memo," from R. D. Plowman, Administrator, Agricultural Research Service, to Stanley Mickaliger, PIADC, 17 September 1991.
Memorandum "Dedicated Efforts During Hurricane Bob," from Walter Sosnoski, USRO Leader "B" Crew, to Drs. Crawford, Breeze, Mebus, and Mr. E. Escarcega, Mr. Hassildine, and Mr. Heaney, 27 August 1991.

THE HOMELAND-DOCUMENTS
Biosafety for Animal Rooms with Glanders—Biosafety Level Four, 20 July 1989.
General Fire Protection Plan and Risk Assessment, Gage-Babcock & Associates, circa 1990.
Application to Visit Plum Island as an Official Visitor, Meeting re Biological Weapons, 9 July 1992.
Individual Event Report Concerning Power Outage in Lab 257, 26 July 1992.
Handbook, Plum Island Decontamination Project, Plum Island Animal Disease Center, May 1993.
Memorandum "Decontamination of Building 101 West Service Wing," from Claude Crawford to All ARS Personnel, 28 May 1993.
Radioactive Material Spill Response and Mitigation Plan, Plum Island Animal Disease Center, August 1999.
E.P.A. Stipulation and Order of Settlement, The State of New York against the U.S. Department of Agriculture, 15 June 2001.

THE HOMELAND-PERIODICALS
Hadwiger, Don F., "The Green Revolution: Some New Perspectives," *Change,* volume 7, no. 9, November 1975.
"Four Foreign Veterinarians Study at Plum Island," *Suffolk Times,* 19 August 1976.
"Mishap Empties Lab," *Newsday,* 3 June 1993.
"Plum Island on Arms Control Map," *Newsday,* 16 July 1992.
"Lab to Try New Recipe," *Newsday,* 27 May 1993.
"Evacuated Lab Is Reoccupied," *Newsday,* 3 June 1993.
"Plum Island Lab Cleanup Missed Pipes," *Newsday,* 4 August 1993.
"Plum Is the Word in Russian," *Newsday,* 2 March 1994.
"Vacco Suit: Plum Island Polluting," *Newsday,* 3 May 1998.
"Plum Island Sued of Sewage," *Suffolk Times,* 7 May 1998.
Franz, David, "Foreign Animal Disease Agents as Weapons in Biological Warfare," *Food and Agricultural Security,* 1999.
Murphy, Frederick, "The Threat Posed by the Global Emergence of Livestock, Foodborne, and Zoonotic Pathogens," *Food and Agricultural Security,* 1999.
Rimmington, Anthony, "Anti-Livestock and Anti-Crop Offensive Biological Warfare Programmes in Russia and the Newly Independent Republics," University of Birmingham, June 1999.
Kohnen, Anne, "Responding to the Threat of Agroterrorism: Specific Recommendations for the U.S.D.A.," J.F.K. School of Government, Harvard University, October 2000.

Pesek, John, "From a Trail to a Path to Sustainable Agriculture," Iowa State University, 1 March 2001.
"Safe at Home," *Suffolk Times,* 18 October 2001.
"Security Stepped Up on Plum Island," *Long Island Traveler-Watchman,* 11 October 2001.
"Hostages to Terror," *Dan's Papers,* 24 May 2002.

CONSULTED BOOKS

A Centennial Celebration: 100 Years of Creating a Healthier Future for Animals and People. Ithaca, NY: Cornell University College of Veterinary Medicine, 1994.
Alibek, Ken, with Stephen Handelman. *Biohazard: The Chilling True Story of the Largest Covert Biological Weapons Program in the World—Told from Inside by the Man Who Ran It.* New York: Random House, 1999.
Bailey, John W. *Pacifying the Plains: General Alfred Terry and the Decline of the Sioux, 1866–1890.* Westport, CT: Greenwood Press, 1979.
Blanchard, Fessenden S. *Long Island Sound.* Princeton: D. Van Nostrand Company, ca. 1958.
Bower, Tom. *The Paperclip Conspiracy.* London: Michael Joseph, 1987.
Brigham Narins, ed. *Notable Scientists from 1900 to the Present.* Farmington Hills, MI: Gale Group, 2001.
Brophy, Leo P., and George J.B. Fisher. *The Chemical Warfare Service: Organizing for War,* Washington, DC: Office of the Chief of Military History, Department of the Army, 1959.
Caro, Robert A. *The Power Broker: Robert Moses and the Fall of New York.* New York: Random House, 1974.
Clark, Neal. *Birds on the Move: A Guide to New England's Avian Invaders.* Unity, ME: North Country Press, 1988.
Cole, Leonard A. *Clouds of Secrecy: The Army's Germ Warfare Tests over Populated Areas.* Totowa, NJ: Rowman and Littlefield, 1988.
———. *The Eleventh Plague: The Politics of Biological and Chemical Warfare.* New York: W. H. Freeman, 1997.
Committee on Foreign Animal Diseases. *Foreign Animal Diseases* (revised 1998). Richmond, VA: U.S. Animal Health Association, 1998.
———. *Foreign Animal Diseases: Their Prevention, Diagnosis and Control* (revised 1984). Richmond, VA: U.S. Animal Health Association, 1984.
Covert, Norman M. *Cutting Edge: A History of Fort Detrick, Maryland* (4th ed.). U.S. Army, 2000.
Daniel, Pete. *Breaking the Land: The Transformation of Cotton, Tobacco and Rice Cultures since 1880.* Urbana: University of Illinois Press, 1985.
———. *Lost Revolutions: The South in the 1950s.* Chapel Hill: University of North Carolina Press, 2000.

Davidson, Osha Gray. *Broken Heartland: The Rise of America's Rural Ghetto.* Iowa City: University of Iowa Press, 1996.

DeGregorio, William A. *The Complete Book of U.S. Presidents: From George Washington to Bill Clinton.* New York: Wings Books, 1997.

DeMille, Nelson. *Plum Island.* New York: Warner Books, 1997.

Drexler, Madeline. *Secret Agents: The Menace of Emerging Infections.* Washington, DC: Joseph Henry Press, 2002.

Dyson, Verne. *Anecdotes and Events in Long Island History.* Port Washington, NY: Ira J. Friedman, 1969.

Eberlein, Harold Donaldson. *Manor Houses and Historic Homes of Long Island and Staten Island.* Port Washington, NY: Ira J. Friedman, 1928.

Engelman, Rose C., and Robert J.T. Joy. *Two Hundred Years of Military Medicine.* Fort Detrick, MD: U.S. Army Medical Department, 1975.

Elphick, Jonathan, ed. *The Atlas of Bird Migration: Tracing the Great Journeys of the World's Birds.* New York: Random House, 1995.

Frazier, Thomas W., and Drew C. Richardson, eds. *Food and Agricultural Security: Guarding Against Natural Threats and Terrorist Attacks Affecting Health, National Food Supplies, and Agricultural Economics.* New York: New York Academy of Sciences, 1999.

Garrett, Laurie. *The Coming Plague: Newly Emerging Diseases in a World Out of Balance.* New York: Farrar, Straus and Giroux, 1994.

Geissler, Erhard, and John Ellis van Courtland Moon, eds. *Biological and Toxin Weapons: Research, Development and Use from the Middle Ages to 1945.* New York: Oxford University Press and Stockholm International Peace Research Institute, 1999.

Gillott, Cedric. *Entomology.* New York: Plenum Press, 1995.

Gipson, Fred. *"The Cow Killers": With the Aftosa Commission in Mexico.* Austin: University of Texas Press, 1956.

Griffin, Augustus. *Griffin's Journal: First settlers of Southhold.* Orient Point, NY: A. Augustus, 1857.

Hamburg, Eric, ed. *Nixon: An Oliver Stone Film* (annotated screenplay). New York: Hyperion, 1995.

Hamilton, Harlan. *Lights and Legends: A Historical Guide to Lighthouses of Long Island Sound, Fishers Island Sound, and Block Island Sound.* Stamford, CT: Wescott Cove, 1987.

Handbook of Infectious Diseases. Springhouse, PA: Springhouse, 2000.

Hickman, Homer. *Torpedo Junction: U-boat War Off America's East Coast 1942.* Annapolis: Naval Institute Press, 1996.

Hoff, Brent, and Carter Smith III. *Mapping Epidemics: A Historical Atlas of Disease.* New York: Franklin Watts, 2000.

Horowitz, Leonard G. *Emerging Viruses: AIDS and Ebola—Nature, Accident or Intentional?* Rockport, MA: Tetrahedron, 1996.

Hunt, Linda. *Secret Agenda: The United States Government, Nazi Scientists, and Project Paperclip, 1945–1990.* New York: St. Martin's Press, 1991.

Junger, Sebastian. *The Perfect Storm: A True Story of Men Against the Sea.* New York: HarperCollins, 1997.

Karlen, Arno. *Biography of a Germ.* New York: Pantheon Books, 2000.

Leonard, Ellis Pierson. *In the James Law Tradition 1908–1948.* Ithaca, NY: New York State College of Veterinary Medicine, 1982.

Loftus, John. *The Belarus Secret.* New York: Paragon House, 1982.

Manchester, William. *American Caesar: Douglas MacArthur 1880–1964.* Boston: Little, Brown & Co., 1978.

Mangold, Tom, and Jeff Goldberg. *Plague Wars: The Terrifying Reality of Biological Warfare.* New York: St. Martin's Press, 1999.

Medical Management of Biological Casualties. U.S. Army Medical Research Institute of Infectious Diseases. Frederick, MD: Fort Detrick, 1998.

Mermin, Lora, ed. *Lyme Disease 1991: Patient/Physician Perspectives from the US and Canada.* Madison, WI: Lyme Disease Education Project, 1992.

Miller, Judith, Stephen Engelberg, and William Broad. *Germs: Biological Weapons and America's Secret War.* New York: Simon & Schuster, 2001.

Mole, Robert L., and Dale M. Mole. *For God and Country: Operation Whitecoat, 1954–1973.* Brushton, NY: Teach Services, 1998.

Munk, Klaus. *Virologie in Deutschland: Die Entwicklung eines Fachgebietes.* Germany: Karger, 1995.

Necrology of the Faculty of Cornell University 1962–1963. Ithaca, NY: Cornell University Press, 1963.

Peters, C. J., and Mark Olshaker. *Virus Hunter: Thirty Years of Battling Hot Viruses Around the World.* New York: Anchor Books, 1997.

Piller, Charles, and Keith R. Yamamoto. *Gene Wars: Military Control over the New Genetic Technologies.* New York: William Morrow, 1988.

Poole, Alan F. *Ospreys: A Natural and Unnatural History.* New York: Cambridge University Press, 1989.

Preston, Richard. *The Hot Zone.* New York: Random House, 1994.

Rafferty, Pierce, and John Wilton, eds. *Guardian of the Sound.* New York: Fossil Photos, 1998.

Rathbun, Captain Benjamin F. *Capsule Histories of Some Local Island and Lighthouses in the Eastern Part of L.I. Sound.* Niantic, NY: Presley Printing, 1999.

Rattray, Jeannette Edwards. *Ship Ashore!: A Record of Maritime Disasters Off Montauk and Eastern Long Island, 1640–1955.* New York: Coward-McCann, 1955.

Regis, Ed. *The Biology of Doom: The History of America's Secret Germ Warfare Project.* New York: Henry Holt, 1999.

Renneberg, Monika, and Mark Walker, eds. *Science, Technology and National Socialism*. Cambridge, England: Cambridge University Press, 1994.

Schlosser, Eric. *Fast Food Nation: The Dark Side of the All-American Meal*. New York: Houghton-Mifflin, 2001.

Schactman, Tom. *Terrors and Marvels: How Science and Technology Changed the Character and Outcome of World War II*. New York: William Morrow, 2002.

Scientific Reprints: 1955–Present. Plum Island Animal Disease Center: U.S. Department of Agriculture, 2002.

Schuyler, Montgomery. *Notes on the Patroonships, Manors and Seigneuries in Colonial Times*. New York: Order of Colonial Lords of Manors in America, 1953.

Sidell, Frederick R., Ernest T. Takafuji, and David R. Franz, eds. *Medical Aspects of Chemical and Biological Warfare*. Washington, DC: Office of the Surgeon General, 1997.

Stockholm International Peace Research Institute. *The Problem of Chemical and Biological Warfare*. Stockholm: Almquist and Wiksell, 1971–1975.

Thompson, Benjamin. *History of Long Island: Containing an Account of the Discovery and Settlement with Other Important and Interesting Matters to the Present Time*. New York: E. French, 1839.

Whitby, Simon. *Biological Warfare Against Crops*. Hampshire, England: Palgrave, 2002.

Williams, Peter, and David Wallace. *Unit 731: The Japanese Army's Secret of Secrets*. London: Grafton Books, 1989.

Wolfgang, Ewert. *Insel der Forscher: ein Bericht über die Gelehrten vom Riems und ihr Werk*. Berlin: Kongress-Verlag, 1962.

Zajtchuk, R., and R. F. Bellamy, eds. *Textbook of Military Medicine: Medical Aspects of Chemical and Biological Warfare*. Washington, DC: Office of the Surgeon General, Department of the Army, 1997.

NEWSPAPERS, PERIODICALS, AND OTHER MEDIA

Agricultural Research
ARS Plum Island Research News
Brooklyn Eagle
Cable News Network (CNN)
Chicago Tribune
Christopher Street
Corpus Christi Caller
Dan's Papers
East Hampton Star

Farm Journal
Hartford Courant
Long Island Press
Long Island Traveler-Watchman
News 12 Long Island
Newsday
New York Herald-Tribune
New York Observer
New York Post
New York Sun
New York Times
New York World-Telegram
Outside
San Diego Union-Tribune
Science
Scientific American
The Day
The Economist
The New Yorker
The Saturday Evening Post
Southampton Press
Suffolk Sun
Suffolk Times
Wall Street Journal
Washington Post
U.S. Department of Agriculture Films
USDA Internet Website <www.usda.gov>
WABC-TV New York

DOCUMENT ARCHIVES
Library of Congress
National Academy of Sciences
National Agricultural Library
National Archives and Records Administration
National Medical Library
Naval Medical Research Institute Library
New York Public Library
Oyster Ponds Historical Society
Pennypacker Long Island Collection, East Hampton Library
Plum Island Animal Disease Center Library
Simon Wiesenthal Center
Society for the Preservation of Long Island Antiquities

Southold Free Library
Southold Town Archives
Suffolk County Historical Society
U.S. Army Corps of Engineers
U.S. Centers for Disease Control
U.S. Department of Agriculture
U.S. Department of Homeland Security
U.S. General Accounting Office
U.S. Office of Technology Assessment
U.S. House of Representatives
U.S. Senate

FREEDOM OF INFORMATION ACT REQUESTS

Army Intelligence and Security Command
Army Soldier Biological and Chemical Command
Central Intelligence Agency
Department of Agriculture[1]
Department of the Army
Environmental Protection Agency
Federal Bureau of Investigation
Nuclear Regulatory Commission
Occupational Safety and Health Administration
Office of Naval Research
Office of Special Counsel

---

[1]This agency did not comply with my FOIA request for documentation. After two years without a response, in July 2003 I received a single document, which was so heavily redacted it was unintelligible and useless.

Acknowledgments

Lab 257 was a project seven years in the making. Many people are responsible for its fruition, but no one save me is responsible for its shortcomings.

To the current and former scientists and workers of Plum Island, and to the local ombudsmen and lay historians featured in the book—including those who asked for anonymity—I thank you for sharing your memories, experiences, and tribulations with me, and for helping me to understand Plum Island through your eyes. Special thanks to Plum Island librarian Honoré McIlvain; photographer Elizabeth Clark; engineer Charles Wenderoth; Martin Weinmiller; William Schlichtig; former Plum Islanders Phillip Piegari, Ben Robbins, Stanley Mickaliger, and John Patrick Boyle; and Lyme disease activist Steven Nostrum. The full story would not be possible without the initial cooperation of former Plum Island directors Drs. Jerry J. Callis, Roger G. Breeze, and David L. Huxsoll.

I would also like to thank New York State Senator Charles J. Fuschillo, Jr., and Assemblywoman Patricia L. Acampora; David Kerkhof of the Suffolk County Historical Society; Joseph Majid and David Van Tassel of the National Archives and Records Administration; Janice Goldblum of the National Academy of Sciences; Southold town historian Antonia Booth; Courtney Burns of the Oyster Ponds Historical Society; Joelle Yudin of William Morrow/HarperCollins; Cheryl Fields of the Army Soldier Biological and Chemical Command; Suffolk County archivist Sharon Pullen; and Anne Kohnen, Tammy-Jo Ferdula, Kate Stroup, Dr. Robert Shope, Dr. Thomas Mettenleiter, Dr.

Donald Smith, Dr. Simon Whitby, Dr. Kenneth Liegner, Hank Alberelli, Ed Regis, Gregory Koblentz, Fareed Zakaria, Pete Daniel, Gary Stubblefield, Marjorie Tietjen, Jon Rand, Fred Ciporen, Lynn Shepler, Linda Hunt, Jeff Benedict, and John Loftus, all of whom graciously assisted me along the way. Stephanie Reidner and Ute Rodrian helped me translate German books, newspaper articles, and government documents into English. Also, thank you to Dr. Marcia Stone for introducing me to the world of microbiology, and to Dr. Lorrence Green, who tutored me further in the discipline and reviewed and corrected scientific portions of the manuscript. That said, any lingering inaccuracies are entirely my own.

I am indebted to journalist and author Karl Grossman, who spent many hours with me recounting his Plum Island observations and investigations spanning thirty years, and for doing something rare in journalism: sharing a career's worth of reporter's notes, a literal treasure trove crucial to this book, and letting me photocopy it all on his dime. Were it not for Karl's generosity, this book would be something far less, and I thank him for his time, his contacts, and his talent. Former *Newsday* reporters John McDonald and Drew Fetherston described their reportage of Plum Island–related stories, and John was kind enough to provide me with declassified Army documents he had obtained on Lab 257.

People who write books of this genre generally fall into two camps: accomplished writers and investigative journalists. I am neither. To the extent this book reads worthy of those titles, much of that honor is due Claire Curry and David H. Wheeler. Claire read and reread drafts of the manuscript more times than I can remember, while adding insights and subtracting digressions along the way. Dave set aside untold hours, provided innumerable suggestions, and coached me all the way to the finish line. Their contributions are inestimable. Kaya Laterman tracked down helpful information on different topics. Rebecca Lynch pored through files at the National Archives, spent hours researching and fact-checking, commandeered manuscript revisions, and compiled chapter notes. She has a brilliant creative future ahead of her.

David Tonsmeire, James Muscarella, and Sandra Lauterbach shot photographs of Plum Island from Orient Point (and from Dave's boat), and William "Hall" Wheeler created the map illustration. I thank them for their talented contributions. Two individuals who taught me how to write and how to think at a formative period in my life should be recognized: Rick Cuchel and Bernie "The Bear" Stein. Fellow writer Amy Schapiro deftly guided me from idea to proposal to agent and to publication.

To my family: my mother, sister, and brother, my aunt and uncle, and to my grandparents Neil, Mildred, and Dorothy, thank you—for your love, for putting up with me, and for keeping my spirits high. To my friends who supported and encouraged me over the years, who gave freely their homes, offices, cars, opinions, and, while learning more than they ever wanted to about Plum Island, their ears—I salute you. They are, in no particular order: Michael Slater, James and Jean Marie Faherty, Thomas and Joann Weghorst, Brian Cohen, Vanessa France, Scott Greenspan, Mark Guthart, David Levinbook, Matthew and Jason Vishnick, Kristen Goelz, Jason Weiler, Robert Salvatico, Scott Fless, Sharon Hill, Michael Cornell, Steve Schapiro, Kathy Wu, Gideon Berger, Brandon Baer, Alvin Murstein, Andrew Murstein, and the late John Geiger. At the beginning of this journey, Jon Huzarsky took down interview notes, poked holes, and rode ferries with me—and helped me realize Lab 257 was a story that could—and should—be told. Betty Leyva provided valuable feedback and was instrumental in the homestretch and I am grateful for her support. And finally, hats off to John DeSimone, a selfless friend who gives of himself tirelessly and deserves nothing but the very best out of life.

A debt of gratitude to my two mentors at my old law firm Willkie Farr & Gallagher, partners Steven J. Gartner and Christopher E. Manno, both of whom taught me the ropes of law practice. Steve cheered me up and fed me periodically while on my sabbatical, and Chris helped resurrect my legal career once the manuscript was complete. The firm itself deserves my sincere appreciation for granting me a leave of absence to chase this story.

I'm grateful to my friend, former New York State Governor Mario M. Cuomo, for his sincere interest in my nascent legal and writing careers, and for sharing his philosophical ponderings with me. I am also indebted to the U.S. Congress for enacting the Freedom of Information Act in 1966, which mandates that federal agencies disclose specific information upon request, so that the citizenry can properly evaluate their government's actions (and inactions).

Alicka Pistek saw fit to pluck my query letter out of the pile, meticulously shape the book proposal, and successfully market it. She is a talented literary agent and I thank her for her faith in me and in the story. From the outset, Mauro DiPreta counseled this rookie writer with great skill and aplomb. He helped me focus and define the story, and refined the manuscript from line to theme. A writer with an editor like Mauro in his corner is a writer truly blessed.

Finally, I must thank the New York Public Library for allowing me to use its Frederick Lewis Allen Room, where I wrote most of the book. The Allen Room is a sanctuary for writers, and it has been an honor and a privilege to research and write there.

Mattituck-Laurel Library